A NEW ECOLOGICAL ORDER

INTERSECTIONS

Histories of Environment, Science, and Technology
in the Anthropocene

Sarah Elkind and Finn Arne Jørgensen, Editors

A NEW ECOLOGICAL ORDER

DEVELOPMENT AND THE TRANSFORMATION OF NATURE IN EASTERN EUROPE

EDITED BY STEFAN DORONDEL AND STELU ŞERBAN

UNIVERSITY OF PITTSBURGH PRESS

Published by the University of Pittsburgh Press, Pittsburgh, Pa., 15260

Manufactured in the United States of America
Printed on acid-free paper
10 9 8 7 6 5 4 3 2 1

Cataloging-in-Publication data is available from the Library of Congress

ISBN 13: 978-0-8229-4717-2
ISBN 10: 0-8229-4717-X

Cover art: Agricultural infrastructure, built on the levees along the Danube River, in a village lying on the Lower Danube's bank. Photo courtesy of the editors.
Cover design: Joel W. Coggins

To Luminiţa, Alexia, and Sofia, who light up my life
—Stefan Dorondel

and

To Tudor, my son
—Stelu Şerban

CONTENTS

Acknowledgments ix

INTRODUCTION
Planning Development, Changing Nature in Eastern Europe 3
Stefan Dorondel and Stelu Şerban

PART I. Planning Territory

1 The Quest for a New Urban Landscape: Spatial Transformation in the Nineteenth-Century Belgrade Environment 29
Dragana Ćorović

2 From Slavic Swamp to Promised Land: Social and Environmental Engineering in a Southern Macedonian Swamp, 1913–1936 46
George L. Vlachos

3 In Quest of Development: Territorialization and the Transformation of the Southern Ukrainian Wetlands, 1880–1960 65
Stefan Dorondel and Anna Olenenko

PART II. Nature, Economy and Experts

4 From Weeds to Commodities: The Translation of Plants into Medicines in Early Twentieth-Century Transylvania 89
Ágota Ábrán

5 The Economy of a Leashed River: State, Experts, and Politics along the Lower Danube, 1900–1940 109
Stelu Şerban and Stefan Dorondel

6 From Nature to National Networks: Hydraulic Bureaucracy and the Modernization of Waters in Czechia, 1890s–1960s 130
Jiří Janáč

7 Big Dam Biographies in Central Asia: Tracing Goals, Actors, and Impacts from World War II to the Present Day 156
Jeanne Féaux de la Croix and Flora Roberts

PART III. Imaging New Nature

8 Goats, Axes, and Uncertain Narratives: Representations of Nature and People in Asia Minor 179
Hande Özkan

9 Constructing Forest Expertise: Foresters in the Białowieża Forest 200
Eunice Blavascunas

10 The Political Ecology of Scientific Innovation in Russia: A Study of a Muskox Domestication Experiment in Siberia 220
Vladislava Vladimirova

11 Humans, Predators, and State Projects: A Look at the Lower Danube, Northwest Bulgaria 240
Yulian Konstantinov

Epilogue
A New Ecological Order at a European Margin 259
Stefan Dorondel and Helmuth Trischler

Contributors 265

Index 269

ACKNOWLEDGMENTS

The idea for this book was born in 2016 during one of our weekly meetings in a tavern in a marginal neighborhood in Bucharest, Romania. The name of the tavern itself—The Red Rooster—invites reflection on the relation between nature (the bird) and its cultural appropriation. Agreeing that Eastern Europe has been neglected in scholarly accounts of the global movement of technology and of development and nature, we decided to put out a call for contributions. The number and quality of abstracts far exceeded our expectations. One might think that our initial idea in that Romanian tavern and the subsequent call for papers would be the entire history of this volume. But of course we could not have envisaged what would eventually emerge from the editors' discussions, debates, and interactions with their peers—an intellectual debt that should be clearly spelled out here. In a fickle academic world, we have received support from wonderful people who spent time reading different chapters and making thorough observations. They are more numerous than we can acknowledge here and we apologize to those we may have forgotten.

Marie Louise Aastrup, George Andrei, Liviu Giosan, Alexandra Ion, Michal Kopeček, and Hande Özkan read and improved the introduction and our contributed chapter. David Moon and Katja Bruisch's crystal-clear suggestions proved invaluable for the construction of the chapter on the transformations of the Ukrainian wetlands. The comments and suggestions of two enthusiastic groups have massively contributed to an improved final version of the volume. First, the summer 2019 cohort of the Imre Kertész Kolleg Jena, where Stefan Dorondel introduced the manuscript for the first time to a public audience within his fellowship presentation. His gratitude goes to the directors of the institution for the opportunity and to his former fellows for their thorough discussion. Slavomíra Ferenčuhová,

Jan Mervart, and Tomasz Rakowski offered great comments and support. Special thanks to Rima Praspaliauskiene, who accompanied Stefan through Jena's startling surroundings and offered thoughtful and significant criticism of the manuscript.

Second, we have to thank the 2020–2021 cohort of the Rachel Carson Center for Environment and Society (RCC) of Ludwig Maximilian University in Munich. Surrounded by a unique and brilliant intellectual and social community that offered many helpful suggestions, Stefan was able to complete the manuscript and submit it to the publisher. He owes a great intellectual debt to Antoine Acker, Anna Antonova, Marcela Cely-Santos, Jean Chamel, Rae Young Choi, Johanna Conterio, Alfonso Donoso, Stephen Halsey, Ute Hasenöhrl, Huiying Ng, Milica Prokić, Kate Rigby, William San Martin, Astrid Schrader, Nicole Seymour, Kirk Sides, and Tracie Wilson. Their good-spirited company during the long lockdowns and restrictions kept him sane and substituted for his home family. Their comments prompted us to reinforce our arguments and contributed to the restructuring of various chapters. Without them this volume would have been much poorer. Roger Norum generously spent quite some time assisting Stefan in finding a catchier title for the volume, and deserves much gratitude. Often unfairly overlooked in academia, the student assistants merit equal credit, and the unparalleled luxury of having books delivered directly to one's desk is an invaluable time saver. The RCC's Sophia Hörl and librarian Laura Mann are ideal partners in assisting with research. They simply do not know the word "impossible" with regard to finding a book or an article. Special thanks to Sally Fenn and Jemma Deer, our language guardians, for their thorough work. They made the texts more readable.

Several institutions also contributed to the existence of the volume. Besides the two institutions mentioned above, we are also indebted to the Romanian Ministry of National Education. The grant received by Stefan Dorondel opened up new avenues for research in the environmental history and anthropology of Eastern Europe (CNCS—UEFISCDI, no. PN II-ID-PCE-2012–4-0587 [2013–2016]). This volume would not have been possible without the grant. Stelu Şerban also thanks the personnel of the State Archive of Pleven for their courtesy and assistance. Equally helpful was Petko Hristov, the former director of the Institute of Ethnology and Folklore Studies with the Ethnographic Museum, Sofia, who introduced us to the National Library in Sofia, Bulgaria. We enjoyed illuminating discussions with him and his colleagues, which contributed to further clarifying our ideas about technology and nature transformation in Bulgarian modern history. Finally, the Institute for Southeast European Studies of the Romanian Academy—where both editors work—showed patience and support on the long road toward the publication of this volume.

A NEW ECOLOGICAL ORDER

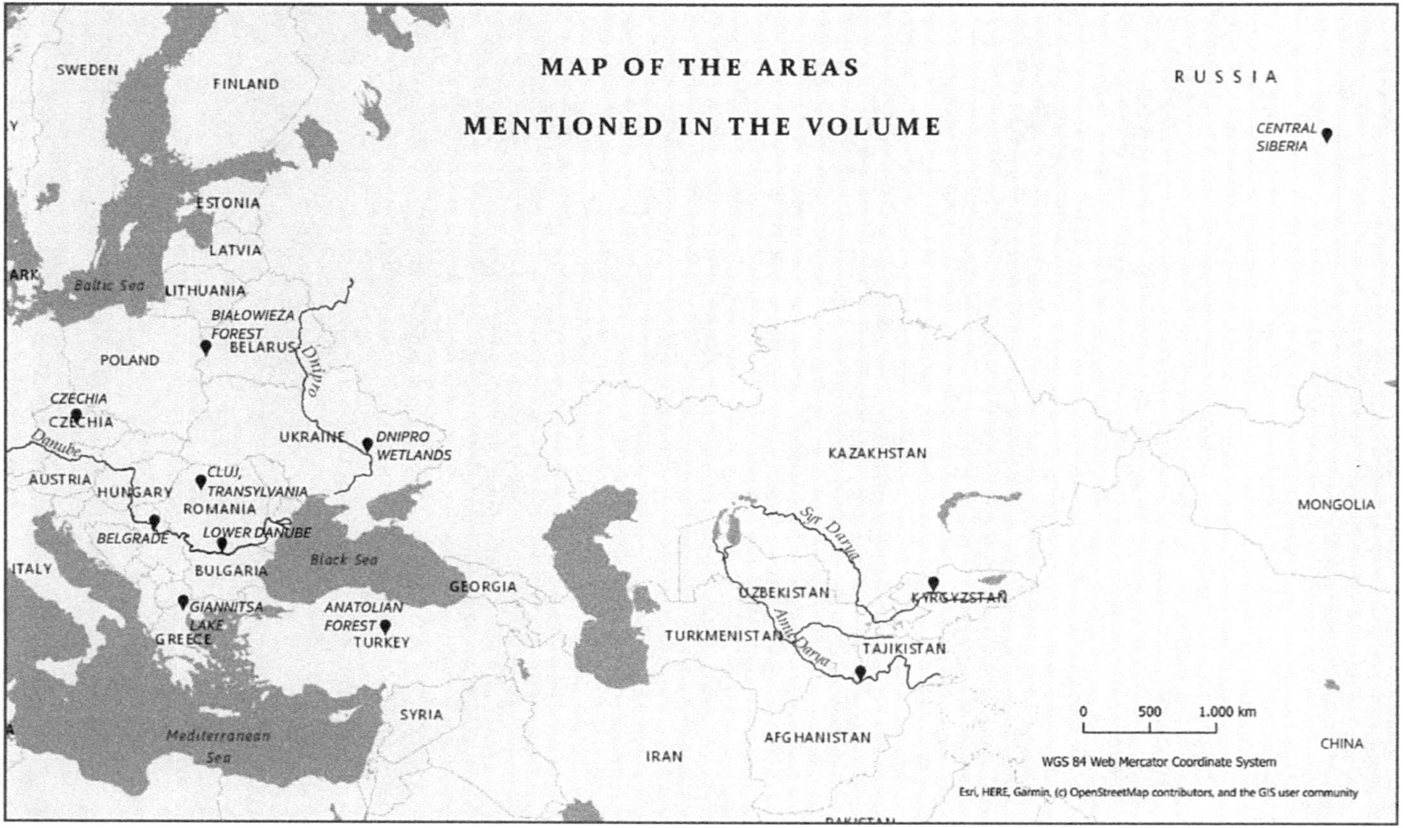

Frontispiece. Map by Marina Virghileanu.

INTRODUCTION

Planning Development, Changing Nature in Eastern Europe

Stefan Dorondel and Stelu Şerban

\\

As an engineering student at the University of Copenhagen in the second half of the nineteenth century, Per, the main character in the Danish novel *Lucky Per*, dreams of a new world that will channel the wasted power of waves, wind, water, and sun—dominated until then by nature—via the network of pipes spread throughout Denmark to improve the condition of human life (Pontoppidan 2013, vol. 1). In a Denmark that found itself at the economic and scientific periphery of Europe at that time, Per advocates a resumption of European Enlightenment philosophy: that the power of nature should be subdued for the purpose of human progress through the use of science and technology (Ferkiss 1993). According to this logic, uncontrolled and unused nature is a "waste" (Cioc 2002; Worster 1973). Rivers, forests, agricultural land, and animals—none escaped state-led and private initiatives to "improve" them through technological and infrastructural interventions.

In the historical period that *Lucky Per* is set, rivers in their natural state were still unpredictable: their shallow sandy waters with meandering features and islands that appeared and disappeared were landscapes that humans attempted to tame (Dorondel, Şerban, and Cain 2019; Lahiri-Dutt and Samanta 2013; Skelton 2017). The surge of industrial capitalism forced rivers to wear narrow jackets of dams and levees, "corrected" the water flow, and used water energy for mechanized production (Barca and Bridge 2015). Wetlands—which were deemed as fetid territories, a habitat of mosquitoes in which malaria reigned—were repugnant to the state as unproductive landscapes that could not contribute to the capitalist economy. Such land was therefore drained and reclaimed for agriculture. Through technological assault marshlands and rivers were transformed into thriving economies based on industrialized agriculture, fish farms, and hydropower production (Blackbourn 2006; Johnston 2007; Pritchard 2011).[1]

Cities also benefited from the exploitation of rivers and wetlands. Globally, many large cities have thrived at the expense of marshes and rivers; the two geographically and culturally distinct kinds of landscapes have had a historically intertwined relationship. Urbanization and industrialization were factors that drastically rearranged the spatial, economic, and biophysical relations between people, rivers, and wetlands (see, for instance, Giblett 2016; Clifford 2017; Knoll, Lübken, and Schott 2017; Way 2018). As Stefania Barca and Gavin Bridge (2015: 368) put it, across North America and Europe (including Russia) rivers and their valleys—and, we would add, marshlands—became places where "industrial capitalism was born and raised" and a new ecological order forged.

The introduction of scientific methods and technological development had radically changed forests and agriculture (Muir 2014). The now famous German School of Forestry began to calculate the volume of produced wood mass in order to maintain a balance between wood exploitation and forest plantation, and implemented a utilitarian vision of the forest with same-age, evenly aligned monoculture trees (Lowood 1990; Peluso 1992; Scott 1998). The forest turned from unregulated, wild "nature" to an "artificial work of forestry" (Radkau 1996: 69).[2] Equally important, forestry started to become institutionalized and bureaucratized and these new trends spread all over the globe as part of colonial domination (Vandergeest and Peluso 2006a, 2006b). In fact, colonies often served as experimental spaces for these new practices. In agriculture, the steam engine and mechanized equipment such as mechanical threshers, sowers, and grain binders, to name a few, radically changed the production and labor process. The terrestrial and maritime transport revolution completely changed the agricultural markets as products were more easily transported over great distances (Mazoyer and Roudart 2006).

All the developments outlined above are well-known to environmental historians. Numerous scholars have explored transformations of the environment in North America and Western Europe since the nineteenth century. We know far less about how Eastern Europe, a kaleidoscope of nation-states, histories, and cultures, went through the same historical processes. With the exception of Russia, which was under closer scrutiny from this perspective (Weiner 1988,1999; Josephson et al. 2013; Moon 2013; Zeisler-Vralsted 2015), the specific regional history of Eastern Europe in the modern period has been less explored in terms of the many connections between the state, the transformation of the economy, and nature.

This volume not only fills a gap in the literature but also makes two interlinked claims. First, transforming nature in Eastern Europe was a significant part of national projects. We understand "national projects" as concerned with building a modern state and a national economy that reflect contemporary developments. We contend that planners, experts, and bureaucrats were the agents of both change and the edification of state infrastructure in the territory. The second claim of the book is that the quest for economic development was carried out by means of controlling and transforming nature. Each historical period—from roughly the 1850s to the post-1989 period—had its own apostles of development, with new economic visions and projections, new plans and new ways to implement them. The national project carriers borrowed discourses about nature, ideas, and technologies of nature governance from those countries that had demonstrated previous experience in putting nature to work in improving the national economy. We show that East European states have not perceived a conflict between the national projects and taking inspiration from neighboring countries and the western world (which had been more successful in taming nature). On the contrary, the construction of a state and a national economy would have been lost without an inspirational model.

Our analysis suggests a historical model in which the "West" had not ended up in the "East" in order to exploit its natural resources in a colonial manner. We show that Eastern states' interest in developing their economy converged with the interest of Western states in transferring experts, technologies, and ideas about nature. This allowed Western institutions, state or private, to invest in various infrastructural projects for transforming nature and gaining financially in the process. The point where the two interests meet is the amalgamation of ideas about nature and developing local economies. The meeting point also reflects the unequal power relations between Eastern and Western countries throughout the twentieth century, which were enacted both directly and indirectly. In the field of economic development, Western technology and knowledge were seen as the most advanced, which created a strong motivation among Eastern countries to reject ideas that originated elsewhere. Disentangling this process and highlighting its main drivers, mechanisms, and outcomes is the aim of this book.

To demonstrate how these two claims come together, we explore the role of state planners, experts, and bureaucrats as carriers and implementers of changes in the natural world, their professional trajectories and actions, and their visions and ideals for a modern state and technologized nature in Eastern Europe. We show how ideas about the capitalist economy and nature, and the scientific knowledge and technologies necessary to implement new ideas, were transferred from Western Europe and blended with local ideas, plans, projects, and policies. Most states in Eastern Europe, even the empires that dominated the zone, became obsessed with the project of the domestication of the natural world, which they saw as the only way to modernize their economies. Eastern states' interest in economic development coincided with Western European interests to find new markets for their products and financial investments. The new nation-states that appeared in the second half of the nineteenth century were greedy for development after centuries of imperial economic and political domination. Furthermore, the Central Eastern European empires—Ottoman, tsarist, and Habsburg (from 1867 Austro-Hungarian)—were themselves in quest of economic, technological, and military development (Grandits, Judson, and Rolf 2020).

To include both nation-states and empires in their mutual pursuit of economic development and the domestication of nature, we engage the concept of K. Sivaramakrishnan and Arun Agrawal's (2003) "regional modernities." They show that a region—which could be a nation-state or a multinational formation—is the place where developmental policies and associated projects occur, as we also see in Eastern Europe.

A New Ecological Order deals with a wide region that encompasses both nation-states that appeared in the second half of the nineteenth century and the above-mentioned regional empires. There are thus chapters that look at travelers' discourses that dehumanize populations and anthropomorphize forests in Republican Turkey, and explore scientific innovations that attempt to domesticate muskox in Siberia or construct dams in the former Soviet Republics of Tajikistan and Kyrgyzstan. These chapters communicate perfectly with those focusing on countries considered indisputably Eastern European for at least two reasons. First, historically it would be impossible to separate the three empires from nation-states such as Bulgaria, Serbia, Poland, Romania, and Ukraine. At least the modern history of these states, as we will briefly point out in the following section, is intrinsically linked to the evolution and fate of the three empires, and for long periods their histories were intertwined to the point of merging. The justification for including Far North Russian territories, Central Asia, and Asia Minor as part of "Eastern Europe" may go even further back, all the way to prehistory, as Chris Hann (2012, 2015, 2016) and Jack Goody (2015) have posited. They refer to the exchange of ideas, technologies, goods, and people across a vast landmass they call Eurasia, which encompassed both the agrarian empires and the pastoral nomads of Central Asia. Second, socialism created a civilization bridge that linked the social, economic, political, and aesthetic similarities of Central and Eastern European countries to Central Asian states (Hann 2016). Throughout this vast area, over large parts of the twentieth century, socialism was the modernizing ideology and political economy that changed both the economy and society (Pine 2007) of places that had previously been quite different. For countries that experienced socialism, current social, economic, and political processes cannot be explained without recourse to socialist economic, political, and ecological conditions (Stenning and Hörschelmann 2008; Feaux de la Croix and Roberts, chapter 7 in this volume; Vladimirova, chapter 10 in this volume).

The concept of regional modernities also allows us to consider developmental policies, discourses, and practices not only found at different levels of government and scales but also as promoted by both institutions and individuals (Sivaramakrishnan and Agrawal 2003: 14). The way developmental discourses and practices were experienced in Eastern Europe shows that they were not imposed by Western centers of power but by the nation-state itself, which was interested in promoting such development as a way to strengthen its power and to legitimize its presence at the local level and on the international stage. In order to realize the modernization metanoia, the nation-states and empires of Eastern Europe commissioned experts (such as engineers, agricultural engineers, geographers, biologists, foresters, and architects) to implement changes and promote ideas about the use of nature for economic development. The relationship between state and experts was not unidirectional. As Ágotá Ábran (chapter 4 in this volume) shows, when the state does not realize the potential of a certain natural resource, it is the experts who recommend political and economic intervention (see also Kohlrausch and Trischler 2014).

The reader may rightfully ask what good another regional perspective on environmental history would do when we know that environmental issues are intrinsically transnational and that "we live in one world in an ecological—environmental—sense" (Vogt 1948: 14–15). While this is true, different regional histories still matter. Does the technoeconomic power—that is, the ability of the state to rearrange nature and society in order to demonstrate its power—of, say, the (semi)colonial and postcolonial Egyptian state function in the same way as in a country that was never a colony?[3] As Timothy Mitchell (2002) brilliantly shows, the hydraulic works and damming of the Nile in Egypt, designed and implemented by British experts, was carried out to suit the Egyptian state, the local elite, and the metropole. In Eastern Europe, the new nation-states liberated from the embrace of the Ottoman, Russian, or Austro-Hungarian Empires were the sole commissioner of such works. In some cases, experts who had worked in African colonies were invited to use their scientific and technological knowledge to build agricultural infrastructure in Eastern European countries, as we will show below. In other cases, local planners and experts who studied abroad brought practices, ideas, technical expertise and scientific knowledge from Western European institutions and used them to reshape the local environment and society. This book delves into the various ways in which national planners, experts, and bureaucrats—all representatives of the state—contributed to transforming nature in the interest of development.

We show that despite radical changes in political regimes from the end of the nineteenth century to the present, these plans share common features and reflect the obsession with "development"—perhaps the buzzword that characterizes modern and contemporary Eastern Europe.

The next section provides a short historical view of Eastern Europe, a sort of descriptive historical map of the region showing the shattering of empires and the rise of nation-states, and the installation and fall of socialist regimes in most Eastern European countries. Each historical stage cried out for urgent development and used its own language to portray the need for the transformation of nature and the installation of a new ecological order.

A FLUID REGION

The book guides the reader through territories that were either engulfed by ruling empires (the Habsburg, Ottoman, or tsarist/Soviet/Russian) or that belonged to disintegrating empires. The end of the nineteenth century marked powerful political movements against the three empires that still remained in place. The unequal development of a capitalist economy and political autonomy in different regions led to the emergence of an autochthonous political and technocratic elite (Ash and Surman 2012; Ábrán in this volume). It was precisely this elite that would constitute the roots of future nation-states.

The nation-state political model emerged in Southeast Europe under several favorable auspices. First, these political ideas were brought from Western Europe, generally by young elite who had been exposed to them in universities. The future Turkish, Bulgarian, and Romanian political elites studied at universities in France, Germany, and Austria during the nineteenth century. When they returned home they had completely changed: from the clothes they wore to the sociopolitical ideas they promoted, the young students entered into conflict with the previous generation who had financially supported them. The local political elite, the Moldo-Walach boyars or the *chorbadzii* in Bulgaria (winners of the economic reforms imposed by the Ottoman Empire) never thought that their own sons would turn the knowledge and experience acquired in Western Europe against their own politically-conservative visions. In Poland, the political elite living abroad elaborated plans for freeing their country and unifying it following the Western European model. By "Western European model" we mean the liberalization of political life, an economic model based on industrialization, the rapid development of science and technology, and a type of societal behavior much less constrained by traditional and conservative ideas.[4] Second, the Ottoman Empire went through a process of economic and military uncertainty in the nineteenth century, which ultimately led to its disintegration after World War I. The empire's weak political and military position allowed the countries of this region (including Greece, Bulgaria, Serbia, and Romania) to become first autonomous and then independent states, and to enlarge their territories. Serbia, for instance, shaken by two major riots in 1804 and 1815, demarcated an autonomous territory within the Ottoman Empire around Belgrade city.[5] Greece gained autonomy over a small territory—which included the Peloponnese, Corinth, Athens, and the Aegean islands—after a violent revolution in 1821, and became an independent kingdom in 1830. After the Russo-Turkish War (1877–1878) that culminated in the peace treaties of San Stefano and Berlin, Romania and Serbia became independent whereas Bulgaria became an autonomous principality. World War I put an end to the Ottoman and Austro-Hungarian Empires, out of which several countries emerged—Turkey, Austria, Hungary, Poland, and Czechoslovakia—while Transylvania become part of the Romanian state (Wandycz 2001; Bideleux and Jeffries 2007). Different regions of the now defunct Austro-Hungarian Empire were included in the new Kingdom of the Serbians, Croatians, and Slovenes (which became the Kingdom of Yugoslavia from 1929) or became part of the Soviet Union (Magocsi 2010).

The elated political elite of these new countries adopted political and administrative reform programs that were inspired by Western European countries. At the same time, the technocratic establishment engaged more and more in a nationalist discourse and actions meant to consolidate the nation-state in the interwar period. The economic development, expansion of a national economy, and construction of the state administration, bureaucracy, and large infrastructure within the national territory were all part of this elite's vocabulary and political action (Grandits, Judson, and Rolf 2020). Inevitably, the natural environment was targeted for change in the name of the nation-state's prosperity (Özkan 2013, 2018).

After World War II the geographical map of Eastern Europe was split between the socialist regimes, under the heavy influence of the Soviet Union, and the capitalist economies of Turkey and Greece. However, Turkey borrowed its first foreign aid from the Soviet Union in the 1920s and also certain ideological aspects of the Soviet model such as the emergence of the "new man" or the use of the elites as vanguards. Turkey's leaders defined Kemalism as a third way between capitalism and socialism, and in many ways the closed, protected economy (which lasted until the 1980 military coup) was very similar to socialism in terms of citizens' everyday experiences (B. Jelavich 1999: 406–12; Zürcher 2017: 223–80; Ahmad 2014: 95–148). The historian Paul Josephson (2016) shows how the socialist countries in Central and Southeast Europe followed the blueprint of the Soviet Union to convert nature into a more productive landscape through building new infrastructure, taming rivers, building thousands of kilometers of levees and irrigation systems, and organizing intensive agriculture and the introduction of new crops.[6]

Giant dams were constructed, such as the Iron Gates dam in 1971 between Romania and Serbia, or were planned to be built, such as Gabčíkovo between Hungary and Czechoslovakia. After Stalin's death (1953) most Southeast European socialist states removed themselves from Soviet control, and some adopted more liberal policies.

Like dominoes, the socialist regimes collapsed one after another between 1989 and 1991. A new economic model was adopted and as the end of socialism coincided with the peak of global neoliberalism, the latter became the most persuasive model (Ban 2016). Some countries, such as Romania and Bulgaria, were more sluggish in adopting new economic and capitalist political models whereas Hungary, Poland, and the Czech Republic implemented the "reforms" more promptly. Rapidly or slowly, these transformations unfolded under the supervision of international institutions such as the International Monetary Fund (IMF) or the World Bank, which deployed "econolobbyists" (Wedel 1998)—fly-in design policies, fly-out international experts—who contributed massively to the economic reforms (Creed and Wedel 1997).

TRAVELING EXPERTISE, TECHNOLOGIES, AND IDEAS

We claim that transferring knowledge, experts, and ideas about nature from more developed countries was part of the national project since its inception. The interwar period produced local and international entanglements of ideas and plans that put nature to work for economic development. Socialism as an "infrastructural modernization" project (Dorondel and Posner forthcoming) had complicated the exchange of expertise, technologies, and ideas of a developmental nature. We further examine the contact points and negotiations between Western and Eastern projections of nature and economy, as well as the avenues on which such expertise traveled.

In Eastern Europe, the state played a major role in creating expertise and commissioning work that was meant to develop a modest, ruralized economy and to build much-needed infrastructure (Gătejel and Kochanowski 2020: 131). The reforms envisaged by the newly emerged nation-state in the late nineteenth and early twentieth centuries had multiple targets. First, as with any other "high modernist state" (Scott 1998)—and we should say that the states of Southeast Europe all emerged as high modernist—control over the territory was the first step (Ćorović, chapter 1 in this volume). As James Scott (1998) has shown for other areas, multiple local property arrangements, a lack of maps and clearly delineated property boundaries, and various historical practices of cultivating land, exploiting forests, fishing, and husbanding animals were not at all pleasing to the newly emerged nation-state. If the state is, as Scott (2017: 139) has defined it, "a recording, registering, and measuring machine" then the myriad of local practices had to be unified and made legible. A second reason for reform was to transform rural economies into capitalist economies and to radically transform production. Integration into the world economic system, or at least the European one, was an immediate target of the state. The Danubian Principalities had already experienced a first "internationalization" and boost to its agriculture in 1829, when, following the Russo-Turkish War, the Treaty of Adrianople allowed the liberalization of the cereals trade. Until then, Walachia and Moldova had been obliged to sell most of their cereal production at a preferential price to the Ottoman market (Palairet 1997; Demeter 2017). The 1829 treaty not only gave wings to the agricultural economy in the two principalities but also led to rapid urbanization along the Lower Danube. New cities appeared on the banks of the lower part of the river, accompanied by an entire infrastructure (including small ports and roads linking the Danube to the interior of the countries) for export of cereals to the world (Hardi 2013). The nation-state was interested in developing agriculture and in making loyal citizens of the peasantry, which was more often under the control of the local elite than the state (Dorondel and Şerban 2014). To accomplish these urgent tasks, the states of Eastern Europe needed technologies for "adapting" nature to a modern economy, and a corps of planners and experts to plan and implement the changes.

Western European expertise traveled east via two main avenues. One ran between western economic stakes in the juggernaut development projects of both states and private companies and the eastern states' interests in such projects. A good example here is engineering works that commenced on the Lower Danube. The mighty gorges of the Iron Gates—which created a bottleneck on the Danube with its rapids, reefs, narrow channels, and shallows "no more than 18 inches of water, with a rush like the race of a mill-stream" (Ardeleanu 2009: 194)—were only possible to navigate when the waters were high enough. This was a nightmare both for the captains of the steamers and for the Austrian company that from 1829 organized the passage between Vienna and Constantinople (Ardeleanu 2009). These new economic opportunities, including the transportation of people and commodities along the Danube, motivated Austrian private investors and state high officials to start the engineering work necessary to regulate the Iron Gates in 1834 (Gătejel 2016). Taming the river to allow the steamers' safe passage was seen as a victory for Western civilization (Quin 1836: 106). Engineering at the mouth of the Danube by the European Commission of the Danube (ECD) was another example of technology transfer and ideas of how to change natural features that were poorly suited to capitalist economic relations. Founded in 1856 by the Habsburg empire, England, France, Germany, the Ottoman and Russian Empires, the ECD aimed at rectifying the Lower Danube and its mouth through hydrotechnical works (Ardeleanu 2020). A tamed Danube would facilitate the cereals trade and represent the export of the Western civilization model to the European periphery, commonly perceived as backward (Yao 2019). After 1880 the new riparian nation-states steadily replaced the Western powers as prime actors in governing the Lower Danube and voiced their own economic and developmental interests (Ashcraft 2011; Hajnal 1920).

Inviting foreign experts and private companies to plan and implement infrastructural projects was another facet of this strategy. For instance, British companies were invited to finance the railways which linked Cernavodă to Constanţa and Ruse to Varna (cities in the Ottoman province Dobroudja). Part of the deal was that they should provide their own technical knowledge and engineers (Karpat 1986: 290; Martykánová 2010: 25). Another project, the Baghdad Railway, which represented one of the most important projects in the Ottoman Empire, was built mostly with German expertise (Martykánová 2010: 221). In 1910 the Romanian government invited William Willcocks, a British engineer who worked in different parts of the world, to draw up plans for an irrigation canal to water the arid yet potentially fertile plains of southern Romania. As a true predecessor of contemporary international experts from the World Bank or the IMF, Willcocks had previously contributed to the construction of the Aswan Dam in colonial Egypt at the end of the nineteenth century, and then came to the Ottoman Empire to help build irrigation systems. The interwar period found him in India assisting in the construction of irrigation systems.

Through their knowledge and expertise these international experts linked areas that were worlds apart by spreading technologies and ideas about economic development and the need to transform nature for development purposes. States seeking to transform urban areas took great pride in importing modern architectural styles, modern planning, and sewage and electricity infrastructure (Kaika 2006: 281; Inal 2011; Ćorović in this volume). The aim was to transform an amalgam of constructions, roads, and land uses into a modern city, thus contributing to the symbolic power of the state (Ferhadbegović 2020). Ferenc Reitter (1813–1874), an engineer, urbanist, and planner, is another example of a traveling expert. After carrying out successful projects in London, Paris, and Vienna he was hired in the mid-1860s to systematize the urban infrastructure of Budapest (Sisa 2013: 120; Tamáska 2018: 568–69). Greece and Bulgaria asked for the assistance of western planners to reshape Athens and Sofia. Western companies such as Lindley Brothers pioneered the construction of urban sewage systems from western cities to cities in Poland, Romania, and even Moscow (Kohlrausch and Trischler 2014: 41). In the interwar period functionalism and modernism were imported and adapted to local needs: with the assistance of Western urbanists, besides the large Greek cities of Piraeus, Thessaloniki, and Athens, 135 small cities of around 5,000 inhabitants were redesigned in the 1920s–1930s. These small cities needed refurbishing after the Balkan Wars (1912–1913) and the devastation of World War I, but also had to cope with the displacement of large masses after the exchange of territories in the Balkans and Asia Minor (Yerolympos 1993: 241).[7]

Other changes in the natural environment such as the construction of levees along the Lower Danube in Bulgaria and various rivers in Greece and the drainage of the marshlands served the same purpose: settling the surge of immigrants after the Balkan Wars and World War I (Dragostinova 2006; Vlachos, chapter 2 in this volume). These projects were supported technically and financially by the League of Nations (Şerban and Dorondel, chapter 5 in this volume).

The guest experts were not always successful in their endeavors. In 1860 the Ottoman central officials granted permission to exploit 100,000 hectares of forest on the empire's territory, in what is now southern Bulgaria, in order to finance a giant infrastructural governmental program: the construction of a railway to link Istanbul and Vienna (Dursun 2007: 306–26). The project failed due to local peasant resistance and the lack of adaptation of new management practices to local customs. In Greece, after Otto from the dynasty of Bavaria was crowned king in 1832, an ambitious plan of forest management and exploitation was conceived following the German model, with all personnel, from engineers to foresters, of German origin (Seirinidou 2017). For many reasons, including peasant resistance to the imposition of new foreign forest exploitation practices, all personnel were withdrawn after the mid-1840s and replaced with local foresters who had limited experience (Kostov 2016).

A second avenue for transferring expertise from Western Europe was through the formation of local technical cadres. In the second part of the nineteenth century, the countries of Eastern Europe began developing native schools of engineers, urbanists, architects, agronomists, and foresters—most of the founders having being trained in western academia. This was a steady but essential process. The new universities and schools developed not only in the newly emerged nation-states but also in the empires that dominated the region. For instance, in Ottoman Egypt beginning in 1820, French and English experts contributed to the development of civil schools for the training of water engineers (Martykánová 2014: 17–21). The Ottoman Imperial School of Naval Engineering was established in 1773 (Kohlrausch and Trischler 2014: 30), whereas in 1809 in tsarist Russia an institute for training road construction engineers—inspired by the Parisian École nationale des ponts et chaussées—was established. Also, in the western peripheries of the tsarist empire, from Finland to Ukraine, a network of polytechnic schools was founded to provide desperately needed experts and technical cadres (Gouzévitch,

Cardoso de Matos, and Martykánová 2017: 248, 270–73). In Poland, the Polytechnic University in Warsaw was opened in 1915 after only three months of German occupation of the city. The new establishment was immediately perceived not only as compensation for the long Russian domination but embodied hope for national construction through technical expertise (Kohlrausch and Trischler 2014). In what was still Ottoman Bulgaria a corps of local agronomists trained in Western and Central Europe emerged in 1860–1870, just before the appearance of the modern Bulgarian state (Angelova 2019).

During the interwar period the nation-states "nationalized" and strengthened their technical and scientific educational institutions. This period represented the finest "hour of the experts" as Martin Kohlrausch (2015) puts it. Experts and scientists were the arms of the state in establishing political control, organizing territory, developing local economies, and shaping the national landscape.

This trend continued after 1945 when, except for Greece and Turkey, all other countries of the region came under the political and economic control of the Soviet Union.[8] This means that they entered a political-economic system dubbed "the greatest state planning scheme the world has ever seen" (Dodd 1933: 34; Cook, Ward, and Ward 2014). Planning profound economic, political, and social changes had radical repercussions for the natural world as well. Again, the state—this time socialist—actively sought fresh technologies that could reinforce the new directions of economic development. In the decade after 1945 the Soviet Union exported technical expertise to other countries of the socialist bloc. Yet the relationship and technological flow between the socialist and capitalist blocs was not totally interrupted. There were contacts and transfers of knowledge, experts, and development ideas between the Soviet Union and Western Europe (Cook, Ward, and Ward 2014). When the socialist bloc experts lacked knowledge in a certain domain they looked to Western countries; the Soviet Union appealed to Finland's experience, expertise, and technology to manage forest harvesting, which was poorly managed in Russia, resulting in a great loss of timber (Kochetkova 2018). Nevertheless, the socialist countries tried to limit the import of western technologies and ideas as they attempted to decolonize their economies and societies (Mark, Kalinovsky, and Marung 2020). Conversely, the socialist countries exported technology to capitalist countries such as West Germany, Italy, the United States, and France. The United States, for instance, imported key technologies in industries such as metallurgy, mining, and technomedicine; in the 1970s Great Britain imported open-end spinning-machine technology from Czechoslovakia; and West Germany, Italy, and Spain imported technologies for the sugar industry and the processing of sulfuric acid from Poland (Kiser 1976; Freeze 2007; Maciejewicz and Monkiewicz 1982). Socialist countries thus represented serious competitors for Western countries' economic interests. For instance, the Soviet Union exported dam-building expertise to Egypt, and countries including Romania, Bulgaria, and Hungary played a significant role in industries such as petrochemistry and urban architecture in Northern Africa and the Middle East (Reynolds 2017; Mark, Kalinovsky, and Marung 2020; Ghettas 2018: 79; US Congress, Office of Technology Assessment 1984: 17, 474; Stanek 2020).

The crushing of socialist regimes in Eastern Europe created new states that emerged from the ruins of their socialist predecessors, to twist David Stark's (1996: 995) famous expression, and new policies of redefining the relationship between state-society-economy and nature appeared.[9] Within this context, a shift has occurred: whereas before 1989 the professional bodies and branches of government dealt with policy transfer, now consultancies, think tanks, and nongovernmental organizations (NGOs) tend to play a more central role (Pojani and Stead 2018). One domain in which knowledge and ideas about environment were transferred through NGOs is the relations between nature, society, and the state. The socialist technological havoc linked to intensive industrialization and brutal exploitation of natural resources produced appalling environmental consequences visible even under socialism (Pál 2017). Thus, the West provided conservation technologies, ideas about nature, and expertise to improve the state of nature. For instance, all attempts to restore areas of the Lower Danube in Bulgaria and Romania after the 1990s were made with financial support, knowledge, and experts from the World Wildlife Fund, Germany or the Netherlands (Vinke-de Kruijf et al. 2012).

Jürgen Kocka and Heinz-Gerhard Haupt(2018) reflect on the ways in which transfers of all sorts—economic, political, cultural, and ideological from one country to another, from one continent to another—unfold. They show that this flow of knowledge, experts, technologies, and ideas is not a simple movement from the West to the rest of the world, but was also negotiated and shaped by the receiving countries. Sometimes knowledge and experience did not even originate in the West but in a neighboring country, as Şerban and Dorondel show in this volume. Thus, the regional modernities—which in our view include the complex relationship between local and international experts, traveling ideas about nature and infrastructural projects, and technologies geared to appropriating nature—challenge the simplified "West created, others received" logic of development.[10]

PLANNERS, EXPERTS, AND BUREAUCRATS AS AGENTS OF CHANGE

We distinguish between planners, experts, and bureaucrats only for analytical purposes. The planners are those experts who are also intricately linked to political power. Their authority is not just scientific but also political, and they often occupy high positions within the state (see, for instance, Dostalík 2017). They are able to circulate ideas and plans regarding a domain of activity, a space or a territory. Sometimes these plans can be traced to a particular person or a group but there are cases in which planning is an anonymous activity and no individual can be tracked down as the mastermind behind the plan. Experts—and we specifically refer here to technoscientific experts—are both trained professionals in a science, from biology and geography to agronomy and engineering of all sorts, but also professionals who are not necessarily scientists but draw on scientific principles in their work (Kohlrausch and Trischler 2014: 8).[11] There are similarities and differences between planners and experts although often scholars do not discriminate between them (see, for instance, Kohlrausch and Trischler 2014; Kohlrausch 2015; Dostalík 2017). For instance, both categories are endowed with political or administrative power and work mostly for state institutions that give them the authority to make decisions in specific domains. They possess knowledge on matters that are either inaccessible or hard to grasp by laypersons (Carr 2010). Both categories solve problems that are both theoretical and practical (Dostalík 2017). They not only *know* things but they *act* based on their knowledge and authority (Carr 2010; Vandendriessche, Peeters, and Wils 2016). Yet there are differences as well. Often experts are involved in international networks and attend international scientific conferences, while planners are more usually state-oriented (Kohlrausch and Trischler 2014; Kohlrausch 2015; Dostalík 2017). Finally, by bureaucrats we mean state employees who have expert knowledge and authority granted them by the state but are expected to maintain the status quo in a certain domain rather than act to change it, as experts would do. Foresters, water technocrats, levee and agricultural technicians, and so on, are the type of bureaucrats we discuss in this book as well as their actions and outcomes (Blavascunas, chapter 9 in this volume, Janáč, chapter 6 in this volume). They are part of an administrative body, usually managing a natural resource (such as forests) at the local level, making it legible to higher offices. They are the contact point between people and the state. Bureaucrats have power that they wield in certain domains through authority bestowed on them by the state (Heyman 2004).[12] Similar to planners and experts, bureaucrats are part of the state's mechanisms of control, organization, and implementation of projects—they are one of the faces of the state and apostles of a new ecological order.

ORGANIZATION OF THE BOOK

The first section, "Planning Territory," explores the way that nation-states try to consolidate power in the territory by planning and organizing it. Dragana Ćorović (chapter 1) analyzes the transformation of Belgrade's changing landscape from the end of the nineteenth century up to World War I. This process comprised various actions, such as the establishment of a land registry, city planning, educational reform, and the construction of modern infrastructure systems. George Vlachos (chapter 2) examines the monumental engineering project carried out in a southern Macedonian wetland at the beginning of the twentieth century in relation to its ethnic and national significance. By eradicating the local population's agricultural practices that were considered deeply rooted in Ottoman and Slavic systems of production, the Greek agronomists transformed the swamps into agricultural fields more suited to modern agriculture and a modern state. The aim was to replace a Slavic population with a Greek-speaking one resettled from Asia Minor. This chapter also makes an excellent link between the engineering of nature and the population exchange between two Balkan countries—a historical process characteristic of the modern Balkans. Stefan Dorondel and Anna Olenenko (chapter 3) explore the history of a Dnipro wetland in Southern Ukraine under two political regimes: tsarist and Soviet. They show that both regimes perceived wetlands as a wasteland that required permanent improvements. The chapter looks explicitly at political and economic ideas concerning the definition of productive land, the scientific knowledge involved in transforming the wetland, and the technical methods required to organize a territory.

The second section, "Nature, Economy, and Experts," reflects on the role of experts in transforming nature for economic development purposes. Their mission is to assist the state not only in rearranging nature to follow an economic rationale—whether capitalist or socialist—but also in escaping the straitjacket of the national market and connecting with regional and international markets. Ágota Ábrán (chapter 4) presents an account of scientists who advocate the conversion of hitherto unused plants into productive ones and encourage the state, the market, and the villagers to view certain weeds as profitable medicinal plants. Alongside the conversion of weeds into commodities it was necessary to convert villagers into economically conscious gatherers and cultivators of medicinal plants. Stelu Şerban and Stefan Dorondel (chapter 5) explore the politics of damming the Lower Danube in Bulgaria and Romania and the role of experts in designing and implementing the state's plans for developing the region between 1900 and 1940. The chapter highlights different strategies to drain the floodplain in order to build an irrigation system and national grid and describes the way the two states incorporated European experts, knowledge, and technologies into these projects. Jiří Janáč (chapter 6) looks at environmental change in Czechoslovakia during the twentieth century from the perspective of a specific group of national technocratic elites—the so-called hydraulic bureaucracy, or hydrocracy. As water came to be regarded primarily as a national natural resource and a vital economic domain, hydraulic expertise became a strategic instrument in the hands of the nation-state. The consolidation of the state was related to the construction of its water infrastructures. Jeanne Féaux de la Croix and Flora Roberts (chapter 7) show the connections between environmental and social engineering through the prism of large-scale water infrastructure projects in Tajikistan and Kyrgyzstan from 1940 to the present day. Drawing on archival research, oral histories, and ethnographic fieldwork, the chapter traces the "biographies" of five large dam-building projects across manifold economic and political changes. Appraising the legacy of late Soviet environmental and social engineering, it traces changes in the assemblage of actors, types of expertise, finance, and political decisionmakers associated with water infrastructure.

The third section, "Imaging New Nature," establishes the consolidation of a technostate apparatus that not only organizes and manages the natural resources of a country or region but also imagines nature and projects its role onto local or regional development. Hande Özkan (chapter 8) analyzes representations of Asia Minor's natural landscapes by exploring the distinct discourses of European experts and the Turkish bureaucracy. She reflects on the processes of anthropomorphization of forests and of naturalization of the local population and shows that the young republic's forestry regime was central to the simultaneous projects of state building, subject formation, and nature making. Why and how a Polish forester became a champion of the Polish homeland through his logging activities in the primeval Białowieża Forest is the topic of Eunice Blavascunas's contribution (chapter 9). The transhistorical narrative that unfolds here demonstrates that the state in Eastern Europe can facilitate and extend its expertise regarding natural forests. In Poland, the forester in Białowieża is more than just a member of the bureaucracy whose contribution helps to fulfill the expectations of the state; he is a figure who believes that Polish forests and forestry are national constructions. Vladislava Vladimirova (chapter 10) engages with an ethnographic description of an experiment to domesticate muskoxen within a reindeer herding enterprise in the Evenki village of Surinda. She shows that the scientific process of muskox domestication becomes a field of negotiation between state administration and the Evenki people and an issue of the region's political economy. Finally, Yulian Konstantinov (chapter 11) presents an account in which humans, animals—the golden jackal (*Canis aureus*)—and infrastructural projects are all implicated in the postsocialist political transformations of a Bulgarian village. Konstantinov gives equal weight to local practitioners' and experts' interpretations of local history and to the jackal's attempt to build a relationship of trust and solidarity with the author.

Stefan Dorondel and Helmuth Trischler's epilogue shows that despite various changes in state ideologies over the past century and a half, state practices regarding the "corrections" made to nature remained the same. These corrections brought to light the existence of what were perceived as "good" and "bad" natural environments, which are also linked to transformations of the economy and political subjects.

A New Ecological Order discusses a "new nature" created by the state's intervention through its agents of change as part of regional modernities in Eastern Europe. It points to the intricate link that exists between global ideas, technologies, and experts, all of which aim to "improve" nature and local politics, economic interests, and natural environments. This link continued unabated throughout three distinct historical periods: the period of nation-state formation at the end of the nineteenth century, the socialist period, and the postsocialist period. In all three periods state planners, experts, and bureaucrats have attempted to impose global technologies and ideas on nature to serve local economic interests.

NOTES

1. For an exhaustive literature on rivers and wetlands, see Schönach (2017).

2. Although forest exploitation has been regulated since premodern times, the complexity of management techniques has reached a completely new level since the eighteenth century.

3. Egypt was not a colony per se but during various periods it was under Ottoman and British rule. Its special status, with the Egyptian state "retaining great control over internal affairs," led Hanan Kholoussy (2010: 688) to call Egypt a "semi-colony."

4. For instance, in the nineteenth century, tsarist Russia forbade the construction of industrial factories, which it feared would bring revolutionary ideas (Acemoglu and Robinson 2012).

5. This paragraph is based on Jelavich and Jelavich (2000). See also Livezeanu and Klimó (2017).

6. For further details concerning agriculture, see Olšáková (2016). For nature conservation policies, see Brain (2010). For building dams, see Cretan and Vesalon (2017) and Fitzmaurice (2018).

7. The dramatic and sudden growth of urban populations in the interwar period obliged Eastern European governments to develop urban and relocative policies in order to cope with the new situation (Müller 2020: 181–85).

8. However, Greece and Turkey are not completely disconnected from the socialist political economic model. For instance, until the opening of markets in 1980, Turkey's state-led economy was organized around five-year plans that were quite similar to the socialist economies (Celâsun and Rodrik 1989).

9. David Stark (1996: 995) argues that in the postsocialist context institutions rebuilt "*not on the ruins* but *with the ruins* of communism."

10. We thank Hande Özkan for this description.

11. For various other types of expertise, see Carr (2010).

12. For an anthropology of bureaucracy, see Hoag (2011).

REFERENCES

Acemoglu, Daron, and James A. Robinson. 2012. *Why Nations Fail: The Origins of Power, Prosperity and Poverty.* New York: Crown.

Ahmad, Feroz. 2014. *Turkey: The Quest For Identity.* London: Oneworld.

Angelova, Milena. 2019. "*The Transfer of Modern Agricultural Knowledge among the Bulgarians in the Danube Province (1860s–1870s).*" In I. D. Michailidis and G. Antoniou, eds., *Institution Building and Research under Foreign Domination: Europe and the Black Sea Region (Early 19th–Early 20th Centuries).* Thessaloniki: Epikentro, 93–106.

Ardeleanu, Constantin. 2009. "From Vienna to Constantinople on Board the Vessels of the Austrian Danube Steam-Navigation Company (1834–1842)." *Historical Yearbook* 6: 187–202.

Ardeleanu, Constantin. 2020. *The European Commission of the Danube, 1856–1948: An Experiment in International Administration.* Leiden: Brill.

Ash, Mitchell G., and Jan Surman. 2012. "The Nationalization of Scientific Knowledge in Nineteenth-Century Central Europe: An Introduction." In M. G. Ash and J. Surman, eds., *The Nationalization of Scientific Knowledge in the Habsburg Empire, 1848–1918.* Basingstoke: Palgrave, 1–29.

Ashcraft, Catherine Marie. 2011. "Adaptive Governance of Contested Rivers: A Political Journey into the Uncertain." PhD diss., Massachusetts Institute of Technology.

Ban, Cornel. 2016. *Ruling Ideas: How Global Neoliberalism Goes Local.* New York: Oxford University Press.

Barca, Stefania, and Gavin Bridge. 2015. "Industrialization and Environmental Change." In T. Perreault, G. Bridge, and J. McCarthy, eds., *The Routledge Handbook of Political Ecology.* London: Routledge, 366–77.

Bideleux, Robert, and Jan Jeffries. 2007. *A History of Eastern Europe: Crisis and Change.* London: Routledge.

Blackbourn, David. 2006. *The Conquest of Nature: Water, Landscape and the Making of Modern Germany.* New York: W. W. Norton.

Brain, Stephen. 2010. "The Great Stalin Plan for the Transformation of Nature." *Environmental History* 15(4): 670–700.

Carr, Summerson E. 2010. "Enactments of Expertise." *Annual Review of Anthropology* 39: 17–32.

Celâsun, Merih, and Dani Rodrik. 1989. "Turkish Economic Development: An Overview." In J. D. Sachs and S. M. Collins, eds., *Developing Country Debt and Economic Performance, Volume 3: Country Studies—Indonesia, Korea, Philippines, Turkey.* Chicago: University of Chicago Press, 617–29.

Cioc, Mark. 2002. *The Rhine: An Eco-biography, 1851–2000.* Seattle: University of Washington Press.

Clifford, Jim. 2017. *West Ham and the River Lea: A Social and Environmental History of London's Industrialized Marshlands.* Vancouver: University of British Columbia Press.

Cook, Ian R., Stephen V. Ward, and Kevin Ward. 2014. "A Springtime Journey to the Soviet Union: Postwar Planning and Policy Mobilities through the Iron Curtain." *International Journal of Urban and Regional Research* 38(3): 805–22.

Creed, Gerald, and Janine R. Wedel. 1997. "Second Thoughts from the Second World: Interpreting Aid in Post-Communist Eastern Europe." *Human Organization* 56(3): 253–64.

Cretan, R., and L. Vesalon. 2017. "The Political Economy of Hydropower in the Communist Space: Iron Gates Revisited." *Tijdschrift voor Economischeen Sociale Geografie* 108(5): 688–701.

Demeter, Gábor 2017. *Agrarian Transformations on the Balkan Peninsula (from the Late 18th Century to WWII)*. Accessed January 29, 2017, https://www.academia.edu/30708102.

Dodd, K. S. 1933. "Planning in the USSR." *Journal of the Town Planning Institute* 20(1): 34–53.

Dorondel, Stefan. 2016. *Disrupted Landscapes: State, Peasants and the Politics of Land in Postsocialist Romania*. Oxford: Berghahn Books.

Dorondel, Stefan, and Cristina Posner. Forthcoming. "Infrastructure, Feral Waters and Social Inequality in Rural Romania." *Critique of Anthropology*.

Dorondel, Stefan, and Stelu Şerban, eds. 2014. "At the Margins of History: The Agrarian Question in Southeast Europe." Special issue of *Martor. The Museum of the Romanian Peasant Anthropology Review* 14.

Dorondel, Stefan, Stelu Şerban, and Daniel Cain. 2019. "The Play of Islands: Danube Dynamics and Border Establishment in Modern Southeast Europe (1830–1900)." *Environment and History* 25(4): 521–47.

Dostalík, Jan. 2017. "The Organicists: Planners, Planning, and the Environment in Czechoslovakia (1914–1949)." *Planning Perspectives* 32(2): 147–73.

Dragostinova, Theodora. 2006. "Competing Priorities, Ambiguous Loyalties: Challenges of Socioeconomic Adaptation and National Inclusion of the Interwar Bulgarian Refugees." *Nationalities Papers: The Journal of Nationalism and Ethnicity* 34(5): 549–74.

Dursun, Selçuk. 2007. "Forest and the State: History of Forestry and Forest Administration in the Ottoman Empire." PhD diss., Sabanci University.

Ferhadbegović, Sabina. 2020. "Projections and Representation of Statehood." In W. Borodziej, S. Ferhadbegović, and J. von Puttkamer, eds., *The Routledge History Handbook of Central and Eastern Europe in the Twentieth Century*. Vol. 2: *Statehood*. London: Routledge,1–40.

Ferkiss, Victor. 1993. *Nature,Technology and Society: Cultural Roots of the Current Environmental Crisis*. New York: New York University Press.

Fitzmaurice, John. 2018. *Damming the Danube: Gabčikovo/Nagymaros and Post-Communist Politics in Europe*. New York: Routledge.

Freeze, Karen Johnson. 2007. "Innovation and Technology Transfer during the Cold War: The Case of the Open-End Spinning Machine from Communist Czechoslovakia." *Technology and Culture* 48(2): 249–85, https://doi.org/10.1353/tech.2007.0068.

Gătejel, Luminiţa. 2016. "Overcoming the Iron Gates: Austrian Transport and River Regulation on the Lower Danube, 1830s–1840s." *Central European History* 49: 162–80.

Gătejel, Luminiţa, and Jerzy Kochanowski. 2020. "Transport, Infrastructure and Communication." In W. Borodziej, S. Ferhadbegović, and J. von Puttkamer, eds., *The Routledge History Handbook of Central and Eastern Europe in the Twentieth Century*. Vol. 1: *Challenges of Modernity*. London: Routledge, 129–82.

Ghettas, Mohammed Lakhdar. 2018. *Algeria and the Cold War: International Relations and the Struggle for Autonomy*. London: I. B. Tauris.

Giblett, Rod. 2016. *Cities and Wetlands: The Return of the Repressed in Nature and Culture*. London: Bloomsbury Academic.

Goody, Jack. 2015. "Asia and Europe." *History and Anthropology* 26(3): 263–307.

Gouzévitch, Irina, Ana Cardoso de Matos, and Darina Martykánová. 2017. "La Russie, l'Espagne, le Portugal et l'Empire ottoman: Deux siècles de politiques technoscientifiques à l'épreuve des approches comparatistes." In M. Kleiche-Dray, ed., *Les ancrages nationaux de la science mondiale XVIIIe–XXIe siècles*. Paris: Éditions des archives contemporaines, 239–86.

Grandits, Hannes, Pieter Judson, and Malte Rolf. 2020. "Towards a New Quality of Statehood: Bureaucratization and State-building in Empires and Nation States before 1914." In

Borodziej, Ferhadbegović, and von Puttkamer, *Routledge History Handbook of Central and Eastern Europe*, 2:41–116.

Hajnal Henry, 1920. *The Danube: Its Historical, Political, and Economical Importance*. Hague: Martinus Nijhoff.

Hann, Chris. 2012. "Europe in Eurasia." In U. Kickell, M. N. Craith, and J. Frykman, eds., *A Companion to the Anthropology of Europe*. Malden, MA: Blackwell, 88–102.

Hann, Chris. 2015. "Goody, Polanyi and Eurasia: An Unfinished Project in Comparative Historical Economic Anthropology." *History and Anthropology* 26(3): 308–20.

Hann, Chris. 2016. "A Concept of Eurasia." *Current Anthropology* 57(1): 1–27.

Hardi, Tamás. 2013. "Spatial Structure and Urban Types of the Danubian Areas." *Problems of Geography* 38(1–2): 59–73.

Kocka, Jürgen, and Heinz-Gerhard Haupt. 2018. "Comparison and Beyond: Traditions, Scope, and Perspective of Comparative History." In H-G. Haupt and J. Kocka, eds., *Comparative and Transnational History: Central European Approaches and New Perspectives*. New York: Berghahn Books, 1–30.

Heyman, Josiah McC. 2004. "The Anthropology of Power-Wielding Bureaucracies." *Human Organization* 63(4): 487–500.

Hoag, Colin. 2011. "Assembling Partial Perspectives: Thoughts on the Anthropology of Bureaucracy." *Political and Legal Anthropology Review* 34(1): 81–94.

Inal, Onur. 2011. "Environmental History as an Emerging Field in Ottoman Studies: An Historiographical Overview." *Osmanlı Araştırmaları / Journal of Ottoman Studies* 38: 1–25.

Jelavich, Charles, and Barbara Jelavich. 2000. *The Establishment of the Balkan National States 1804–1920*. Seattle: University of Washington Press.

Jelavich, Barbara. 1999. *History of the Balkans:Twentieth Century*. Cambridge: Cambridge University Press.

Johnston, J. B. 2007. "Défricheur d'eau: An Introduction to Acadian Land Reclamation in a Comparative Context." *Material Culture Review* 66 (Fall): 32–41.

Josephson, Paul R. et al. 2013. *An Environmental History of Russia*. Cambridge: Cambridge University Press.

Josephson, Paul R. 2016. "Introduction: The Stalin Plan for the Transformation of Nature andthe East European Experience." In Olšáková, *In the Name of the Great Work*, 1–41.

Kaika, Maria. 2006. "Dams as Symbols of Modernization: The Urbanization of Nature Between Geographical Imagination and Materiality." *Annals of the Association of American Geographers* 96(2): 276–301.

Karpat, Kemal H. 1986. "The Crimean Emigration 1856–1862 and the Settlement and Urban Development in Dobruca." In C. Lemercier-Quelquejay, G. Veinstein, and S. E. Wimbush, eds., *Passé turco-tatar: Présent soviétique*. Paris: Editions Peeters, EHESS, 275–306.

Kholoussy, Hanan. 2010. "Monitoring and Medicalizing Male Sexuality in Semi-Colonial Egypt." *Gender and History* 22(3): 677–91.

Kiser, John III. 1976. "Technology Is Not a One-Way Street." *Foreign Policy* 23(3): 131–48.

Knoll, Martin, Uwe Lübken, and Dieter Schott, eds. 2017. *Rivers Lost, Rivers Regained: Rethinking City-River Relations*. Pittsburgh, PA: University of Pittsburgh Press.

Kochetkova, Elena. 2018. "Industry and Forests: Alternative Raw Materials in the Soviet Forestry Industry from the mid-1950s to the 1960s." *Environment and History* 24(3): 323–47.

Kohlrausch, Martin. 2015. "The Hour of the Experts? Reflections on the Rise of Experts in Interbellum Europe." In Vandendriesske, Peeters, and Wils, *Scientists' Expertise as Performance*, 67–79.

Kohlrausch, Martin, and Helmuth Trischler. 2014. *Building Europe on Expertise: Innovators, Organizers, Networkers*. Basingstoke: Palgrave.

Kostov, Alexandr. 2016. "Nachaloto na modernoto gorsko delo na Balkanite: gorski sluzhbi i lesovudno obrazovanie." *Spisanie na BAN* 2: 30–38.

Lahiri-Dutt, Kuntala, and Gopa Samanta. 2013. *Dancing with the River: People and Life on the Chars of South Asia*. New Haven, CT: Yale University Press.

Livezeanu, Irina, and Árpád von Klimo, eds. 2017. *The Routledge History of East Central Europe since 1700*. Milton Park: Routledge.

Lowood, Henri E. 1990. "The Calculating Forester: Quantification, Cameral Science, and the Emergence of Scientific Forestry Management in Germany." In T. Frangsmyr, J. L. Heilbron, and Robin E. Rider, eds., *The Quantifying Spirit in the Eighteenth Century*. Berkeley: University of California Press, 315–43.

Maciejewicz Jan, and Jan Monkiewicz. 1982. "Technology Export from Poland." *Economics of Planning* 18(1): 22–42.

Magocsi, Paul Robert. 2010. *A History of Ukraine: The Land and Its Peoples*. Toronto: University of Toronto Press.

Mark, James, Artemy M. Kalinovsky, and Steffi Marung, eds. 2020. *Alternative Globalization: Eastern Europe and the Postcolonial World*. Bloomington: Indiana University Press.

Martykánová, Darina. 2010. *Reconstructing Ottoman Engineers: Archaeology of a Profession (1789–1914)*. Pisa: Plus-Pisa University Press.

Mazoyer, Marcel, and Laurence Roudart. 2006. *A History of World Agriculture: From the Neolithic Age to the Current Crisis*. Translated by J. H. Membrez. London: Earthscan.

Mitchell, Timothy. 2002. *Rule of Experts: Egypt, Techno-Politics, Modernity*. Berkeley: University of California Press.

Monkiewicz, Jan, and Jan Maciejewicz. 1986. *Technology Export from the Socialist Countries*. Boulder, CO: Westview Press.

Moon, David. 2013. *The Plough That Broke the Steppes: Agriculture and Environment on Russia's Grasslands, 1700–1914*. Oxford: Oxford University Press.

Muir, Cameron. 2014. *The Broken Promise of Agricultural Progress: An Environmental History*. London: Routledge.

Müller, Dietmar. 2020. "Statehood in Central Eastern and Southeast Europe: The Interwar Period." In Borodziej, Ferhadbegović, and von Puttkamer, *Routledge History Handbook of Central and Eastern Europe*, 2:148–93.

Olšáková, Doubravka, ed. 2016. *In the Name of the Great Work: Stalin's Plan for the Transformation of Nature and Its Impact in Eastern Europe*. New York: Berghahn Books.

Özkan, Hande. 2013. "Cultivating the Nation in Nature: Forestry and Nation-Building in Turkey." PhD diss., Yale University.

Özkan, Hande. 2018. "Remembering Zingal: State, Citizens, and Forests in Turkey." *International Journal of Middle East Studies* 50: 493–511.

Pál, Viktor. 2017. *Technology and the Environment in State-Socialist Hungary: An Economic History*. Cham: Palgrave Macmillan.

Palairet, Michael. 1997. *The Balkan Economies c. 1800–1914: Evolution without Development*. Cambridge: Cambridge University Press

Peluso, Nancy Lee. 1992. *Rich Forest, Poor People: Resource Control and Resistance in Java*. Berkeley: University of California Press.

Pine, Frances. 2007. "Dangerous Modernity? Innovative Technologies and the Unsettling of Agriculture in Rural Poland." *Critique of Anthropology* 27(2): 183–201.

Pojani, Dorina, and Dominic Stead. 2018. "When West–East Planning Policy Advice Fails to Gain Traction." *Journal of Environmental Planning and Management*. https://doi.org/10.1080/09640568.2018.1497586.

Pritchard, Sara B. 2011. *Confluence: The Nature of Technology and the Remaking of the Rhône*. Cambridge, MA: Harvard University Press.

Pontoppidan, Henrik. 2013. *Per cel norocos* [The Lucky Per]. Vols. 1–2. Romanian translation from Danish by I. Soare and Al. Budişteanu. Bucharest: Litera.

Quin, Michael. 1836. *A Steam Voyage down the Danube: With Sketches of Hungary, Wallachia, Servia, Turkey, etc.* 3rd. ed. Paris.

Radkau, Joachim. 1996. "Wood and Forestry in German History: In Quest of an Environmental Approach." *Environment and History* 2(1): 63–76.

Reynolds, Nancy Y. 2017. "City of the High Dam: Aswan and the Promise of Postcolonialism in Egypt." *City and Society* 29(1): 213–35.

Schönach, Paula. 2017. "River Histories: A Thematic Review." *Water History* 9(3): 233–357.

Scott, James C. 1998. *Seeing Like a State: How Certain Schemes to Improve the Human Condition Have Failed*. New Haven, CT: Yale University Press.

Scott, James C. 2017. *Against the Grain: A Deep History of the Earliest States*. New Haven, CT: Yale University Press.

Seirinidou, Vaso. 2017. "Environmentalism Narratives and Sociopolitical Agendas in Greece in the Eighteenth and Nineteenth Centuries." In H. Petrić and I. ŽebecŠilj, eds., *Environmentalism in Central and Southeastern Europe: Historical Perspectives*. Lanham, MD:Lexington Books, 91–110.

Sisa, József. 2013. "Budapest in der Gründerzeit: Architektur und Städtebau." *Gründerzeit:. Jahrbuch des Bundesinstituts für Kultur und Geschichte der Deutschen im östlichen Europa* 21: 119–44.

Sivaramakrishnan, K., and Arun Agrawal. 2003. "Regional Modernities in Stories and Practices of Development." In K. Sivaramakrishnan and A. Agrawal, eds., *Regional Modernities. The Cultural Politics of Development in India*. Stanford, CA: Stanford University Press,1–62.

Skelton, Leona J. 2017. *Tyne after Tyne: An Environmental History of a River's Battle for Protection, 1529–2015*. Cambridge: White Horse Press.

Stanek, Łukasz. 2020. *Architecture in Global Socialism Eastern Europe, West Africa, and the Middle East in the Cold War*. Princeton, NJ: Princeton University Press.

Stark, David. 1996." Recombinant Property in East European Capitalism." *American Journal of Sociology* 101(4): 993–1027.

Stenning, Alison, and Kathrin Hörschelmann. 2008. "History, Geography and Difference in the Post-socialist World: Or, Do We Still Need Post-socialism?" *Antipode* 40(2): 312–35.

Tamáska, Máté. 2018. "The Industrialization of the Danube Cityscape: Danube Ports during the Concurrent Urban Development of Vienna and Budapest 1829–1918." *Zeitschrift für Ostmitteleuropa-Forschung* 67(4): 552–77.

US Congress, Office of Technology Assessment. 1984. Technology Transfer to the Middle East. Washington, DC: Government Printing Office.

Vandendriessche, Joris, Evert Peeters, and Kaat Wils, eds. 2016. *Scientists' Expertise as Performance: Between State and Society, 1860–1960*. London: Routledge.

Vandergeest, Peter, and Nancy Lee Peluso. 2006a. "Empires of Forestry: Professional Forestry and State Power in Southeast Asia, Part 1." *Environment and History* 12(1): 31–64.

Vandergeest, Peter, and Nancy Lee Peluso. 2006b. "Empires of Forestry: Professional Forestry and State Power in Southeast Asia, Part 2." *Environment and History* 12(4): 359–93.

Vinke-de Kruijf, Joanne et al. 2012. "The Challenges of Transferring the Knowledge: Experiences from Romania." *Water Governance* 1: 63–70.

Vogt, William. 1948. *Road to Survival*. New York: William Sloane Associates.

Wandycz, Piotr S. 2001. *The Price of Freedom: A History of East Central Europe from the Middle Ages to the Present*. London: Routledge.

Way, Thaïsa, ed. 2018. *River Cities, City Rivers*. Cambridge, MA: Harvard University Press.

Wedel, Janine R. 1998. *Collision and Collusion: The Strange Case of Western Aid to Eastern Europe 1989–1998*. New York: St. Martin's Press.

Weiner, Douglas R. 1988. *Models of Nature: Ecology, Conservation and Cultural Revolution in Soviet Russia*. Bloomington: Indiana University Press.

Weiner, Douglas R. 1999. *A Little Corner of Freedom: Russian Nature Protection from Stalin to Gorbachëv*. Berkeley: University of California Press.

Worster, Donald, ed. 1973. *American Environmentalism, the Formative Period (1860–1915)*. New York: Wiley.

Yao, Joanne. 2019. "'Conquest from Barbarism': The Danube Commission, International Order and the Control of Nature as a Standard of Civilization." *European Journal of International Relations* 25(2): 335–59.

Yerolympos, Alexandra. 1993. "A New City for a New State: City Planning and the Formation of National Identity in the Balkans (1820s–1920s)." *Planning Perspectives* 8(3): 233–57.

Zeisler-Vralsted, Dorothy. 2015. *River, Memory and Nation Building: A History of the Volga and Mississippi Rivers*. New York: Berghahn Books.

Zürcher, Erik J. 2017. *Turkey: A Modern History*. London: I. B. Tauris.

PART I

Planning Territory

1

The Quest for a New Urban Landscape

Spatial Transformation in the Nineteenth-Century Belgrade Environment

Dragana Ćorović

\\\

This chapter explores the transformation of the city of Belgrade's environment, using research that considers societal, political, economic, and cultural changes during the nineteenth century. The various phases in Belgrade's urban environmental transformation took place as part of the Pashalik of Belgrade (1459–1815), the Principality (1815–1882), and then the Kingdom of Serbia (1882–1914). The initial period from 1791 to 1815, bore the first indications of transformation; the second lasted until 1867, and the third, main phase of the process, continued to the beginning of World War I (Ćorović 2015). The First and Second Serbian Uprisings, in 1804–1813 and 1815–1817, the declaration of partial autonomy in 1830, the withdrawal of the Ottoman military from fortified cities of Serbia in 1862–1867, and full independence in 1878, all influenced the transformation of the society, state, and economy of Serbia. The context of the new, capitalist system, the adoption of new ideas regarding cultural, technological, and economic development, hand in hand with national emancipation, also contributed to the transformation of Belgrade's urban environment. Questions of landownership and the establishment of a land registry across the country were prominent during almost the entire century. At the historical core of Belgrade, the transformation of the Ottoman city unfolded in tandem with the process of organizing the Belgrade Cadastre, but also with the implementation of a new urban plan and the undertaking of important communal works.

ENVIRONMENT, LANDSCAPE, MODERNIZATION, EXPERTS

The end of the eighteenth and beginning of the nineteenth century marked the end of the classical period and the start of a new era in science (Foucault 1972: 215–37). This was also a turning point in the creation of the modern capitalist world, which would, in the long term, produce the novel landscape of the Industrial Revolution (Wallerstein 1990). The human role in the environment differs from that of other species, precisely due to the imprint of cultural patterns that characterize human activity. The understanding of an environment is similar to the understanding of a landscape, of the way it transforms, but also of the manner in which we perceive and present it. On the other hand, "every landscape is a symbolic environment," as Thomas Greider and Lorraine Garkovich (1994: 1) believe. They see the connection between an environment and a landscape in the way that, by changing the natural environment, people create the landscape through a symbolic transformation of nature, specifically in the manner in which "landscapes are reconstructed in response to people's changing definitions of themselves" (1994: 1). Numerous definitions of a landscape reveal the inherent contradiction of the concept: at the same time, landscape is a set of physical, geographical, and geological characteristics of the Earth's surface, and, on the other hand, landscape is also the subject of perception, reflection, and representation (Wylie 2007: 1–16). A landscape receives its meaning from actors, those who inhabit it and work in it, and also those who perceive it: different meanings of landscape are expressed through its visual presentations and numerous theoretical interpretations and interdisciplinary studies.[1]

In his book on the political, social, and economic aspects of the modernization of Egypt in the twentieth century, Timothy Mitchell studies, among other things, the specific role of experts as actors in a complex process: "Political power now had a new form: the knowledge and command of space" (2002: 90). In the past ten years, infrastructure has been the current topic of study in social theory, anthropology, ethnography, technical sciences, and so forth. In scientific research, infrastructure appears, as does landscape, as the context and mediator in which various social and economic relationships and characteristics of cultural systems are reflected: "So many cares, feelings and dispositions are arraigned through infrastructural interactions" (Amin 2014: 145).[2] The concept of infrastructure is represented in a wide range of theoretical discourses: the interdisciplinary study of infrastructure is important in the search for means to overcome environmental problems (Grabowski et al. 2017); the study of infrastructure is the context for defining a new constellation of spatial relations, introduced by new technologies

(Dourish and Bell 2007); and together with the complete network of social and political relationships it comprises, nowadays, infrastructure marks a dividing line between war and peace, wealth and poverty, and it represents a critical element of modern society's complexity (Harvey, Jensen, and Morita 2017). The process of the construction of modern infrastructure systems is an excellent illustration of the manifestation of these divisions, which, nevertheless, defined the new society and, thereby, the new landscape as well.

The modernization of Belgrade in the late nineteenth century and in the first decades of the twentieth century was, in fact, the Europeanization of the city, implemented in the spirit of the cultural and ideological characteristics of modern capitalism. At the time, Serbia's modernization included a wide range of changes and innovations of both the physical and social environment, which "did not necessarily imply the existence of modern society elements: defined property relationships, the rule of law, human rights, and civil society" (Perović 2006: 52), but it certainly included aspirations toward their achievement, which gradually became an "obsession with development" (Dorondel and Şerban, Introduction in this volume). In Serbia during the nineteenth century, state institutions, professional and management bodies gradually undertook more and more responsibility for decision making on interventions in Belgrade's urban space. Although they belonged to the social elite, and some of them had significant political roles in the life of the city, and even of the state, we cannot conclude that the engineers in Belgrade in the nineteenth century belonged to the technocracy, because it did not exist in Serbia at that time, considering that "technocracy has more to do with organization than with machines, more with systematic approaches to objectives than with specific techniques, and more with efficiency and rational decision making than with technological determinism" (Radaelli 1999: 12).

LAND, ENVIRONMENT, AND BORDERS

On his return from Constantinople and the Middle East, the French poet and politician, Alphonse de Lamartine, approached Belgrade along the Constantinople Road (Lamartine 1835). He saw the fortress and city as the limit of the Orient, and its surroundings, in contrast to the landscape he was leaving behind, as bare. Lamartine admired the harmony between the people, culture, and landscape of Šumadija (forest land), the geographic region of central Serbia. Traditional Serbian villages were in fact economic and kin communities with joint ownership of fields and forests (Zundhausen 2008: 31–72).[3] However, during the Ottoman rule, Belgrade was an important hub for trade routes going both east–west and north–south. Due to its position, Belgrade was of great importance to the Ottoman Empire as both a military and trade port, as a combat encampment and a food and armaments store, all of which later influenced the form and speed of its metamorphosis from a medieval European fortress into an Ottoman city.[4]

Belgrade became part of the Ottoman Empire in 1521, at a time of the "centralized quasi-feudal [Ottoman] state" (1421–1596), a well-connected network of urban centers with the following key characteristics: mobility, circulation, and exchange.[5] In this network, cities, including Belgrade, were seats of bureaucracy, the military, religion, and trade (Zundhausen 2008: 31–72). Toward the end of the seventeenth century, during the Great Turkish War of 1683–1699 (against the Habsburgs), the border between Turkey and Austria stabilized at Belgrade, along the Sava and Danube Rivers.[6] During the unstable period from 1788 to 1815, Belgrade was completely evacuated several times as the inhabitants fled due to wars, uprisings, and rebellions, as well as a plague outbreak in 1794–1798 (Popović 1964: 301–28). From the beginning of the nineteenth century to World War I, Belgrade's surface area increased about ninefold (120–1,051 ha) (Dervichévitch 1939; Gavrilović 1974). The greatest expansion occurred in the period 1889–1907, when the Belgrade district grew some 90.0 percent (Ćorović 2015). Turning the surrounding land into urban building land represented one of the starting points for the transformation of the urban environment and invites an analysis of a series of various questions regarding the agents and actors of the change, scope and nature of the process, and the essence of the newly established relationships between the city and its surroundings in the given period.

The transformation of urban surroundings is extremely important for the transition from a feudal to a modern industrial society because it implies providing the town a position within the state territory and the opening of towns toward a broader environment, namely, their reestablishment in the "space of circulation": "The sovereign will be someone who will have to exercise power at that point of connection where nature, in the sense of physical elements, interferes with nature in the sense of the nature of the human species, at that point of articulation where the milieu becomes the determining factor of nature" (Foucault 2007: 23). At the very beginning of the nineteenth century, the structure of Belgrade (see figure 1.1) was characterized by clearly separated wholes: the fortress, the historic city core (known as the City within the Moat, or *Varoš u šancu,* in Serbian), and the part outside of the Moat, that is, rural settlements among the swamps, fields, meadows, and ponds. In the area of Belgrade's historical core, within the Moat, different religious and ethnic groups lived separately, each in its own *mahalle* (in Turkish—neighborhood or residential quarter). The City Field, Kalemegdan, was situated between the Belgrade fortress walls, where the Turkish army was located, and the town, namely, between the Turkish garrison and the civilian population of

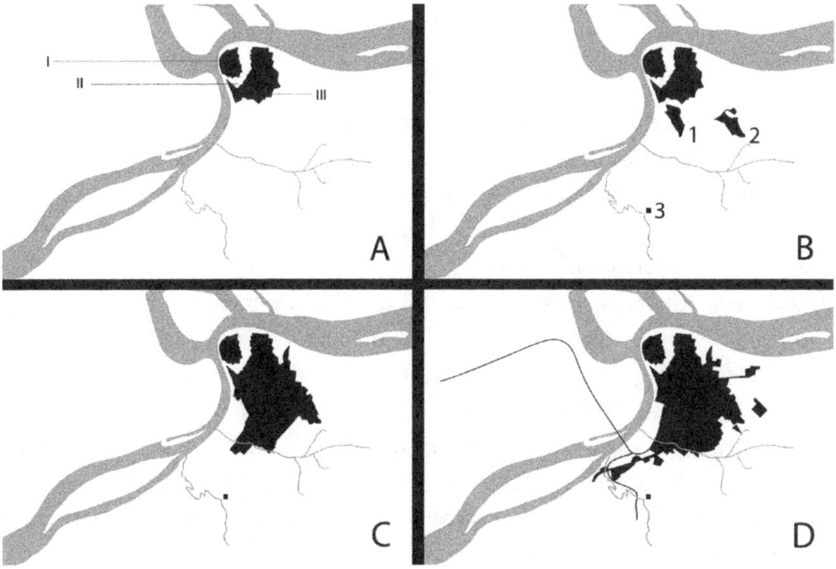

FIGURE 1.1. *The increase in the area of Belgrade's territory, 1806–1898*: (A) 1806—(I) the Belgrade fortress, (II) the City Field, (III) the City within the Moat; (B) 1835—new parts of the city, (1) Savamala, (2) Palilula, (3) Topčider, the residence of Prince Miloš and court economic facilities; (C) 1878—joining of different parts of the city; and (D) 1898—establishment of railway (1884) and tramway network (1892). Author: Dragana Ćorović, design: Slobodan Radosavljević. Source: Ćorović (2015).

different ethnic and religious origin. This was an open area about five hundred meters wide, on which construction, greening, and the formation of a city walkway were forbidden by the Turkish army. The fact that, as previously stated, the Sava and the Danube Rivers had represented the state borders for almost two centuries, resulted in the expansion of the city area toward the south and east, with regard to the City within the Moat during the nineteenth century.

The land beyond Belgrade's core had become a factor of territorial positioning of the city in the Principality of Serbia in the new historical circumstances. In this period, Belgrade was the headquarters of both the Serbian civilian government and the Turkish garrison and, therefore, no conditions existed for undertaking any significant transformations of the urban area. The main decisions regarding the spatial restructuring of Belgrade's vicinity were made by Prince Miloš Obrenović (ruled 1815–1839 and again 1859–1860), who personally determined some of the main directions of expansion of the new city (Tasić et al. 1995: 164). He was a true example of Foucault's *sovereign* (2007), that is, the epitome of political power in its natural environment. Efforts to solve the basic questions of organization within a country, such as the execution of a land survey and the elaboration of property relationships with regard to land, appeared after the Principality of Serbia acquired a degree of autonomy and delimitation from the Ottoman Empire, that is, after the period of 1829–1833.

Belgrade became the capital of the former Principality of Serbia in 1841. The first attempt at delineating the district of Belgrade was in 1842, but the boundary between Belgrade and the surrounding villages was not determined (Maksimović 1955; Ranković 1939: 690). The villages that surrounded a city in nineteenth-century Serbia were closely tied to the urban center, composing an almost indistinguishable whole. This was, naturally, also true of Belgrade, whose relationship to nearby inhabited places was particularly strengthened after 1867. This relationship was manifest in a more intense inclusion of the villages into the city's economy and social life. "Liberated Belgrade saw the influx of new residents, coming from nearby villages, working mostly as agricultural workers, but also craftsmen and merchants, holding on to their property, consisting mostly of cattle and land. In this way, Belgrade's immediate surrounding came to be property owned by city dwellers" (Nikolić-Stojančević 1967: 39).

The political process of solving the issue of the Turks' position in Belgrade and other fortified towns in the Principality of Serbia (Aleksić-Pejković 1967), as well as the issue of the realization of the autonomy of the Principality of Serbia, started with an international conference in Kanlıca, near Constantinople, in 1862. Pursuant to the agreements reached at this conference, which was held after the escalation of Serbian–Turkish conflicts in Belgrade culminating with the bombing of the town by the Turkish garrison, the Turkish civilian population started emigrating from Serbia, thus initiating the sale and purchase of Turkish property. This was followed by the Turkish army's retreat from fortified Serbian towns, except for Belgrade, Kladovo, Smederevo, and Šabac, which were abandoned by Turkish troops in 1867.[7]

The situation in the Belgrade environment was complicated by the fact that ever since the proclamation of Serbia's partial autonomy from the Ottoman Empire in 1830, the boundaries of Belgrade toward the hinterland represented an international border between the Principality of Serbia and the Turkish Empire, and extended along the line of the City Moat. In 1867 Belgrade was given its first modern urban plan as presented by the engineer Emilijan Josimović (1823–1897), who wrote a comprehensive study of the city at the time (1867).[8] Josimović was born in Moldova Veche, Romania. He studied philosophy and natural and technical sciences in Vienna, and was also a trained mason (Medaković 1997). In 1845 Josimović came to Belgrade, where he taught mathematics, mechanical engineering, and geodesy at the Military Academy. He later became a professor at the Higher School, where he taught mathematics, civil architecture, and practical geometry. Josimović was the author of textbooks in the field of technical education, the first president of the Technicians' Group, the Association of Engineers of Serbia, founded in 1868, and also president of the First Belgrade Singing Society, as well as member of the Theatre Construction Board, the Board for Support and Education of Poor and Abandoned Children, and a member of various municipal boards (Medaković 1997). From 1864, Josimović surveyed, measured, and estimated the quality of existing buildings and streets of the City within the Moat, namely, the area between the City Moat and the City Field, Kalemegdan. In an urban plan published at his own cost, together with a comprehensive explanation, Josimović proposed forming a rectangular network and 119 regular urban blocks instead of the former Ottoman urban matrix and the existing network of narrow streets. The adoption of this plan represented the beginning of the gradual disappearance of the Ottoman city landscape, together with the urban system that had reflected compliance with unwritten rules of construction and life in the Ottoman culture. Although the new urban plan for Belgrade was created as the idea of a single man, even before Josimović published his book the state had "embraced" this initiative. Two years after the publication of the plan, as recorded in archival documents, the Ministry of Construction was authorized to initiate the process of its implementation, even though, as has been recorded, the Municipality of Belgrade was economically too weak to accomplish this assignment (Ćorović 2018; Peruničić 1970). The same document also recorded the hopes that the state had for accomplishing this plan: "Let the development of freedom and progress in Serbia reflect in development of its capital, in its spacious, wide, clear, and straight streets and its beautiful buildings, so that the passengers arriving on steamboats that stop at our riverbank, feel the urge to stay in our environment for a while" (Peruničić 1970: 738). The process of implementing Josimović's plan started in 1867, and included the cutting of new streets and the broadening of existing ones, the leveling of ground, and the preparation of the communal infrastructure. In addition, some of today's most important green areas of the city were formed at that time: Kalemegdan, the former City Field, the neglected space around the Belgrade fortress, which became a significant public park. Also, according to Josimović's plan, a "green belt," which was supposed to include six public parks, was designed to replace the City Moat. The idea of the green belt around the city was never realized, but the Moat was removed, in a process that started in 1880 and ended in 1883.

Control over the city limits meant control over the city's economy, land, productive forces, and capital (Peterson 1981: 17–40). The question of the city borders was, therefore, a fundamental issue of Belgrade's urban regulation, but Josimović's plan for the City within the Moat did not resolve the city's problematic economic relationships with its surrounding environment (Maksimović 1968; Nedić 1976).

The issue of the city borders came under debate again in 1885, and it manifested discord on various levels of decision making—that is, between the relevant ministerial cabinets and the Belgrade City Assembly (Gavrilović 1974: 256). This situation was the result of, among other things, the 1866 Law on Municipalities, which, even when abolished, left a legacy by which the previous degree of independence of municipalities in Serbia was significantly decreased, whereby governmental authorities were entitled to control the decisions and finances of municipalities and to sanction or dismiss presidents of municipalities and councillors. Municipalities were under the constant supervision of the district authorities, the government, and the police, and the same situation was also true in Belgrade, where the municipality, although with a certain degree of autonomy, was, to a greater or lesser extent, supervised by the police throughout the nineteenth century. The implementation of Josimović's new urban plan came again under the jurisdiction of the municipality of the city of Belgrade in 1892. Following on from Josimović's plan, Belgrade received several more regulatory and situational plans (1878, 1886, 1893, and 1903), which corresponded to the expansion of the city area (in 1878 it was 463 hectares; in 1889, 573 hectares; and in 1907, 1,091 hectares). Over the course of the nineteenth century, the population of Belgrade and Serbia increased constantly.[9] The 1890 census shows the population of Belgrade was more than double that of 1866, at 55,000. In 1900, this had risen to 70,000, in a country of nearly 2.5 million (Kostić 1927: 62). There was disagreement regarding the establishment of the city area limits between the councillors themselves as well as between the councillors and the Ministry of Construction. The disagreement was, in part, over the inclusion of the poorer peripheral neighborhoods, and also over the municipality's desire to establish a distinction between the City within the Moat and the space outside this area (Gavrilović 1974). After a series of municipal debates, the line of demarcation of the city was set in 1890, but would be moved several times in the following decades (see figure 1.2). The question of city limits was posed periodically in 1901, 1904, and 1906, as well as in the period between the world wars (Lapčević 1929: 6). The city limits were determined by the limits of the areas where "communal order" had been established according to the European model—that is, the modern industrial society model.

After the declaration of full independence of the Principality of Serbia in 1878, and the Law of Industry in 1880, the country underwent a period of more intensive capitalist development. The 1885 land survey in Belgrade was the basis for communal development from around the turn of the century until World War I. For a long time the question of establishing surveying offices was one of the topics that incited debates about the various urban plans for the city.[10] The first precise recording of Belgrade, which was supposed to provide exact data on the borders

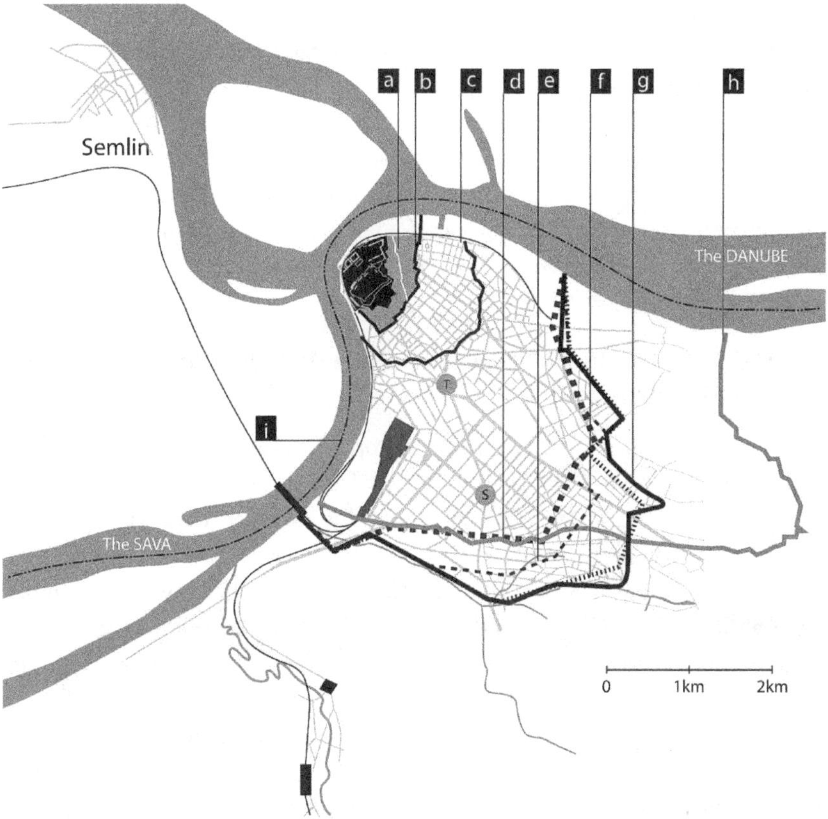

FIGURE 1.2. *Belgrade boundaries, from the eighteenth to twentieth century, compared to Belgrade before World War I*: (a) the City Field limit before 1862, (b) the City Field limit after 1862, (c) the spread of the line of the Moat in the nineteenth century, (d) the district boundary in 1893, (e) the district boundary in 1904, (f) the district boundary in 1906, (g) district street in 1909, (h) Eugen von Savoyen's Moat line in the eighteenth century, and (i) the state border, 1791–1918; central squares of Belgrade today—(T) Terazije, (S) Slavija. Author: Dragana Ćorović, design: Slobodan Radosavljević. Source: Ćorović (2015).

of plots, objects, and terrain, occurred in 1885 and was supposed to serve the realization of the contemporary regulation plan: to provide the basis for technical works (Obradović 1929). This endeavor ran parallel to the surveying done in the whole country. The municipality engaged surveyors from Budapest for work that was to take ten months, "to survey all the streets, markets and public plots, and perform the leveling in the town of Belgrade [as well as to produce] plans and reports regarding their findings" (Pajević 1923: 30). This plan did not fulfill its mission, as the data recorded were incomplete, and also because it was not revisited, checked, or completed in sufficient time. It was a missed opportunity to create a basis for the organization of the Belgrade Cadastre; rather, during the following twenty years this question was resolved only in fragments. As in the broader case of the Kingdom of Serbia, the problem was considered for the first time only in 1904, when the head of the Belgrade Cadastre organized the service and prepared the laws and regulations relevant to the cadastre. An intensive effort to create surveying plans did not even begin before 1911, and it represented the basis for the creation of regulatory and leveling plans for Belgrade in the interwar period, creating yet another reason for again raising the "endless question of surveying the city" (Obradović 1929: 9).

ENGINEERS, EDUCATION, AND COMMUNAL INFRASTRUCTURE FOR TRANSFORMING BELGRADE'S ENVIRONMENT

Besides Emilijan Josimović, approximately one thousand engineers also worked in Serbia during 1834–1919. They can be classified in several groups, including foreign experts, Serbs from the Habsburg monarchy and other countries, the citizens of Serbia who had been educated abroad as state scholars or at their own expense, and, of course, those engineers who had acquired their technical education in Belgrade at the Technical Faculty of the Higher School (Šolaja and Magdić 1994: 4). Together with experts in other fields (physicians, bankers, and lawyers), the engineers took part in making political decisions in Belgrade, in the capacity of councillors and presidents of the municipality and members of municipal commissions dealing with communal issues, although their authority, as already stated, was limited, and their work was constantly supervised by the government and the police.

With regard to building activity in Serbia, in the first half of the nineteenth century it took place amid conditions of inadequate procedures, poor organization, and a lack of building materials (Nestorović 2006: 43). Later, in the nineteenth century, there was a qualitative change in the technical and engineering culture in Serbia, and gradually, decisions about intervening in a space began to fall under the responsibility of the state, its institutions, various expert and administrative bodies, and, thus, indirectly engineers, as one layer of a societal elite. Given that the institutionalization of higher architectural education was not completed until the end of the nineteenth century, the design and execution of objects as well as any other work falling within the architectural profession were conducted either by architects educated abroad or people educated in closely related technical or artistic fields. The official gazette of the Society of Serbian Engineers and Architects, *Srpski tehnički list* (the Serbian technical journal), appeared in 1890, the same year the society itself was founded. By the 1890s, experts born in Serbia in the seventh and eighth decade of the nineteenth century gradually took over positions from Serbs born in the Habsburg empire and other foreign experts, who had previously worked on these jobs. The last quarter of the nineteenth century represented the period in which the number of foreign engineers in Serbia was reduced, a process that was also influenced by the establishment of the Technical Faculty in 1863 (Nestorović et al. 1996: 7–9). By 1891, educated experts, some of whom had been recipients of state scholarships, were returning from various European capitals.

Serbian students in the nineteenth century gravitated toward three cultural circles: Central European, French, and Russian. From 1882 to 1914, the Kingdom of Serbia sent 853 scholarship recipients to study in Europe, of whom 51 out of a total of 77 recipients of scholarships specifically from the Ministry of Construction went to German engineering schools (Trgovčević 2003: 55), as was the case of the famous engineer and public worker, Konstantin Kosta Glavinić (1858–1939). Born in Belgrade, into a wealthy merchant's family, Glavinić became a state scholar after the death of his father and studied at the Technical Faculty in Belgrade. As he later wrote, most engineers in Serbia of that time were Germans, Czechs, Poles, and Serbs from the Habsburg monarchy (Glavinić 1937), that is, foreign experts, which was not exclusively characteristic for Serbia at the time (Vlachos, chapter 2 in this volume). Having graduated from the Technical Faculty in Belgrade, Kosta Glavinić continued his education as a state scholar at the Technical University, Berlin (Die Technische Hochschule Berlin).[11] Glavinić started his career as a civil engineer in Belgrade in 1884 in the Railway Department of the Ministry of Construction section for iron bridges.[12] In the same year he became a professor at the Technical Faculty, where he taught the science of building bridges and tunnels. He was a municipal councillor (1886–1903) and a member of numerous municipal commissions, formed on the occasions of: the establishment of large-scale construction and infrastructure projects, the first official naming of Belgrade streets, street paving, the preparation of the Belgrade regulatory and leveling plan, and the construction of water supply and sewerage systems. Glavinić was president of the City Municipality of Belgrade in 1903–1907 and 1910 (Glavinić n.d.). During his time in the City Municipality of Belgrade, very important utility works were done: the construction and reconstruction of the city water supply and sewerage systems, the construction of several schools and two gymnasiums, the creation and landscaping of several city parks, land survey and elaboration of cadastral plans, and the introduction of the first public toilets.

Of all the members of the society, and not the least important, the engineers were most engaged in the construction of the modern city infrastructure, the physical technical system. At the turn of the century, engineers had an important role in the creation of the physical urban environment—Belgrade's new landscape—that would "influence and cultivate people and society" (Janáč, chapter 6 in this volume). One of them, the famous Serbian politician who served as prime minister several times, Nikola Pašić, studied as a state scholar at the Swiss Federal Institute of Technology (Eidgenössische Technische Hochschule) in Zurich. He was president of the Belgrade City Municipality 1889–1890, when he started negotiations with the Serbian-French Society (*Société serbe-française*) on the organization of utility works. These negotiations were concluded in 1891 by his successor as president of the municipality, by the signing of a contract for the construction of a water supply system, a power plant, public electric lighting, and a public tramway network (Govedarević 2008: 37–129). The new water supply system started operations on June 29, 1892. On that occasion a celebration was organized in one of the main city squares and, on the same evening, a concert and fireworks took place in Kalemegdan Park. The fireworks "sent messages in all directions that Belgrade had chosen the track of progress, development, and modern life in every respect!" (Bunjac et al. 1992: 31). Soon rule books were published prescribing the method by which households were to be connected to the new water supply system as was a guide for calculating the cost of this utility service (*Zbornik* 1906). The municipality collected the fee for maintenance of the water supply system, and water consumption was calculated based on the number of rooms in a household. The price for drinking water in Belgrade was subsidized by the municipality. In 1904 the Municipality of Belgrade authorized an expert board to start the execution of the design and construction of the sewerage system and to engage experts from Berlin for supervision of the main design, which was adopted in March 1905, and the sewerage system in the very center of Belgrade was finished in 1910. It was planned that two-thirds of the city at the time would have street channels and collection bins by 1914 (Bunjac et al. 1992: 40). Before World War I, the sewerage network was seventy-one kilometers long.

In the first decades of the nineteenth century, the urban development of Belgrade was at the level of preindustrial cities, where streets were used mostly for the transport of people, animals, and goods (Sjoberg 1960). In other words, Belgrade had not yet entered the phase of creation of a democratic sphere of public space. In 1867 Emilijan Josimović wrote that in the City within the Moat, certain streets were so narrow that even horse-drawn carts could not pass through. For the purpose of improving the street network's functionality, in 1891, an important document was produced, *Regulation of Belgrade Town*. It contained a list of all streets that needed to be reconstructed (223), their position, direction, and prescribed width (*Regulacija* 1891). Carriageways were cobblestoned with various materials: crushed stone, stone cubes, and wooden cubes (Gavrilović 1974). Just before World War I, Belgrade had about three hundred streets, a number ten times that at the end of the eighteenth century (Popović 1964: 302). The first electric power plant in Belgrade started operations in autumn 1893. The first lamp, a streetlighting test, was illuminated in February 1893, on the occasion of the public event of a merchants' ball: "The crowd admired the first electric *Bogenlampen*. Citizens were particularly impressed with the brightness of the light which they considered to be stronger than the sunlight. There were many who covered their eyes with their hands to protect them from the strong light" (Jovanović 1979: 101). The streetlighting network, installed in major streets of the city, and 7 kilometers long, including 65 *Bogenlampen*, and 1,000 *Glühlampen*, with a total power of 81,000 candela, was installed along a number of streets.

EMERGING ECOLOGY IN THE TRANSFORMED LANDSCAPE

To a great extent, the transformation of the urban environment of Belgrade from an Ottoman to a European city unfolded through the conversion of the natural setting—the space surrounding the city core—into a new urban environment. Up to 1867, the transformation of Belgrade's environment was the consequence of the city's repositioning into a "space of circulation." The main phase of the transformation of Belgrade's urban landscape started with the crucial year of 1867 and the termination of the simultaneous presence of two authorities in the city. At the transition from the nineteenth to the twentieth century, according to numerous indicators, such as the existence of modern utility infrastructure, public green areas, modern urban planning, and institutionalized public health care, Belgrade was already a modern European city. In the period up to 1914, together with the historical core, Belgrade's environment became the stage of the creation of "the machinic city," which "becomes a kind of clock and map writ large" (Amin and Thrift 2002: 78, 98). However, the radical transformation of the urban environment was achieved by the introduction of modern infrastructural systems, which had a direct influence on all areas of life in the city, especially on the modern arrangement of space and the new experience of time, and corresponded to new social and economic conditions.

The transformation of Belgrade from 1867 to 1914 produced a landscape that followed a mental matrix that was the result of the Europeanization and modernization of society. The urban history of Belgrade shows that the main issue of its development until 1914 had been the question of the boundaries of Belgrade's territory, namely, of determining the line to which the "modern city" would spread. From the condensed history in which Serbian society had accomplished the progress that societies with a less turbulent history would have achieved in several centuries of gradual development (Zundhausen 2008), Belgrade also entered a period of intense modernization and urbanization in which the existing capitalistic social relationships provided the impetus for specific challenges and differences between the new and the old, nature and culture, knowledge and control, uniqueness and uniformity, and city and state—between the different logic and languages inherent in systems of government at different levels (Scott 1998: 53–83), which in time has produced yet unresolved issues and unanswered questions.

NOTES

This chapter was written as a part of the project "Studying Climate Change and Its Influence on the Environment: Impacts, Adaptation, and Mitigation" (43007) and realized as a part of the agreement on financing the scientific-research work of the scientific-research organization in 2021, registration number 451-03-9/2021-14/200169 datec 5/2/2021, supported by the Ministry of Education and Science of the Republic of Serbia, within the framework of integrated and interdisciplinary research.

1. See also Roca, Claval, and Agnew (2011); Dorondel (2016); Berrizbeitia (2015); and others.
2. See also Schnitzler (2013); Chalfin (2014); and Jensen (2017).
3. These communities were called *zadruga*, that is, a form of a cooperative unit characteristic of traditional Slavic rural settlements (see also Vlachos, chapter 2 in this volume).
4. On medieval Belgrade, see Popović and Bikić 2004.
5. Serbia was under Ottoman rule in the period from 1459 to 1878. See also Vaten (2002).
6. This border stood until 1918, with brief interruptions in 1717–1739 and 1788–1791.
7. See also Jovanović (2013).
8. According to the first modern census in 1866, Belgrade had a population of 25,000 (Kostić 1927: 59–69; *Stanovništvo* 1953). On Josimović's plan, see also Vuksanović-Macura (2018).
9. The population growth of Belgrade was due to migration, as the birthrate was lower than the death rate throughout almost the entire nineteenth century (Radovanović 1974).
10. On the cadastre and urban development of Belgrade up to the interwar period, see Obradović (1929).
11. Die Technische Universität Berlin, after 1946.
12. The first railway line in Serbia (Belgrade–Niš) was put into operation in 1884.

REFERENCES

Aleksić-Pejković, Ljiljana. 1967. "Beograd između velikih sila i Turske od Jedrenskog mira (1829) do 1867. godine" [Belgrade between the Great Powers and Turkey from the Treaty of Adrianople (1829) to 1867]. *Godišnjak grada Beograda* 14: 5–21. [Cyrillic]

Amin, Ash. 2014. "Lively Infrastructure." *Theory, Culture and Society* 31(7–8): 137–61.

Amin, Ash, and Nigel Thrift. 2002. *Cities: Reimagining the Urban*. Cambridge: Polity.

Berrizbeitia, A., ed. 2015. *Urban Landscape*. Oxon: Routledge.

Bunjac, Vladimir et al. 1992. *Jedan vek Beogradskog vodovoda i kanalizacije 1892–1912* [One century of Belgrade water supply and sewerage system 1892–1912]. Belgrade: Beogradski vodovod i kanalizacija. [Cyrillic]

Chalfin, Brenda. 2014. "Public Things, Excremental Politics, and the Infrastructure of Bare Life in Ghana's City of Tema." *American Ethnologist* 41(1): 92–109.

Ćorović, Dragana. 2015. "Beograd kao evropski grad u devetnaestom veku: transformacija urbanog pejzaža" [Belgrade as a European city in the nineteenth century: urban landscape transformation]. PhD diss., University of Belgrade. [Cyrillic]

Ćorović, Dragana. 2018. "Plan Emilijana Josimovića u kontekstu transformacija urbanog pejzaža Beograda u devetnaestom veku" [Emilijan Josimović's plan in the context of Belgrade urban landscape transformation in the nineteenth century]. *Izgradnja* 72(4–6) (April–June): 151–63. [Cyrillic]

Dervichévitch, Chemso. 1939. *Évolution de Belgrade*. Paris: Jouve.

Dorondel, Stefan. 2016. *Disrupted Landscapes: State, Peasants, and the Politics of Land in Postsocialist Romania*. Oxford: Berghahn Books.

Dourish, Paul, and Genevieve Bell. 2007. "The Infrastructure of Experience and the Experience of Infrastructure: Meaning and Structure in Everyday Encounters with Space." *Environment and Planning B: Planning and Design* 34: 414–30.

Foucault, Michel. 1972. *The Archaeology of Knowledge*. Translated by A. M. Sheridan Smith. New York: Pantheon Books.

Foucault, Michel. 2007. *Security, Territory, Population: Lectures at the Collège de France, 1977–78*. Edited by Michel Senellart. Translated by Graham Burchell. New York: Palgrave Macmillan.

Gavrilović, Danica. 1974. "Razvoj vlasti u Beogradu u XIX veku: Uprava i sudstvo od 1867. do 1914 godine" [Development of governance in Belgrade in the nineteenth century: administration and judiciary from 1867 to 1914]. In V. Čubrilović, ed., *Istorija Beograda, 2: Devetnaesti vek* [History of Belgrade, vol. 2: Nineteenth century]. Belgrade: Prosveta, 249–63. [Cyrillic]

Glavinić, Kosta. 1937. "Malo uspomena iz vremena moga đakovanja u Beogradu i Berlinu od 1874–1884. god" [Some memories from the time of my student days in Belgrade and Berlin from 1874 to 1884]. Porodični fond Glavinić, Historical Archives of Belgrade, signature: 1119 K 4. 3.4.7.3. [Cyrillic]

Glavinić, Kosta. n.d. "Podaci za biografiju Koste D. Glavinića" [Data for the biography of Kosta D. Glavinić]. Porodični fond Glavinić, Historical Archives of Belgrade, signature: 1119 K 1. 1.13. [Cyrillic]

Govedarević, Nenad. 2008. *Tramvaji u Srbiji 1892–2008* [Tramways in Serbia 1892–2008]. Niš: Mašinski fakultet. [Cyrillic]

Grabowski, Zbigniew J. et al. 2017. "Infrastructures as Socio-Eco-Technical Systems: Five Considerations for Interdisciplinary Dialogue." *Journal of Infrastructure Systems* 23(4) (December). https://www.doi.org/10.1061/(ASCE)IS.1943-555X.0000383.

Greider, Thomas, and Lorraine Garkovich. 1994. "Landscapes: Social Constructions of Nature and the Environment." *Rural Sociology* 59(1) (March): 1–24.

Harvey, Penny, Casper Bruun Jensen, and Atsuro Morita. 2017. "Introduction: Infrastructural Complications." In P. Harvey, C. Bruun Jensen, and A. Morita, eds., *Infrastructures and Social Complexity: A Companion*. Oxon: Routledge, 1–22.

Jensen, Casper Bruun. 2017. "Pipe Dreams: Sewage Infrastructure and Activity Trails in Phnom Penh." *Ethnos* 82(4): 627–47.

Josimović, Emilijan. 1867. *Objasnenje predloga za regulisanje onoga dela varoši Beograda, što leži u šancu: sa jednim litografisanim planom u razmeri 1/3000 / od E. Josimovića* [Providing explanations of the proposals meant for regulating the part of Belgrade that is situated in the Moat, by a lithographic plan scaled 1/3000, E. Josimović]. Belgrade: E. Josimović. [Cyrillic]

Jovanović, Milan M. 1979. "Elektrifikacija pojedinih elektrodistributivnih područja u Srbiji, 1. Područje Elektrodistribucije Beograd" [Electrification of certain electric power distribution areas in Serbia, 1. Electric power distribution in the Belgrade region]. In B. Cvetković, ed., *Od Đetinje do Đerdapa, Elektrifikacija Srbije 1893–1941* [From Đetinja to Đerdap, the Electrification of Serbia 1893–1941]. Belgrade: Poslovna zajednica Elektrodistribucije Srbije, Združena Elektroprivreda–Beograd, 99–113. [Cyrillic].

Jovanović, Miloš. 2013. "'The City in Our Hands': Urban Management and Contested Modernity in Nineteenth-Century Belgrade." *Urban Studies* 40(1) (February): 32–50.

Kostić, Lazar M. 1927. *Beograd u prošlosti i sadašnjosti* [Belgrade, past and present]. Belgrade: Tucović. [Cyrillic]

Lamartine, Alphonse de. 1835. *Souvenirs, impressions, pensées et paysages pendant un voyage en Orient, 1832–1833 où, Notes d'un voyageur*. Brussels: L. Hauman.

Lapčević, Dragiša. 1929. "Reonsko pitanje" [The question of districting]. *Opštinske novine* 33(8) (June 1): 6. [Cyrillic]

Maksimović, Branko. 1955. "Prilozi za proučavanje arhitekture starog Beograda" [Contributions to the study of the architecture of old Belgrade]. *Godišnjak muzeja grada Beograda* 2: 231–36. [Cyrillic]

Maksimović, Branko. 1968. "Razvoj Beograda van građevinskog rejona krajem XIX veka" [Belgrade's development outside of its district at the end of the nineteenth century]. *Godišnjak grada Beograda* 15: 129–42. [Cyrillic]

Medaković, Dejan. 1997. "O Emilijanu Josimoviću" [On Emilijan Josimović]. In Dejan Medaković et al., *O Emilijanu Josimoviću*. Belgrade: Društvo urbanista, 5–16. [Cyrillic]

Mitchell, Timothy. 2002. *The Rule of Experts: Egypt, Techno-Politics, Modernity*. Berkeley: University of California Press.

Nedić, Svetlana. 1976. "Urbanističko uređenje Beograda od 1886. do 1914. godine" [Urban plan of Belgrade from 1886 to 1914]. *Godišnjak grada Beograda* 23: 175–217. [Cyrillic]

Nestorović, Bogdan. 2006. *Arhitektura Srbije u XIX veku* [Serbian architecture in the nineteenth century]. Belgrade: Art Press. [Cyrillic]

Nestorović, Bogdan et al. 1996. *Visokoškolska nastava arhitekture u Srbiji: 1846–1971* [Teaching architecture at the university in Serbia: 1846–1971]. Belgrade: Arhitektonski fakultet / Ministarstvo za nauku i tehnologiju Srbije / Plato. [Cyrillic]

Nikolić-Stojančević, Vidosava. 1967. "Etnička, demografska i socijalno-ekonomska struktura Beograda 1867. g." [Ethnic, demographic and socioeconomic structure of Belgrade in 1867]. *Godišnjak grada Beograda* 14: 23–44. [Cyrillic]

Obradović, Jovan I. 1929. "Katastar Beograda" [Belgrade Cadastre]. *Opštinske novine* 47(19) (November): 7–9. [Cyrillic]

Pajević, Bora M. 1923. *Regulacija Beograda, 1867–1923. god* [Regulation of Belgrade, 1867–1923]. Belgrade: Grafički institut Narodna misao. [Cyrillic]

Perović, Latinka. 2006. *Između anarhije i autokratije: Srpsko društvo na prelazima vekova (XIX–XXI)* [Between anarchy and autocracy: Serbian society between the centuries (nineteenth–twenty-first)]. Belgrade: Helsinški odbor za ljudska prava u Srbiji.

Peruničić, Branko. 1970. *Uprava varoši Beograda 1820–1912* [Administration of the city of Belgrade 1820–1912]. Belgrade: Muzej grada. [Cyrillic]

Peterson, Paul E. 1981. *City Limits*. Chicago: University of Chicago Press.

Popović, Dušan J. 1964. *Beograd kroz vekove* [Belgrade through the centuries]. Belgrade: Turistička štampa. [Cyrillic]

Popović, Marko, and Vesna Bikić. 2004. *Kompleks srednjovekovne Mitropolije u Beogradu* [Complex of the Diocese of Belgrade]. Belgrade: Arheološki institut. [Cyrillic]

Radaelli, Claudio M. 1999. *Technocracy in the European Union*. Oxon: Routledge.

Radovanović, Milovan. 1974. "Grad i njegovo stanovništvo: Demografski odnosi 1815–1914" [The city and its population: demography in the period 1815–1914]. In Čubrilović, *Istorija Beograda*, 2: 267–98.

Ranković, Dragomir J. 1939. "Šetnja kroz stari Beograd" [A stroll through old Belgrade]. *Beogradske opštinske novine* 57(11): 687–92. [Cyrillic]

Regulacija varoši Beograda odobrena Opštinskim odborom od 13. i 14. februara 1891. godine [Regulation of Belgrade city approved by the Municipal Board dated February 13–14]. 1891. Belgrade: Parna štamparija Narodne radikalne stranke. [Cyrillic]

Roca, Z., P. Claval, and J. Agnew, eds. 2011. *Landscapes, Identities, and Development*. Surrey: Ashgate.

Schnitzler, Antina von. 2013. "Traveling Technologies: Infrastructure, Ethical Regimes, and the Materiality of Politics in South Africa." *Cultural Anthropology* 28(4): 670–93.

Scott, James C. 1998. *Seeing Like a State: How Certain Schemes to Improve the Human Condition Have Failed*. New Haven, CT: Yale University Press.

Sjoberg, Gideon. 1960. *The Preindustrial City: Past and Present*. Glencoe, IL: Free Press.

Šolaja, Vladimir B., and Adela Magdić. 1994. *Inženjeri u Knjaževstvu / Kraljevini Srbiji od 1834. godine do završetka Prvog svetskog rata* [Engineers in the Principality / Kingdom of Serbia from 1834 to the end of World War I]. Belgrade: Zajednica tehničkih fakulteta Univerziteta u Beogradu / Muzej nauke i tehnike / LOLA Institut.

Stanovništvo Narodne Republike Srbije od 1834–1953 [Population of the People's Republic of Serbia 1834–1953]. 1953. Belgrade: Zavod za statistiku i evidenciju N. R. Srbije.

Tasić, Nikola et al. 1995. *Istorija Beograda* [History of Belgrade]. Belgrade: Srpska akademija nauka i umetnosti / Draganić. [Cyrillic]

Trgovčević, Ljubinka. 2003. *Planirana elita* [Planned elite]. Belgrade: Službeni glasnik. [Cyrillic]

Vaten, Nikola. 2002. "Uspon Osmanlija (1362–1451)" [The Ottoman Rise (1362–1451)]. In Robert Mantran, ed., *Istorija Osmanskog carstva* [The History of the Ottoman Empire]. Translated by Ema Miljković-Bojanić. Belgrade: Clio, 38–92. [Cyrillic]

Vuksanović-Macura, Zlata. 2018. "Inženjer i knez: modernizacija i evropeizacija Beograda" [Engineer and prince: modernization and Europeanization of Belgrade]. *Izgradnja* 72(4–6) (April–June): 140–50. [Cyrillic]

Wallerstein, Immanuel. 1990. "Culture as the Ideological Battleground of the Modern World-System." In Mike Featherstone, ed., *Global Culture. Nationalism, Globalization, and Modernity: A Theory, Culture and Society Special Issue*. London: Sage, 31–55.

Wylie, John. 2007. *Landscape*. London: Routledge.

Zbornik opštine Beogradske [Codes of the Municipality of Belgrade]. 1906. Belgrade: Opština grada, 178–84, 185–90.

Zundhausen, Holm. 2008. *Istorija Srbije od 19. do 21. veka* [The History of Serbia from the Nineteenth to the Twenty-First Century]. Translated by Tomislav Bekić. Belgrade: Clio. [Cyrillic].

2

From Slavic Swamp to Promised Land

Social and Environmental Engineering in a Southern Macedonian Swamp, 1913–1936

George L. Vlachos

Many scholars have rightfully argued that modernity has been largely characterized by the domination of mankind upon the environment (Leiss 1994; Merchant 1979; McNeill and Engelke 2016). Science and technology have been the primary pavers of this path; the elusive notions of progress and development have been its instigators, and state-employed experts and technocrats have often been the "neutral" pawns realizing this subjugation on the ground.

Among all the lands, soils, and landscapes these state experts considered dispensable, marshlands were certainly the most commonly unwanted. Guided by their manifest of efficiency and efficacy, modernists regarded swamps not merely as nuisances that had to be avoided, but as nature's pathologies that should be tamed and utilized (Dorondel and Olenenko, chapter 3 in this volume). Technological advancements in environmental engineering during the late nineteenth and early twentieth centuries (in some cases even earlier) provided the necessary know-how to achieve this.[1] Dredges and ditches started appearing all over the world, continuously altering the surface of the earth one bed and one bank at a time, as such "wastelands" turned into arable and habitable land.

They were not about to stop there though. The environment was not the only element that needed a drastic makeover. People(s) did as well. Aided or employed by the state, these representatives of modernity put certain communities in their crosshairs. Rural populations, in particular, were seen as the most backward and thus in dire need of all-around refinement.[2] Interestingly, in cases where technocrats and state experts were not just impartial cogs facilitating the function of the state apparatus but also acted on behalf of a nation-state, the backward peasantry was not seen only as primitive but as a disgrace to the nation.

By narrating the story of a lake and its eventual demise, this chapter aims to show how deeply entangled environmental and social historiographical schools can become, to the point that excluding one or the other from a certain case study can result in an incomplete history. This case study will take us to Giannitsa Lake, which is the main focus of this chapter. Despite its name, it was hardly a lake. Its irregular banks and seasonally unstable depth make it easier to classify it as a swamp rather than a lake. Located in southern Macedonia almost fifty kilometers west of Thessaloniki, the capital port city of the province, the lake dominated the region in terms of economic activity and environmental impact. Around its wetlands extended the Giannitsa or Thessaloniki plain, devoid of any vegetation and fed by four rivers. Scattered across the plain, were forty or more settlements as well as the town of Giannitsa. Most of these places had vigorous Slavic-speaking communities that, as we shall see, came to be regarded as potential trouble by the Greek authorities after Macedonia was ceded to Greece in 1913.

AMPHIBIOUS NATIONALISMS

Many Greek nationalists, from the beginning of the Macedonian Question in the late nineteenth century to the present, have sworn by the allegedly indisputable "Greekness" of southern Macedonia. Historical demographics prove them wrong though. By the late nineteenth century, when communities and individuals in the area started being labeled as nationalities, a large part of Macedonia was inhabited by Slavic-speaking populations that in Ottoman times were merely defined as Christians. This was particularly pronounced in the northern, eastern, and central parts of the province where, as Konstantinos Karavidas—a man who will be mentioned several times in this chapter—wrote sometime between 1917 and 1922: "Beyond Loudias [a river located in the central-western part of the province] the Greek language stops and the Bulgarian starts, with a few exceptions. The prefectures of Giannitsa, Kilkis, Gevgeli, Vodena, Karatzova, and even those zealous Muslims of Karatzova, apart from very sparse exceptions, speak Bulgarian."[3]

Karavidas also concluded that this "Slavicness" of Macedonia was even more accentuated in the countryside of the province. According to him, Greeks had a natural tendency to be drawn to cities because they preferred the pursuit of commerce rather than working the soil. This left rural Macedonia open for migrating Slavs through the centuries.[4] Even though his racial argument today seems far-fetched, the fact remained. Across the fields and valleys of Macedonia, Slavic dialects and languages were far more common while Greek and Turkish were used mainly as lingua francas of commerce and administrative matters, respectively. Vassilis Demetriadis suggested that the consolidation of the *chiflik* system in the province during the eighteenth and nineteenth centuries accelerated the consolidation of Slavic communities in the area. More precisely, the Ottoman owners of the large agricultural estates facilitated the internal migration of Slavic populations in order for them to work as agricultural laborers (Demetriades 1981).

Whatever the reason that led to this situation was, the potent Slavic population of Macedonia substantially facilitated the advance of nationalism, not because the locals became "enlightened" by ideals of ethnic superiority, but because both Bulgarian and Serbian nationalists could easily claim Macedonia as their irredenta based on the ethnic composition of its inhabitants. While Serbian governments remained inconspicuous, valuing their alliance with Greece, Bulgarian governments sought to reinstate the 1878 Treaty of San Stefano, according to which Macedonia was to be ceded to Bulgaria. It was not long before nationalist agitators infiltrated several provinces of the European part of the Ottoman Empire in an effort to achieve this goal.

Giannitsa Lake was one of the focal point of nationalist interest. Before its involuntary national awakening, the whole area around the lake had sustained a stable ecosystem that included people and animals alike. The vitality of this system was defined and dictated by the swamp. On the one hand, local fauna, from eels to boars and from waterfowl to mosquitoes, used the swamp as breeding grounds and shelter from predators. On the other hand, the local communities had managed to set up stable economies that made use of the resources that were offered. Fishing and hunting were naturally popular activities, especially as a means of complementing the regional grain-based diet; mud, reeds, and straw provided the building materials for their houses (Didaskalou 1993: 42; Kottaris and Xatzibrettas 1993).

These days of relative tranquility came to an end, however. At the dawn of the twentieth century Giannitsa Lake came into play for the first time. In 1904 a Greek journalist reported that the swamp was overtaken for the first time by nationalist activists. On that occasion, the Bulgarian government-led nationalists of the Internal Macedonian Revolutionary Organization (IMRO) had occupied the swamp in an effort to continue their activities in secrecy, after the Porte had crushed the Ilinden Uprising of 1903. This renewed interest in the area—which the Greek state considered well within its sphere of interest—awakened even the most dormant nationalist circles in Athens. Quite quickly, those intellectual and military heads demanded action against the Bulgarian guerrillas (known to Greeks as *comitadjis*).

The Greek government reacted in an assertive manner. Bands of armed paramilitaries were formed, drafted mostly from the national army, their minds inflamed with stories of savage Bulgarians and enslaved brothers, dozens of Macedonian Strugglers (*Makedonomachoi*), as they came to be known, crossed the northern borders and entered Ottoman Macedonia in 1905, with the sole purpose of rooting the *comitadjis* out of the swamp. What ensued was a brutal four-year undercover war that took place primarily in and around Giannitsa Lake, between the Greek and Bulgarian nationalists, who fought under the not-so-watchful eye of the Ottoman gendarmerie.[5] Amid all this, the residents of the surrounding settlements were caught in the crossfire, as armed men from both camps casually entered their villages in order to plunder their livestock or terrorize them into declaring one or the other national identity to any demographer who might have traveled through the area.[6]

Only the Young Turk Revolution of 1908 put an end to this war, before kindling a new one. The Turkish nationalist movement demanded the removal of rival like-minded agents and activists of both camps from Macedonia. After only a few years the Balkans were once again in turmoil due to the two Balkan Wars of 1912 and 1913, the first of which was conducted by an uneasy alliance between Bulgaria, Serbia, Greece, and Montenegro against the Ottoman Empire, and the second was between Greece, Serbia, Romania, Montenegro, and the Ottoman Empire this time against Bulgaria. By the end of 1913, everything was already over. Greece had annexed the apple of discord, Macedonia and the European part of the Ottoman Empire had practically dissolved. With the wars over, all Balkan governments were now free to deal with internal enemies, namely, those who did not abide by their nation-building agendas.

OF AGRONOMISTS, SOCIETIES, AND PERSISTENT PEASANTS

Freshly arrived officials entered Giannitsa as the Greek army was occupying the town in 1912. Most of them had come from "Old Greece," as experienced engineers and agronomists.[7] They had been instructed to survey the new land and estimate the profit margin that the development of agriculture would bring. During their stay, however, they ran short of positive comments to offer about the status of the Giannitsa plain. In their reports back to the Ministry of Economy, they explained in technocratic terms rather than nationalist pomp, all the elements that stalled the progress of agricultural production in the area. According to them, Macedonia had the potential for acquiring true economic momentum. Also according to them, agricultural intensification and commodification had to be applied under their vigilant eyes and intellects. It is worth briefly exploring who those agronomists were.

Although the Agricultural School of Athens would not be founded until 1920 (later it was renamed the Agricultural University of Athens), the youth who wished to obtain secondary agricultural education had some alternatives. The most popular was enrollment in one of the three agricultural schools that had been available from 1887 to 1914, either in Athens, Peloponnesus, or Thessaly. However, the overall impact of those schools could only be described as ineffective, as during their operating period they managed to provide Greek agriculture with a disproportionately small number of experts. Undoubtedly, some of them ended up in Macedonia, settling in promising rural areas, trying to set up the agricultural infrastructure of the land. As low-level civil servants though, little is known about them. Their deeds might have been salvaged in local oral histories, but for the most part, their names have been wiped clean. On the contrary, those who guided their actions and trained them in the science of agronomy—a group of people who constituted the first golden generation of Greek agronomists—have enjoyed the recognition of history. These agronomists, unlike their lowly counterparts, had been brought up with Western European standards. Most of them graduated from famous European universities during the first decade of the twentieth century, and they all irrevocably represented both progress and their employer: the Greek state (Kandila 2001).

One of them, a prestigious ministry agronomist named Pilavios Papageorgiou, published a lengthy article in August 1913, in which he gave a concise outline of how the Greek state should proceed in exploiting Macedonia. All that was needed, he insisted, was "good management and well-handled agricultural propaganda," and at the same time recommended a long list of crops that would presumably be marketable (Papageorgiou 1913). To his disdain, it turned out to be much more complicated than that. As months passed, Greek state agronomists, who by then had surveyed the villages around the swamp, became aware that they were not going for just an easy tidying up but for a complete reshuffling of the cards in regard to the production mentality followed in the area.

A report written the following year, in 1914, demonstrates that grim realization. The disheartened author and agronomist Georgios Palamiotis tried to pinpoint all the impediments that had rendered the Giannitsa plain a "desolate" land. He underlined three issues. The first one was the people. When the author inspected the villages around the swamp his first impression was: "[The locals] lack any idea of place, time, and [market] price; and due to having spent many years under slavery and distress, all of them, with no exceptions, are suspicious and cunning, not to mention that they bear characteristic slowness [*yavaşyavaş*] in terms of thinking, working, walking, and trading" (Palamiotis 1914a: 305).[8]

According to Palamiotis, it was, moreover, the peasants' tools that had hindered their progress, a complaint that indeed occurred very often in many similar agronomic reports. Western Europe's hardware and technology had not penetrated the Macedonian countryside at all. In fact, it was constantly confirmed that little to nothing had changed in regard to the agricultural equipment since antiquity. As Palamiotis (1914a: 321) put it, "only the Hesiod plow [a simple wooden plow with a metal tip] is used to cultivate the field, while the iron plow, very rare, is used for plowing only in the plain. Neither threshing nor any other work is done by machine."

Finally, the marshy environment of the area presented perhaps the biggest challenge for the Greek government. Palamiotis commented that at one time Giannitsa Lake was a breeding ground for millions of crows that used the dense vegetation of the lake to "raid," as he called it, the nearby sown fields. A state-issued cull, he suggested, would solve this problem. The numerous floods on this land made agricultural production and goods transportation between markets absolutely unpredictable. To make things even worse, the malaria morbidity statistics around the lake were so high that agronomists and government officials realized over the course of the next years that the feeble bodies of these peasants could not possibly evolve into the agricultural power house that Macedonia needed. At this point in history, however, reclaiming the swamp was not yet discussed. Instead, to benefit the locals, engineers proposed general adjustments capable of controlling the flow and floods, at the same time admitting that the fishers of the lake caught "quite tasty eels," which could potentially become a commodity (Palamiotis 1914b). Furthermore, they engaged in an intense campaign to "fix" the first two defects: the work ethics of the local communities and their flawed equipment.

This campaign was spearheaded by a private association, of which Palamiotis was a member, the Greek Agricultural Society (GAS). Interestingly, being private did not mean being independent. On the contrary, the GAS was often given assistance by the state and the crown. The society was founded in 1901 by royal decree and the king appointed himself as president for life. Pursuing total control of the society, His Majesty asserted his right to appoint eighteen out of the thirty-six members of the General Board. Below him were three vice presidents, who were almost always influential political or entrepreneurial personalities.[9] In light of this, it seems naive to claim that the GAS was merely a philanthropic association brought together for the greater good of the Macedonian people. It is far more likely that the society functioned as a well-connected lobby interested in rendering agriculture more productive and profitable.

Members of the GAS did not go to Macedonia untrained. Before attempting to modernize the new lands, the society had presented excellent samples of its efficiency in Old Greece. As of 1909 it had formed 162 agricultural associations, a number that rose by 1911 to 720 all over Greece. In addition to these remarkable achievements, the GAS introduced what came to be known as "model orchards" in every province of the country. These were farms run by GAS agronomists using progressive techniques in accordance with the specialized needs of the province they had been appointed to (Kallivretakis 1990). When the time came, the GAS tried to show the same degree of dedication in Macedonia. As we shall see, however, it was not as easy a target as southern Greece.

In order to initiate the transition to modernity in the province, the GAS first decided to deal with the material aspect. Priority was given to the introduction of advanced tools and fertilizers for the farmers of Macedonia. In the society's annually published bulletin for the year 1915, we come across the first actual data regarding this task. The GAS had established an "agricultural station" in Giannitsa, which presumably sold needed materials to local farmers at lower-than-market prices (Anonymous 1915). Moreover, the society had also undertaken the distribution of booklets that contained practical information for designing more productive fields. The texts of these brochures could possibly minimize their effectiveness in rural areas, but they also included illustrated instructions to help illiterate farmers adopt the correct practices. As a bulletin suggested, in 1914 at least thirty thousand copies of such booklets were distributed across Greece. Some were destined for Macedonia to be used in schools as a means of "indoctrinating pupils and farmers in agricultural matters" (Anonymous 1915: 170–71). Despite the organizational ability of both the GAS and the state in planning and executing these attempts, the campaign to turn locals from peasants into farmers did not bring the desired outcome.[10] Based on the official agricultural reports of the ministry from 1914 to 1919, the extent of cultivated land was almost the same as before and, in fact, as Mark Mazower claims, the index of crop production per hectare in Macedonia followed a downward path beginning in 1914 (Anonymous 1914: 6; Mazower 1991: 76).

Speculating as to why that was the case is not hard. Although Macedonia had been ceded to Greece, the local population was largely indifferent toward the Greek administration as well as disinterested in abandoning their way of life and adopting a new one that would not necessarily improve their standing. The atrocities committed by the Greek army during the Balkan Wars had not been forgotten.[11] Furthermore, even after the wars had ended, the representatives of the Greek state in Macedonia tended to treat the Slavophone locals as potentially hostile to the state and therefore to them personally. It is thus no wonder why mutual suspicion had halted any attempt at convergence between natives and the Greek government, and it had no prospect of being restored.

To add to this, another key reason that technocrats and agronomists were failing in Macedonia can be traced to the *chiflik* ownership model, which did not reward increased the productivity and commodification of agricultural products. More precisely, *chiflik* estates, predominant in the area around Giannitsa Lake, were large hereditary estates that required of their laborers a bare minimum of agricultural production, enough to cover the needs of the owner and to sustain themselves.[12] Due to this noncompetitive modus operandi, once the state got hold of the *chiflik* estates, the new production guidelines set by the Greek agronomists were not met with enthusiastically by the peasants, who resisted by remaining as idle as possible. Under such circumstances, no wonder the nationalizing project in Macedonia stalled. The solution came in the form of a catastrophe.

THE UNLIKELY COLONISTS

In 1913 in Giannitsa plain, a colonizing process began that would deeply change Macedonia. In the most well-read newspaper of the province we find a passage, meaningless at the time, informing its readership about the resettlement of two thousand refugees from Bulgaria on the outskirts of Koulakia (Chalastra), a small community close to Giannitsa (Anonymous 1913a: 3). A few days later the article's author wrote again: "Nice Idea!... We can only congratulate those who initiated such action since the countryside is in need of working hands, and the refugees who resettle there, being the most capable and deftest of farmers, will contribute as much as possible to the improvement and refinement of agriculture in these extensive and fertile lands" (Anonymous 1913b: 1).

Resettling populations loyal to the state in crucial regions was not something new. In fact, it had been a sound practice of states that wanted to assert dominance over a certain geographical area. However, it was not common to resettle uprooted and impoverished refugees in lands that already had an established community, but not the desired one. In Macedonia, this took place erratically until 1923 whenever the government had displaced populations at its disposal and expropriated or abandoned lands to settle them on. Greece's defeat in the Greco-Turkish War of 1919–1923 changed this drastically. More than one million Orthodox Christians were driven out of Asia Minor, Eastern Thrace, and the Caucasus region and came to Greece as part of the massive population exchange that ensued (Özkan, chapter 8 in this volume). They immediately became the best candidates the state had in its pursuit to replace unwanted populations.

Some journalists saw the war-torn people as an opportunity. To them, these refugees were an "agricultural capital" that Greece had to exploit. In early January 1923 a journalist wondered: "Therefore, today we have in the bosom of our country more than 300,000 agricultural hands capable of multiplying our productive output and rejuvenating our economy. How did Greece put this colossal economic and agricultural force to work? How did it distribute this wealth, and in what proportions to all the anemic plains of our country? ... What kind of efforts did it put into turning this entire dead mass [the refugees] into robust entities of economic vigor? And finally, how did it utilize this agricultural capital?" (Zografos 1923: 1)

The author also posed for the first time the question of what should be done about the Giannitsa swamp, asserting that "it is turning out to be an insurmountable obstacle to the safe resettlement and living conditions of thousands of refugees" (Zografos 1923: 1). Besides, it was no secret that the majority of the refugees would be resettled in the "desolate" and "backward" plains of Macedonia. It was even eloquently illustrated on a French map published after 1926 marking the density of agricultural colonies in Macedonia, where a dark cloud of countless settlement names covers the province, as opposed to the very few dots located across the rest of Greece.[13]

The land reclamation project in the swamp became more pressing during the following months when the technocrats running the land distribution affairs in the region realized that the appropriated *chiflik* estates were not nearly sufficient to host the huge number of refugees that had come to the area (Janáč, chapter 6 in this volume). For example, one prefecture of the Giannitsa plain alone accepted more than 40,000 individuals, which probably meant that combined with the other two prefectures of the plain, more than 100,000 new residents overwhelmed the plain and its settlements overall (Koliadimou 2014: 111). In fact, even a major progovernmental newspaper openly criticized governmental negligence (Anonymous 1923: 1). At last, after an extended period of uncertainty, the contract regarding the reclamation works was signed between the Greek government, at the time led by the dictator General Theodoros Pagalos, and the Foundation Company (henceforth, Foundation), an American New York-based company, specialized in hydraulic engineering (Anonymous 1925a: 1).

As the contract stated, the great "productive works" would be completed in three stages within a decade. The first included "complete detailed survey of the whole area west to Thessaloniki" as well as the eradication of two smaller lakes north to the Giannitsa plain, lakes Aimatovo and Artzan, which fed the rivers responsible for flooding the Giannitsa plain. The second stage was to divert three of the four rivers crossing the plain, at the same time erecting flood barriers along their banks. More importantly, this stage would also include the draining of Giannitsa Lake. The water volume of the lake would initially be collected in a central basin and then released to the nearby Thermaic Gulf aided by the main drainage ditch. The whole project would conclude with the diversion of the mouth of the Axios River away from the port of Thessaloniki, as its alluvia had been threatening the undisturbed coming and going of ships (see figure 2.1).What was interesting though, was what the Greek government would gain from these works in overall extent of land reclaimed or secured from floods. Foundation had estimated that it

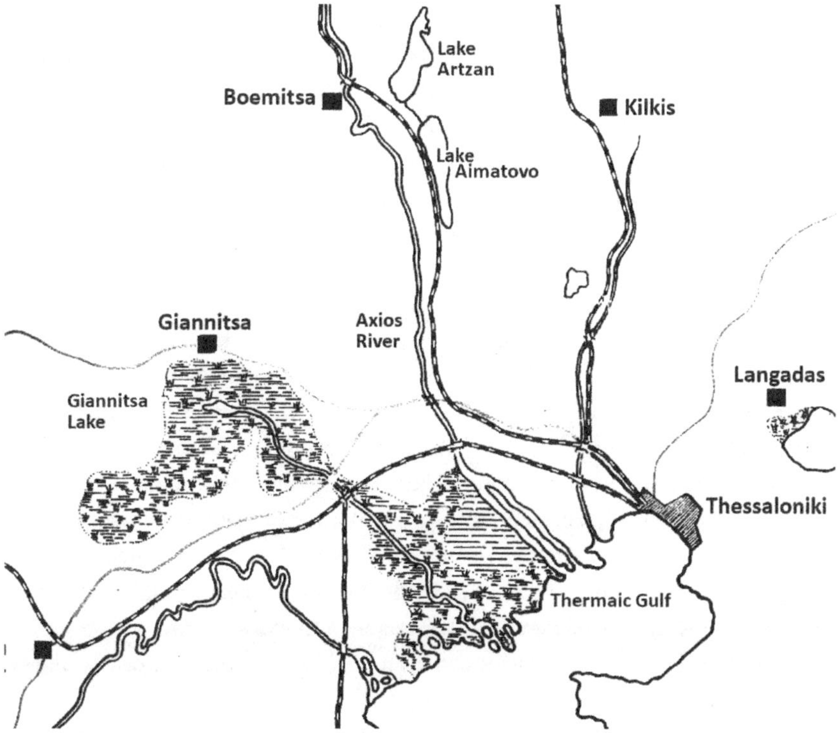

FIGURE 2.1. The general area of the hydraulic works.

would amount to or even surpass 130,000 hectares. Such improvement came at a high cost, however, as the Greek government would pay almost $27 million for completion of the work.[14]

Remarkably, the scheduled reclamation enjoyed the support of both the general public and the political parties in Greece. A few days before the contract was officially proclaimed in the newspaper *Macedonia,* which always supported the Liberal Party, the paper could not hide its enthusiasm for the finalized agreement between the dictator Pagalos and Foundation Company. Featured on the first page of the paper we find an article titled "Foundation: The First Grand Benefactor of Central Macedonia" as well as a smaller piece suggesting that with this hydraulic project, Pagalos would at last overwhelm the "enemies of the nation." It is unknown whether this ambiguous statement was directed against the peasants of the Giannitsa plain, who were considered by many to be Bulgarian agents, or the natural elements that had condemned Giannitsa plain to misery (Anonymous 1925b: 1).

Furthermore, it would be interesting to explore the efforts that various interwar dictators put into similar engineering projects as a means of sanctifying their image and legitimizing their rule. Besides the example of Mussolini's reclamation of the Pontine Marshes, the Spanish dictator Franco had managed to gain the tolerance, if not the support, of the Spanish masses with his dam-inauguration campaign, which brought water to all areas of the country, at the same time loudly proclaiming its irrevocable unity against separatists. Indeed, it seems that his political maneuvers on this matter were so successful that they earned him the positive nickname "Paco el Rana" (Frankie the Frog) (Swyngedouw 2007). For General Pagalos however, things did not go so well. It seems that the project had already stalled before it began. The dictator clung onto his position for barely one year before being arrested and jailed. For unspecified reasons ranging from administrative obstruction to economic hardship, the project would not start for three more years until 1928. During this period of complete inactivity, pressure mounted in both discursive and material terms. Newspapers were giving weekly updates on when exactly the work would continue, always postponing to a later date (Anonymous 1926: 2). With a large number of the scheduled refugee resettlements already completed and without new lands to be distributed to them, refugees remained landless and idle, thus becoming a desperately impoverished minority in settlements where the locals had land, income, and equipment. This contrast produced two results.

First, the refugees and the locals clashed. In the vast majority of cases, this happened due to the fact that the newly arrived settlers forcefully occupied and cultivated former *chiflik* lands that were claimed by the locals. A plethora of hostilities took place between the two communities. In one case, in the Rachtina settlement, the president of the community complained to the local authorities that refugees had encroached on up to two hundred hectares belonging to the local population (Anonymous 1928a: 2). In another case, in the village of Asvestario, in Pella Prefecture, locals and refugees were involved in a massive brawl that left twenty injured (Katsapis 2007: 316).

The second result was observed by Kostantinos Karavidas, an expert on agrarian matters and a higher civil servant of the Greek state serving in Macedonia. Karavidas unveiled a different perspective regarding the interaction between the landless mass of refugees and the established Slavic communities of Giannitsa plain, one that has been wildly overlooked in modern historiography. While touring on behalf of the Ministry of Agriculture to solve management problems regarding the resettlement of refugees in Macedonia, he noticed something odd about them, which he documented in one of the many essays he wrote. Karavidas asserted that refugees were being led astray by other agricultural production modules. He specifically claimed that a Slavic grassroots cultural and economic institution that was scattered across communities in Macedonia and composed of joint self-sufficient extended families was leaving a positive impression on refugees. The *zadruga*, as Karavidas called it, maintained some key egalitarian characteristics that made it attractive to the impoverished newcomers. First, the whole set of assets of the *zadruga* were common property to all its members, and second, every decision concerning the community was made after egalitarian processes had been followed (Novaković 2005; Damianakos 2002: 213–14; Todorova 1993). Karavidas also claimed that the local *zadrugas*, were starting to absorb the landless refugees possibly by employing them, not only economically but also culturally, as those two aspects went hand in hand in his reasoning. To him, *zadruga* was not merely a system of producing agricultural goods but also a "living covenant of Slavic traditions" that showed refugees an alternative communal way of life, unlike the one intended by the Greek state.[15]

Apart from Karavidas's warning, which is interesting in its own right, it was unique in his confession that the "Greekness" of the Orthodox refugees was not at all taken for granted by state officials, which meant that Greece had to somehow find a way to keep refugees loyal. Based on Karavidas's train of thought, such a result could only be achieved by pulling them away from the grasp of alternate production structures. The best way to handle this was to indoctrinate them into the "righteous" Greek agricultural community (which was suspiciously reminiscent of Western European ones), a precondition of which would be to grant them adequate plots of land. In the face of land shortages, this could only been achieved with the land reclamation of the Giannitsa Lake.[16]

THE THREE STEPS TO SUBJUGATION

At last, the reclamation works began in the summer of 1928. The first improvements quickly became noticeable. Shortly after the process began, Foundation pledged that to hand over eight thousand hectares of arable land to the Greek authorities by the end of 1929 (Anonymous 1928b: 6). Also beginning during the same year were the reclamation and draining works in the two small lakes to the north to the swamp. The event drew the attention of newspapers and was thoroughly covered. In fact, the prime minister himself was present at the opening ceremony, where he triumphantly exclaimed:

> The small Greece, the destroyed Greece of 1922 managed, with her wounds still open, to undertake these colossal works, a part of which we inaugurate today. In five years' time, the works in Thessaloniki's plain [Giannitsa plain] will have been completed, handing over for cultivation 500,000 *stremmata*, securing 750,000 more and irrigating up to 800,000. Here, around us, a small Egypt will arise, fertile and hygienic, capable

of protecting the prosperity of the population in Central Macedonia . . . after the completion [of the works] Macedonia will certainly be the most enviable region in the East. (Anonymous 1928c: 1)[17]

It was only then that the locals realized the negative momentum of the reclamation to them. The more the works progressed, the more they were left with only two choices. Either conform to the new circumstances or immigrate.[18] From 1923 until 1932, according to data collected by the General Directorate of Macedonia, 6,670 Slavophone locals had abandoned the prefecture of Giannitsa for Bulgaria, namely, almost half the local Slavic populace (Ioannidou 1997). Three indirect yet effective mechanisms, all coordinated and supervised by the Greek government, contributed to the final subjugation of all who remained. The first concerns the labor force that was required for the works. Article 11 of the finalized agreement between Foundation and the government stated that "the contractor shall recruit all the remaining technical assistant and labor personnel, preferring those among the Greek subjects—that is, insofar as it is possible and practicable, at least one-fifth of whom shall be employed from men who have taken part in wars to the extent any competent men are available."[19] This piece of information discloses something that it is not clear at first sight. Apart from the obvious preference for war veterans, the other part of the article mentions "Greek subjects." This is not suspicious. Both locals and refugees were naturalized Greek citizens, which theoretically meant that Foundation would draw workers from both groups. Although this might have been the case, however, the possibility that these laborers were almost exclusively refugees is high. In fact, this is not so much a matter of a mere percentage derived based on the number of them residing in proximity to the work as one of organized labor associations on their part. More precisely, once the refugees arrived in Greece, the government sorted them to the extent possible into homogeneous refugee groups, based on their place of origin (Koliadimou 2014: 346). The ultimate goal was to bind together and resettle coherent communities of refugees, consisting of those who already knew and trusted each other. This led to the consolidation of an information network among refugees, which spread news of open manual labor positions to those willing to work. In an area where all productive activity had ceased because of ongoing reclamation work (as specified in the contract with the Greek government), refugees were most likely to absorb the principal source of income, which in turn sowed the seed of their imminent supremacy.[20]

The second mechanism is strongly connected to these refugee groups as well. With the tables turning in their favor, refugees were steadily being transformed from landless rubble to tillers of the reclaimed lands. Guided by the state apparatus, and more precisely by the Ministry of Agriculture, these refugee groups evolved into agricultural cooperatives, which entailed some very crucial advantages for the refugees. Perhaps the main advantage was the ability to obtain loans with low interest and highly favorable conditions of repayment, especially after the Agricultural Bank of Greece was founded in 1929 on precisely these premises. Also cooperatives' members had access to much more advanced equipment, which was bought and shared collectively by the cooperative, thus significantly reducing the cost. Similarly to the first mechanism, locals could participate on absolutely equal terms with the refugees to those cooperatives. However, they did not, probably due to persisting bad blood between the two communities. The data are quite compelling on this. The participation count for the more than one hundred agricultural cooperatives that were active in the area by 1929 was 3,388 refugees, while the participation of locals did not surpass three hundred members (Koliadimou 2014: 341).

The last subjugation mechanism concerned the distribution patterns of the reclaimed lands. As we saw, by 1936 when the reclamation project was completed, the government had more than 100,000 hectares at its disposal, which were to be granted to the residents of the area, both new and old. A priority list had to be followed, however. At the top of the lists were those who had helped the Greek cause in times of war: veterans or the families of the fallen. Agricultural expertise was also promoted, as large expanses of land were given to certain agronomists and agriculturalists who originated from Macedonia or were willing to move to Giannitsa plain, in order to pave the way for efficient production. The distribution to them was implemented on a merit-based system, according to the educational degrees they held. Namely, those who had been awarded a university degree, either from Greece or abroad were eligible for twenty-five hectares of reclaimed land, while those with a lower degree were assigned only five to fifteen hectares (Konstantinidis 1989: 95). The resettled refugee population, however, got the lion's share of the reclaimed or secured lands. Interestingly, among the refugees, larger parcels of land were given to those who were married, a clause that apparently pushed many to hastily start new families in order to receive the advantages (Koliadimou 2014: 246). Locals were not left out of the distribution; they were still eligible candidates, but as many of them had already been granted land from appropriated *chiflik* estates, usually not particularly fertile, they received only smaller fields.

The new engineered environment was capable only of sustaining the equally engineered ecosystem. With the lake gone and its aquatic environment destroyed, fishermen, hunters, and reed gatherers were now in distress. In an effort to salvage what they could, these environmentally unemployed workers founded their own association in 1932, numbering more than two hundred members. Similarly to the refugees, they requested land and support. Unlike them, however, what they got was a lukewarm response. Although their pleas were heard and their case was included in the official law concerning the manner in which the reclaimed lands would be distributed, it was specified that they would be given fields "insofar as there are available," placing them at the bottom of the priority list (Koliadimou 2014: 248).

The impact of the reclamation project was also very prominent in aesthetic terms, not that this mattered at all to those directly involved, refugees and locals alike. It did, however, matter to modernists and nationalists, as it constituted proof of a successful political agenda. It would not be an exaggeration to claim that a by-product of the draining was the construction of a landscape admired and venerated by those with nationalist convictions. The wild marshland had been replaced by well-kept farms that reflected the modernist virtues of the state and the rigorousness of the race. The new houses were another proof of that. As the resettlement of refugees in the area intensified, the construction of houses, often out of mudbricks and thatch, gave way to facilities built with concrete and tiles, both being products that needed to be imported from elsewhere by the millions. Furthermore, the organically developed villages with their irregular shapes turned into planned settlements defined by a modern grid. A time lapse between 1926 and 1936 would indeed be spectacular, as it would document the complete metamorphosis of a vast area, both environmentally and socially.

Using the case study of a southern Macedonian swamp and its reclamation, this chapter sought to draw attention to topics that have been similarly addressed in Alan Mikhail's *Nature and Empire in Ottoman Egypt* (2013). There, Mikhail tells the story of how, during the early days of the Ottoman Empire in Egypt, the management of water irrigation from below shaped a quasi-autonomous production system that succeeded in breaking away from the grasp of the imperial administration. The ingenuity in this simple notion hides in the fact that the author focused on a subject of social history, only this time taking the transformation of an ecosystem as the central event, around which he built his narrative (Mikhail 2013).

This chapter strove to abide by this methodology by unveiling the social history of local and refugee communities in relation to the presence and reclamation of Giannitsa Lake. Compared to Mikhail's presentation of a bottom-up view of such processes, in this case we have the definite interference of a centralized nation-state that pushed for national homogenization coupled with agricultural intensification by carrying out hydraulic engineering works.

On that front, it is also crucial to underscore the fact that modern historiography has often presented such projects merely as unbiased projects authorized by equally unbiased state technocrats, obscuring the political agendas that may lie beneath the alteration of an ecosystem. In light of this, the reclamation of Giannitsa Lake might not present a special case at all, meaning that historians should probably revisit and rewrite the history of similar events, but this time with increased alertness toward the hidden parameters of supposedly gleaming public works.

NOTES

1. See Ciriacono (2016) for a specialized but engaging collective volume on several studies regarding land reclamation before the advent of modernity.

2. A list for anyone who would like to see the impact of nationalism on rural populations would definitely include Weber (1976), Judson (2006), and Cole and Wolf (1999).

3. American School of Classical Studies at Athens (ASCSA/GR GL KDK) 049, folder 04, subfolder 01, document 01, "Periepoikismou Makedonias" [On the colonization of Macedonia].

4. ASCSA/GR GL KDK 049, folder 10, subfolder 01, document, "Anakefalaiotikai Paratiriseisepi tis egkatastaseos prosfygon stin Makedonia" [Recapitulating notes on the resettlement of refugees in Macedonia].

5. This particular theater of war of the Macedonian Struggle was immortalized in the novel by Penelope Delta, *Secrets of the Swamp* (2012 [1937]). Note however that it is essentially a propaganda novel openly leaning toward the Greek nationalist cause.

6. For an interesting analysis on adulterated censuses, see Kostopoulos (2003).

7. The term "Old Greece" refers to the continental provinces that made up the initial core of the Greek state, namely, Peloponnesus and Central Greece. It could also been claimed that Thessaly, a province that was acquired later, in 1881, also belongs to the "Old Greece."

8. "Yavaşyavaş" is a Turkish expression that translates into "slowly." The phrase was transferred in Greek as a demeaning comment upon the, presumably, Ottoman mentality toward work.

9. Tellingly, the first three vice presidents were G. Theotokis, A. Zaimis, and P. Stefanovik-Skilitsis; the first two served multiple times as prime ministers of the Greek government, and the third was a prominent landowner and investor.

10. For similar case studies, see Damianakos (2002), Nitsiakos (2015), and Karakasidou (1997).

11. For a detailed list of the horrific war crimes committed by both Bulgarian and Greek nationalists during the Balkan Wars, see Anonymous (1914a).

12. A concise yet thorough analysis on how *chifliks* functioned within the Ottoman Empire is included in Vergopoulos (1975: 65–68).

13. Cartographic Heritage Archive, map 932, in Barbaritis (1926).

14. Hellenic Republic 1925, "On the Ratification of the Contract regarding the Execution of Hydraulic Works on Thessaloniki's Plain," *Efimerida tis Kyverniseos*, October 8, 2020–2022. It is also important to note here that the funds for this project came from two loans obtained by the Greek government from the Hambro credit house, especially for this purpose.

15. ASCSA/GR GL KDK 049, folder 17, subfolder 06, document, "Ipomnima peri tou kathestotos gaioktisias kai tis oikonomikis theseos ton slavikon meionotiton en ti periochi tis

Florinis" [Memorandum concerning landownership and economic position of the Slavic minorities in the Florina area].

16. ASCSA/GR GL KDK 049, folder 17.6, "Peritoukathestotosidioktisias, ton anomalon dikaiopraksion epi akiniton kai tis oikonomikis theseos ton slavofonon meionotiton tis Dytikis Makedonias (1927)" [On landownership, abnormal legal transactions, and economic position of Slavophone minorities in western Macedonia (1927)].

17. One *stremma* equals one-tenth of a hectare.

18. The immigration conditions between Greece and Bulgaria had been facilitated after the noncompulsory population exchange that was signed between the two countries in the Treaty of Neuilly-sur-Seine.

19. Hellenic Republic 1925, "On the Ratification of the Contract."

20. Hellenic Republic 1925, "On the Ratification of the Contract."

REFERENCES

Anonymous. 1913a. "Egkatastasis Prosfigon" [Refugee resettlement]. *Makedonia*, June 13.

Anonymous. 1913b. "Oraia idea!" [Nice idea!]. *Makedonia*, June 16.

Anonymous. 1914a. *Georgiki Apografi 1914* [Agricultural Census 1914]. Athens: Elstat.

Anonymous. 1914b. *Report of the International Commission to Inquire into the Causes and Conduct of the Balkan Wars* (No. 4). Washington, DC: The Carnegie Endowment.

Anonymous. 1915. *Georgiko Deltio* [Agricultural Bulletin]. Athens: Greek Agricultural Society, 170–71.

Anonymous. 1923. "I anaptyksis tis paragogis kai i Epanastasis" [The development of production and the revolution]. *Makedonia*, March 13.

Anonymous. 1925a. "To kinima tou stratou epekratise se olin tin Ellada" [The Military Movement prevailed everywhere in Greece]. *Makedonia*, June 25.

Anonymous. 1925b. "Faoundeison: i proti megali euergetis tis kentrikis Makedonias" [Foundation: the first grand benefactor of central Macedonia]. *Makedonia*, September 20.

Anonymous. 1926. "To provlima aergias sti Thessaloniki" [The idleness problem in Thessaloniki]. *Makedonia*, September 28.

Anonymous. 1928a. "Parapona" [Complaints]. *Makedonia*, September 22.

Anonymous. 1928b. "I statherotis tis Thessalonikis" [Thessaloniki's stability]. *Makedonia*, September 7.

Anonymous. 1928c. "I nea Epochi" [The New Era]. *Makedonia*, November 25.

Barbaritis, K. 1926. *Carte d'etablissement des refugies ruraux et urbains en Grece (1923–1926)*, 1:500000. Paris: Imp. Berger.

Ciriacono, Salvatore, ed. 2016. *Land Drainage and Irrigation*. Abingdon: Routledge.

Cole, John W., and Wolf, Erik R. 1999. *The Hidden Frontier: Ecology and Ethnicity in an Alpine Valley*. Berkeley: University of California Press.

Damianakos, Stathis. 2002. *Apo ton choriko ston agroti* [From peasant to farmer]. Athens: Exadas-Ekke.

Delta, Pinelopi S. 1937. *Sta mystika tou valtou* [The secrets of the swamp]. Athens: Vivliopoleion tis Estias. English translation: Delta, Penelope. 2012. *Secrets of the Swamp*, trans. Ruth Bobick. Portsmouth, NH: Peter E. Randall.

Demetriadis, Vassilis. 1981. "Problems of Land-Owning and Population in the Area of Gazi Evrenos Bey's wakf." *Balkan Studies* 22: 43–57.

Didaskalou, Ifigeneia. 1993. "Oi psathades kai oi psarades ton Giannitson" [The reed-gatherers and fishermen of Giannitsa]. *Istorika kai Laografika Themata* 3: 42–53.
Ioannidou, Alexandra. 1997. "Ta slavika idiomata stin Ellada: Glossologikes sigliseis, politikes apokliseis" [Slavic idioms in Greece: linguistic convocations, political deviations]. In Vasilis Gounaris, Iakovos Michailidis, and Giorgos Aggelopoulos, eds., *Taytotites sti Makedonia* [Identities in Macedonia]. Athens: Papazisis, 123–42.
Judson, Pieter M. 2006. *Guardians of the Nation: Activists on the Language Frontiers of Imperial Austria*. Cambridge, MA: Harvard University Press.
Kallivretakis, Leonidas. 1990. *I dynamiki tou agrotikou esksigchronismou stin Ellada tou 19ou aiona* [The dynamic of agricultural modernization in nineteenth-century Greece]. Athens: Educational Institute of the Agricultural Bank.
Kandila, Ioulia. 2001. "I Averofeios Georgiki Scholi Larisas: symvoli stin istoria tis georgikis ekpaideysis stin Ellada" [The Averoff Agricultural School of Larissa: contribution to the history of agricultural training in Greece]. Phd diss., University of Ioannina.
Karakasidou, Anastasia N. 1997. *Fields of Wheat, Hills of Blood: Passages to Nationhood in Greek Macedonia, 1870–1990*. Chicago: University of Chicago Press.
Katsapis, Kostas. 2007. "Isilthaneis to spiti mas: Koinonikes proipotheseis ton antiparatheseon anamesa se gigeneis kai prosfyges sto Mesopolemo" [They barged into our house: the social conditions of the confrontation between locals and refugees during the interwar period]. Proceedings of the conference of I elliniki agrotiki koinonia, 303–19. Athens, 2007.
Koliadimou, Agni. 2014. "Prosfyges sto nomo Pellas" [Refugees in Pella Prefecture]. Phd diss., Aristotle University of Thessaloniki.
Konstantinidis, Konstantinos. 1989. *Ta eggeioveltiotika erga stin pediada Thessalonikis* [Land-improvement projects in the Thessaloniki plain]. Thessaloniki: ETYEM.
Kostopoulos, Tasos. 2003. "Counting the 'Other': Official Census and Classified Statistics in Greece (1830–2001)." *Jahrbücher für Geschichte und Kultur Südosteuropas* 5: 55–78.
Kottaris Spiros, and Xatzibrettas Dimitrios. 1993. "I limni ton Giannitsonkai I apoksiransi tis" [The lake of Giannitsa and its draining]. *Filippos* 4: 14–27.
Leiss, William. 1994 [1972]. *Domination of Nature*. Montreal: McGill-Queen's Press.
Mazower, Mark. 1991. *Greece and the Inter-war Economic Crisis*. Oxford: Oxford University Monographs.
McNeill, John Robert, and Peter Engelke. 2016. *The Great Acceleration*. Cambridge, MA: Harvard University Press.
Merchant, Carolyn. 1979. *The Death of Nature: Women, Ecology, and the Scientific Revolution*. San Francisco: Harper & Row.
Mikhail, Alan. 2013. *Nature and Empire in Ottoman Egypt: An Environmental History*. Cambridge: Cambridge University Press.
Nitsiakos, Vassilis. 2015. *Peklari*. Ioannina: Isnafi.
Novaković, Nada. 2005. "Theories on the Beginnings and Structure of Joint-Families ('zadruga')." *Stanovništvo* 43: 105–34.
Palamiotis, Georgios. 1914a. "Kazas Giannitson" [Giannitsa Prefecture]. *Oikonomiki Ellas*, July 6.
Palamiotis, Georgios. 1914b. *Agricultural Census of Macedonia* [Georgiki Ereuna Makedonias]. Athens: Greek Agricultural Society, 59–64.
Papageorgiou P. 1913. "The Agricultural Future of New Greece." *Oikonomiki Hellas*, August 10.
Swyngedouw, Erik. 2007. "Technonatural Revolutions: The Scalar Politics of Franco's Hydro-

social Dream for Spain, 1939–1975." *Transactions of the Institute of British Geographers* 32(1): 9–28.

Teichmann, Christian. 2007. "Canals, Cotton, and the Limits of De-colonization in Soviet Uzbekistan, 1924–1941." *Central Asian Survey* 26(4): 499–519.

Todorova, Maria. 1993. "Slava und Zadruga." *Historische Anthropologie* 1(1): 123–30.

Vergopoulos, Kostas. 1975. *To Agrotiko Zitima stin Ellada: i koinoniki ensomatosi tis georgias* [The agricultural question in Greece: the social incorporation of agriculture]. Athens: Exantas.

Weber, Eugen. 1976. *Peasants into Frenchmen: The Modernization of Rural France, 1870–1914*. Redwood City, CA: Stanford University Press.

Zografos, Dimitris. 1923. "To georgikon prosfygikon kefalaion" [The refugee agricultural capital]. *Makedonia*, January 9.

3

In Quest of Development

Territorialization and the Transformation of the Southern Ukrainian Wetlands, 1880–1960

Stefan Dorondel and Anna Olenenko

"I think it is necessary to recall that there are wetland regions with rich silty soil, in conditions favorable for all kinds of amelioration, not only on the main rivers of our republic, the Dnipro and Dniester. Wetlands also exist on the Southern Buh, Ingul, and many other rivers in almost all provinces of Ukraine. We should consider them as 'wasteland' [*neudobnaia zemlia*] that brings no or very insignificant income. The main task of our generation is to reclaim wetlands from nature, make them an object of labor in order to obtain the benefits that they can bring." This statement by Aleksandr Yaroshevsky (1925: 4), a Ukrainian engineer of the Southern Regional Ameliorative Organization (Odessa) who settled in the south in the early 1920s, demonstrates his thinking about Ukrainian wetlands and suggests ways that Soviet power should engage in making them economically profitable. These words could have come from almost any engineer or scientist of the same period elsewhere in Europe. The idea of fixing the "imperfections" of nature by putting rivers and wetlands to work for the benefit of humans dates from the seventeenth century and has Western European origins in the development of hydrology as a "science of rivers" (Cioc 2002).

Wetlands were deemed to be pestilent environments, unproductive landscapes that had to be subjected to radical transformations. Few landscapes, if any, were more negatively perceived than the wetlands (Williams 1991; Adger and Luttrell 2000; Bruisch 2019). Associated with rivers that flooded annually, wetlands were considered enemies of the riparian population and thus subjected to modernization and technologization. For the capitalist economy wetlands were poorly productive, at least compared with agricultural fields, an indeterminate mixture of water and land that escaped clear definition. The name itself, "wetland," was long a pejorative term, "less than ideal for the majority of practical purposes" (Maltby and Barker 2009: ix). Whereas the ideal river was a channel of slowly flowing water, a uniform, tamed, navigable, and predictable environment (Worster 1992) that could be easily harnessed, the ideal wetland was an arable field. These ideas floated around in the West as well as the East, exciting engineers and planners. This is not to say that hydrological works only began in the nineteenth century. Many authors have shown that complex hydraulic systems were built after 1500 in China, India, and Ottoman Egypt, for instance (Tvedt 2016; Mikhail 2011). Nevertheless, the emergence of capitalism and its way of organizing nature (Moore 2016) proved to be radical in terms of its alterations to rivers and wetlands. Transforming nature became a synonym of modernity.

This chapter explores the history of the transformation of Dnipro wetlands in the steppes of Southern Ukraine under two political regimes: the tsarist era and the Soviet period. By exploring the fate of wetlands under these two regimes, we argue that the modern state and its experts linked two radically different environments—wetlands and the southern steppes—through agricultural development. The richness of wetlands, with their tremendous array of aquatic and terrestrial animals, plants, and birds as well as their various amphibious landscapes have proved less attractive to experts and planners; instead Russian authorities prized wetlands' possible contribution to the improved productivity of neighboring spaces—the steppes. This chapter aims to understand the transformation—and, at the end of the 1950s—the disappearance of those wetlands that formed part of the steppes. This process was part of the reorganization of Imperial Russian territory: a politico-economic rearrangement of the physical world with deep implications for local people and their economy. The Russian state considered this rearrangement, the reclamation of land from the wetlands, a "civilizational act" and the only way to shape a modern future for various regions of the empire (Peterson 2019; Bruisch 2019). Covering just 2 percent of the vast steppe, the wetlands have hitherto been ignored by environmental historians.[1] This chapter attempts to analyze wetlands as a distinct part of the southern steppe environment. By focusing on two political regimes the chapter contributes to the understanding of how the wetlands came to be viewed as an economic utility, as a result of political economic decisions, expert evaluations, and actions of state-commissioned engineers. We show the changes induced in the wetlands by the two political regimes on different scales, by different means, and with different scopes. By focusing on the Soviet period we intend to contribute to the still small but burgeoning literature that emphasizes the environmental damage caused by the socialist regimes (see Weiner 1999; Josephson 2002; Stahl 2010; Dorondel 2016; Olšáková 2016; Pál 2017). Comparing the tsarist state's attitude to the Dnipro wetlands with the Soviet state's vision for them allows us to understand the similarities and differences between two radical political regimes and administrative states regarding this "amphibious landscape" (van Dam 2016).

The tsarist state seized this territory (now in southern Ukraine) from the Ottoman Empire after the Russian–Turkish wars (1768–1774 and 1787–1791). Soon after, the Zaporozhian Sich (a central fortress) was liquidated (1775) and the Zaporozhian Cossacks were brought under tsarist rule.[2] Before the Russian Empire conquest, the vast territories of the steppes and the land bordering the Dnipro River were sparsely inhabited by seminomadic pastoral populations and migrants (peasants) who had arrived from other parts of Ukraine. The wetlands themselves belonged to the Zaporozhian Cossacks. It was the last stretch of Cossack land before the border between the Cossacks and Tatars. The colonization of the steppes continued from the end of the eighteenth century and into the nineteenth, not only by Ukrainian peasants but also by Russians, Germans, and other ethnic groups (Staples 2003; Moon 2013). As a result, in Ekaterinoslav province (guberniia) at the end of the nineteenth century—the focus of our study—68.9 percent of inhabitants were Ukrainians and 17.3 percent were of Russian origin (Ia. Boiko 1993).

The Russian state perceived the vast steppe area to be "as wide as the sea"—as "wild and unpossessed" (O'Neill 2017: 4). Thus, one of the most urgent tasks of the Russian state was to organize the territory administratively and economically as a means of civilizing it (O'Neill 2017: 23). This action cried out for knowledge of the southern empire, hence the scientific expeditions that began under Catherine the Great (1762–1796) (Moon 2013). State officials used the weapons of modern scientific knowledge such as maps, inventories, and statistics (Blackbourn 2006; Moon 2010a, 2013) in order to reorganize a territory where the constant negotiation between water and land produced an environment that was considered less appropriate to a modern economy. Imperial Russian scientific knowledge was in the vanguard of economic and political interest, as many authors have pointed out (Moon 2010a, 2013; Fedotova and Loskutova 2015; Bruisch and Gestwa 2016; Peterson 2016a).

Grand ideas about transforming the wetlands were not in short supply during the imperial period. Different authors proposed different projects for increasing the economic profitability of the wetlands. Throughout most of the nineteenth century though, wetlands were discussed in terms of afforestation, land amelioration, or drainage for reclaiming agricultural land. Experts and tsarist officials alike were obsessed with the efficient use of fertile black soil found in the southern steppes—a fertile but drought-prone land—increasing the area of arable land and obtaining higher yields. However, drainage of the wetland was too expensive for the Russian authorities who concentrated instead on making the steppes more suitable for agriculture through afforestation, creation of shelter belts and ponds, and regulation of the Dnipro River. In 1880 the Ministry of State Property organized expeditions in Southern Russia (to the provinces of Kherson, Tauride, and Ekaterinoslav) in order to develop recommendations on combating drought and building infrastructures for irrigation and water supply (Moon 2010a).

However, the Russian state encountered multiple obstacles to organizing the territory—and we refer here especially to wetlands—as they wished. First, the landlords, most of them of Russian origin who received land as the empire advanced into the southern steppes, were not easily convinced of the need for radical change as these lands encompassed valuable woodland (in a completely treeless environment) and pasture, as we will detail below. Yet, as Ekaterina Pravilova (2014) shows, the growing state interest in managing Russian imperial natural resources—especially rivers and forests—came into conflict with the local understanding of ownership and management of the natural environment. Second, the state lacked the necessary resources and technology. As Daron Acemoglu and James Robinson (2012) have pointed out, the nineteenth-century West European revolutionary surge frightened the tsarist officials to the point that they halted the construction of new industrial factories in order to stop the penetration of revolutionary ideas in Russia. Thus Russia was not at the forefront of technological power in the nineteenth and early twentieth centuries. Until the end of the nineteenth century, imperial attempts to change the environment were impeded by the lack of specialists. Not until 1894 did the Ministry of State Property create the new Department of Land Improvement. The opening in 1894 of the Moscow Agricultural Institute created the conditions to train specialists in land draining and land improvement (Kolganov et al. 2016: 11). Multiple congresses on land amelioration issues—in Odessa in 1909, in Kyiv in 1912, in Moscow in 1913, and in Novocherkassk in 1914—were held to strengthen technical expertise in the field.

Finally, 90 percent of Russia's population lived in rural areas: the country had an administrative corps that was neither particularly effective nor sufficient for such a vast territory (Subtelny 2009 [1988]: 25). A corrupt and inefficient bureaucracy proved to be ineffective in transforming the environment so radically. In short, the lack of resources, expertise, and an efficient corps of bureaucrats meant that tsarist officials only intervened when expedient in the wetlands, mostly to carry out projects of afforestation. The aim was to "enhance" the wetlands to the point that they became useful for agricultural purposes, not to transform them into a different environment, as the Soviet state later did.

The Soviet state was a different matter; its gargantuan plan for harnessing the Dnipro River led to the disappearance of the wetlands. The Soviet state had not been hindered in its technological expansion by private landowners—they disappeared in the years following the October Revolution. Collectivization did not play a significant role in the transformation of the wetlands, at least not at the very beginning, as it was more concentrated on arable land. However, as we will show later in this chapter, the pastures in the wetlands steadily became the property of the collective farms. The fast-growing number of collectivized peasants led to the incorporation of the wetland pastures into the state and collective farms. Stalin's plan for the transformation of nature (1948) was a gargantuan attempt to change the physical environment for developmental purposes. The southern steppes were directly targeted: rivers were to be transformed into energy production machines, with plans for 45,000 dams to be constructed along the rivers, 70,000 kilometers of forest shelterbelts planted—all to improve agricultural production and the industrial sector in European parts of Soviet Russia (Josephson 2016: 1; see also Brain 2010a and Olšáková 2016). The Soviet state entered into a fierce economic competition with the Western world, and one way to close the gap was through massive investment in industrial projects regardless of the environmental consequences (Oldfield 2005). The Soviet state deployed technology, resources, and labor to build a huge dam in the south of Ukraine that led to the total destruction of the wetlands. Before the construction of the Kakhovka dam there were plans to increase the productivity of the wetlands within the collective farm. The central Soviet state proved ultimately ruthless concerning the fate of the wetlands, as we will document later in the chapter. The Soviet state had the scientific knowledge, the power, the technological means, and the political determination to erase an unwanted resource.

THE DNIPRO WETLANDS AND THE RUSSIAN IMPERIAL QUEST

At the end of the nineteenth century the southern Ukrainian Dnipro wetlands were, as wetlands elsewhere in the world, an endless source of livelihoods for the local population. The annual floods of the Dnipro River created a unique microclimate with extensive flooded areas in April and May, and small ponds that remained after the waters retreated. A multitude of backwaters and canals intertwined with mounds of land, sand, and small patches of forest created a biodiverse and complex system for animals, plants, birds, fish, and other aquatic species. The shallow, warm waters were the nursery for a variety of fish that in turn attracted other wild animals and birds. The yearly floods brought silt, which created rich fertile soils suitable for gardening (Gerchyk 1900: 1–2; Selezhinskii 1899: 2).

People living next to the marshes had a variety of ecological resources to rely on. Predictably, fishing featured heavily in the family's diet and income. Fish such as crucian carp, trench, pike, and crayfish were abundant. As in other wetlands of the world (Antipa 1910; Bakema, Howard, and Wood 2009; Payne 2012), locals took advantage of the ecology of fish spawning, attracted by warm and shallow waters. The abundance of fish provided riparian villagers with a guaranteed food supply, and the surplus was sold in nearby markets (see figure 3.1).

Riparian communities had access to a diversified ecological system in which wild fowl also played a significant role. Deer, elk, grouse, and black grouse were all part of the wetland peasant diet, and wild birds' eggs represented another significant source of food (Bilivnenko 2012).

Farming was another of the villagers' economic activities. The inhabitants of the surrounding villages cultivated gardens in the wetlands, which they planted after the water of spring floods withdrew. On fertile wetland soils, locals raised tomatoes, cucumbers, and other vegetable crops. The gardens were so important that each homestead had its own path down to the water and their plots (Bosa 2014).[3] Cereals were rarely planted in the wetland—for this there were the vast grasslands on the high steppes of the south (Moon 2013). Animals, especially sheep, grazed in the wetlands from spring to mid-autumn. Woodlands that grew in the wetlands were at the villagers' disposal, especially willow, poplar, and oak. Villagers used wood for domestic needs such as furniture, baskets, and heating. The riparian population created a complex economy that used many of the natural resources of this amphibious landscape (Riabinin 1890). No wonder this wetland cornucopia elicited from locals epithets such as "treasure," "paradise," "beauty," and "wonder" (Bilivnenko 2012).

Yet the urbanite's view of the wetlands was quite different from that of the locals. In a provincial newspaper in 1888, an unknown author portrayed the Dnipro wetlands as a trouble spot in the southern steppe, a hotbed of diseases and epizootics. The author conceded that for the landowners the wetlands was important for the livestock; still, the authorities' main task was, he posited, their draining (Anonymous 1888: 3–4).

The patchwork landscape also created a web of formal and informal tenure systems and a complex economic relationship. A statistic from the end of the nineteenth century shows that in Ekaterinoslav province 27 percent of the wetlands were commons belonging to peasant communities, most of them Ukrainians, 65.6 percent belonged to noble landowners of Russian origin, and only 7.4 percent was state owned (Selezhinskii 1899: 48). Landowners were nobles who had received

FIGURE 3.1. An image of fishing nets, most probably belonging to the fishing team from the collective farm in Kamianka-Dniprovska district, taken in the early 1950s. Copyright Kamianka-Dniprovska Museum.

land in the southern provinces immediately after the Russian Empire conquered them in the last quarter of the eighteenth century. Because wetlands became economically interesting by providing access to water and forest, it is no surprise that large landowners owned most of the wetlands (A. Boiko 2009: 51). Landowners usually leased parts of the ponds to associations of fishers and grazing land to those who did not have pastures.

After the southern steppes became part of the Russian Empire under Catherine II (1762–1796), the imperial authorities initiated scientific research studies of the entire area starting with the last quarter of the eighteenth century. The scientific inquiry would have helped them decide what to do with the new territories, how to colonize them, and how to organize the economic activity and the exploitation of their natural resources (Bekasova 2010; Moon 2010a, 2013; Peterson 2016a).[4] The scientific expeditions sponsored by the imperial state produced various kinds of data, which contained information about the climate, soil, forest, river system, minerals, flora, and fauna. Over several decades a sufficient number of ethnographic and geographical descriptions accumulated, which gave the state a clearer picture of the wetlands in the south (Afanas'ev-Chuzhbinskii 1861; Pallas 1999).[5]

Yet it was not the wetlands that were the immediate target of scientists' interests but rather the woodland that grew there. In the treeless landscape of the steppes, the woodlands were truly valued: in Ekaterinoslav province, for instance, woodland accounted for 4 percent of the whole territory, 63 percent of which was located in the wetlands (Selezhinskii 1899: 63). Despite its poor quality, wetland forest was of paramount importance to the locals for heating and construction.

In the second half of the nineteenth century Russian scientists reached the conclusion that the woodland had been more extensive in the past (see also Fedotova and Loskutova 2015). The forest scientist I. Gerchyk (1900: 5) pointed out that there was black earth underneath the mounds of sand scattered throughout the southern provinces—which he posited had been carried there by the wind. Another scientist, P. Lisitsin, considered that in the late eighteenth and early nineteenth centuries the banks of the Dnipro and of other rivers in the region such as the Donets, Kalmius, Volcha, and Konka had been covered by valuable forest that had recently disappeared. This was the reason, he argued, that these rivers once had a higher water level but they had turned into "puddles." The statistics he quoted indicated that between 1875 and 1899 the forested areas decreased by 10 percent (Lisitsin 1900: 30).

Scientists identified the culprits: locals with their practices of unsustainable forestry and overgrazing. Deforestation and local practices of cultivating hay left behind large wastelands and sandy areas and lands disfigured by deep ravines (Selezhinskii 1899: 3; Moon 2010b). The small plots of land available to peasants meant that they grazed their cattle in the wetlands. The increase in population and cattle, industrial growth, and extensive plowing of the steppes exacerbated the process. Overgrazing, claimed scientists, led to erosion of the soil and desertification (Gerchyk 1900).

Scientists blamed the local population but also the gaps in the Russian Empire's legislation. In the early eighteenth century Peter I issued laws to protect forests. But at the end of the century Catherine II abolished some of these restrictions and allowed all forests to be transferred to the full ownership of the landowners. The unauthorized felling of the forest was flourishing anyway, not only in Southern Ukraine but throughout the tsarist empire (Costlow 2009; Moon 2010b). In an attempt to stop deforestation and manage wood exploitation, Tsar Alexander III passed the Regulation on the Preservation of Forests in 1888. The law regulated the exploitation of forests, including the wetland forests. It stipulated the constitution of a Forest Protection Committee (FPC) in each province of the empire. The committee was charged with monitoring forests, including private ones, and preventing their illegal exploitation. In Ekaterinoslav, the FPC established that the forests in the wetlands and islands of Dnipro played a defensive role against flooding, preventing bank and soil erosion and impeding the formation of snowfalls and fast streams. The FPC established strips of forest 275 meters wide along the

Dnipro and 90 meters along the Konka River (Gerchyk 1900: 7). Having a protective role, the forests were exempted from exploitation.[6] The regulations included significant technical changes such as banning haymaking in the glades between the protective forests and the river's bank. Regulations also imposed bans on locals mowing or bringing cattle into newly exploited forests (Gerchyk 1900: 13). The new impositions met with opposition and discontent among the local population. They circumvented the law by ignoring it and continued their long-standing practices (Selezhinskii 1899: 47).

The need for timber for construction, for economic and military needs, and for the improvement of farming conditions in the steppe encouraged—along with the scientific conclusions of experts—the idea of afforestation (Bark 1880). Thus, in the 1870s afforestation after a clear-cut became compulsory (Lisitsin 1900). However, long before this became law locals had experimented with afforestation. In 1816 V. P. Skarzhynsky founded a tree nursery in Kherson province and from 1818 onward experimented with growing oak, pine, poplar, and other species on 656 hectares. Problems including methods of soil cultivation, planting time, choice of tree species, number of plants, and care requirements needed to be addressed if the experiments were to be successful (Selezhinskii 1899). The experimental character of this enterprise and other similar ones also drew on the experience of German colonists who managed to achieve reasonably good results in reforesting the steppe.[7] The Forestry Society founded in Molochansk district in 1830, and headed by Johann Cornies, promoted experimentation in afforestation. Besides developing a rigorous program of planting trees, Cornies also built a library of books and periodicals in German and Russian (Staples 2003: 116). German forestry was perhaps the most advanced at the time and the first to establish stations for experimental forestry (Scott 1998; Fedotova and Loskutova 2015). Despite many failures (planting pine, birch, aspen, and oak in sandy dunes was rarely successful), scientific methods to combat crop failure and drought gained ground in society at the end of the nineteenth century.

Following this experimental model, the minister of state property, Pavel Kiselev, established the Berdyansk model of forest plantation in 1841 (Staples 2017). This plantation became the main supplier of planting materials for the peasants of the southern provinces. In addition, foresters intended to educate peasants' children in the forestry business in order to encourage further afforestation of the steppes. The Horticulture Society, a branch of the Imperial Russian Society of Horticulture established in 1890, created a tree nursery the aim of which was to provide the local population with planting material (Selezhinskii 1899). In 1893, "The organization for the promotion of the afforestation of the steppes in Ekaterinoslav" (Savchuk 2014) was founded, along with the Imperial Agricultural Society of Southern Russia, which was based in Odessa and engaged in disseminating knowledge about afforestation, renewal, and conservation of forests in the steppes (Anonymous 1897: 3).

Research into wetland forests was not motivated by purely scientific interests but by the active colonization policy of the Russian government in the southern steppes. The Russian state took seriously its mission to "tame" the steppe and transform it in order to create optimal conditions for productive management (Shenrok 1874). For this purpose, the wetlands were of high scientific and political interest not for what they could offer as natural resources to the imperial state but rather as a site for experiments in afforestation and for providing invaluable services to agriculture in the steppes (Doklady 1898: 2).

The attempt to protect and multiply the wetland forests went hand in hand with the development of soil science. Of critical importance was the work of V. V. Dokuchaev, who founded the discipline of scientific soil science and soil geography (McNeill and Winiwarter 2004). He argued that deforestation led to negative consequences for steppe arable farming, such as dust storms and droughts, and caused great damage to productivity despite the richness of the chernozem (Moon 2013). Dokuchaev proposed the creation of forest belts and ponds in the steppes and the afforestation of the rivers' banks. In his understanding, afforestation meant not only planting trees but also planting shrubs such as bog willow especially on the sand dunes that spread in the wetlands (see figure 3.2).

These "natural constructions" helped retain moisture in the soil, lower the temperature of the air near the surface of the earth, and prevent hot winds. Dokuchaev's theory concerning the paramount importance of forests for agriculture and his pleas for afforestation became widespread at the end of the nineteenth century (Dokuchaev 1892). Despite his refusal to consider the wetland vegetation as forest, his work had a positive impact on wetlands. For instance, the FPC in Ekaterinoslav recognized the need for the afforestation of sandy areas of wetlands due to his work. They realized that a lack of forest leaves steppes completely unprotected against the dry east wind (*sukhovey*) (Gerchyk 1900: 14–15).

During the imperial era wetlands were scientifically invisible. The object of scientific inquiry was the woodland associated with wetlands and their paramount importance for the agriculture of the steppes. This is hardly surprising as most "wetland scientists" were forest scientists and administrators. Iosiph Selezhinskii was the head of the Zurin forestry department. Ernest Shenrok, the author of works on the history of foreign forestry and coauthor of the 1884 Forest Management Instruction was vice-director of the Forest Department of the Russian Empire. Ivan Akinfiev (1851–1919) was a scientist and educator from Ekaterinoslav (Savchuk 2014). He was a member of the organization for the promotion of the

FIGURE 3.2. Mounds of sand (dunes) in the wetlands near Kamianka-Dniprovska. Photo taken in the early 1950s. Copyright Kamianka-Dniprovska Museum]

afforestation of the steppes of Ekaterinoslav province and of the Horticultural Society. Some of them, as in the case of Shenrok, were intimately acquainted with the Western models of forestry, especially the German, Austrian, and French ones. He made a research trip to all these countries in order to study their forestry systems (*Obzor Ustroistva lesnago upravleniia* 1889). Experimentation and organizational ideas were debated in international conferences where Russian foresters also attended and contributed their experience (Fedotova and Loskutova 2015).

For the imperial administration, the steppes' wetlands and forests were not just an innocent object of scientific study but a means to an end: the growing productivity of the steppes. The Russian state was interested in taming a territory that was considered wild in a triple sense: its nature was unsettled, and its culture was both "savage" and economically backward. Wetlands and steppes—not only in Ukraine but in many places around the globe—were considered uninhabited and unproductive lands (Dove and Kammen 2015; Cybriwsky 2017). As uninhabited territories they were constantly reclaimed by the state for "development" and reconversion to make them more economically productive. The populations that inhabited these territories when Catherine the Great first took them over—nomadic and settled pastoralists, then Ukrainian peasant emigrants—were perceived as uncivilized and backward (O'Neill 2017). I. V. Riabinin (1890: 41) described the primitive economy of the wetlands in the following terms: "Getting acquainted with wetland economy takes us far back into the history of agriculture when man was still content only with what nature gave him without any efforts on his part; the extensive exploitation of the land and the complete dependence of the economy on spontaneous influences without any introduction of improved methods of exploitation of lands constitute the dominant feature of economic management in the wetlands." Thus, the Russian imperial state acted as a "high modernist state" (Scott 1998). Science was the main driver in transforming the wasteland into a productive land. However, we emphasize two important features in our story. One is that the local elite was also interested in developing the new territory: the local composition of the Horticultural Society or the organization for the promotion of the afforestation of the steppes in Ekaterinoslav shows that clearly. A second feature is that some of the knowledge developed by Russian and Ukrainian scientists was derived from Western knowledge of forestry (especially German forestry), which employed new mathematical methods and developed strict plans for exploitation that maintained a balance between extraction and afforestation. This was a magnet for regional forestries (see, for instance, Radkau 1996; Scott 1998; Vandergeest and Peluso 2006a, 2006b). Its utilitarian scope was a powerful tool in the hands of the state that wielded it in order to control and organize territories and people. By organizing its territory and emphasizing its "productivity," imperial Russia mirrored larger trends taking place in a capitalist context. This exploitative relationship with the environment served the ruling classes and was linked to capitalist principles. Yet often this relationship was established with the assistance of the state.[8]

THE END OF WETLANDS

After the October Revolution in 1917, the Soviets seized power in Russia. They gradually applied a program of nationalizing private enterprises and the collectivization of agricultural land. Large land and forest properties were collectivized or transformed into state farms (Conquest 1986). After the installation of Bolshevik power, as with all Soviet republics, Ukraine was involved in the process of industrial and agrarian transformation of nature. The steppe regions continued to be targeted as their climatic conditions—high temperatures, unstable and low precipitation, drought, and dry winds—impeded high agricultural production. The early Soviet governments considered that the nature of the steppe had to be "corrected" and the wetlands, still considered wasteland, might turn out to be helpful. Aleksandr Yaroshevsky, whose proposal for the radical transformation of the "wasted land" opened this chapter, suggested that the wetlands should supply water to a network of irrigation systems throughout the southern steppes. Wetlands, he posited, could be useful as long as a series of dams and canals were constructed that would allow both reclamation of agricultural land and protection against floods for the inhabited areas. Moreover, he envisaged using the wetlands to grow rice, but only if an irrigation system could be constructed that would enable the water level to be permanently monitored.

In 1925 the All-Ukrainian Commission on the Land Management of Jewish Workers under the Central Executive Committee of the Ukrainian SSR organized a complex survey of the Dnipro wetlands (geodesic, hydrological, botanical, and statistical-economic). The aim of the study was to gain a clear picture of the way locals used the wetlands as a resource and to find more productive ways to use them.[9] At the end of the 1920s, the Dnipro wetlands were gradually allocated to collective and state farms while local villagers received small plots of 0.15 hectares per family. The collective farms used the wetland natural pastures to graze their animals (Titov 1946). The practice was accompanied by the building of infrastructure: roads, bridges, barns and stables, houses for the caretakers, and ferries to cross the Konka River to transport hay and vegetables from individual and auxiliary gardens were all part of a productive wetland plan (Anonymous 1945).

Yet wetlands were not an environment easily adapted to economic plans. Their specific vegetation, shrubs, the thickness of perennial weeds, and stumps interfered with Soviet plans for productive land use. In addition, the annual floods left behind cover vegetation residues that delayed the grass growth; tree stumps prevented the mechanization of hay gathering. Thus, regional authorities tasked some members of the collective farms with finding ways to improve the quality of wetlands meadow grass and increasing their yield. To achieve that, special teams were organized to tame the wetland (Iakovlev and Chalenko 1948) whereas the scientific staff of the Kamensky irrigation pilot field study indicated that the Dnipro wetlands required the implementation of special agrotechnical measures. Clearing the stumps, destruction of cover vegetation, weed control, the collection of high value grass seeds, and land improvements were a few of the measures taken by the task force (Iakovlev and Chalenko 1948).

Between 1920 and 1947 different regional institutions initiated projects that proved to be a continuation of policies from imperial times. The amelioration of wetlands, plantation of forest belts, land reclamation of flooded areas, and irrigation of the steppes were the major plans for the Dnipro and its wetlands. These plans were different from those of the imperial period as they pushed for building dams and levees, having as their model those built on the Po and Mississippi Rivers (Nizhnii Dnepr 1930: 52). This is hardly a surprise. As Maya Peterson (2016b) has shown, US expertise and technical ingenuity helped Soviet Russia to change the Russian landscape and to implement agricultural infrastructures in Russian provinces of Central Asia in the late 1920s and early 1930s. In 1948, 4,615 hectares of wetlands were transferred to the use of collective and state farms of the Kamianka-Dniprovska district. Special teams of geodesists, ameliorators, and agronomists had been tasked to "find the best fields, in order to make full use of their richness" (Skripets', 1948).

There were also economic plans at the central level that superseded regional interventions to "correct" nature. Stalin's plan for the transformation of nature that was approved in 1948 (Josephson 2002) was just such an attempt to correct the "unpleasant" features of the steppe, including wetlands.[10] Thus, in 1950 the government initiated the construction of the Kakhovka hydroelectric power station on the river Dnipro. This dam was part of a cascade of reservoirs on the Dnipro: starting with the one in Zaporizhia in 1927 and finishing at the end of the 1970s with Kakhovka, Kremenchug, Kyiv, Dniprodzerzhynsk, and Kanev hydroelectric power stations (Cybriwsky 2017). The entire Promethean program embodied Lenin's definition of Communism: "Soviet power plus the electrification of the whole country" (Lenin 1970: 159).

The documentation for the construction of the Kakhovka dam and its reservoir was revised several times in an attempt to reduce the costs. The Decree of the Council of Ministers of the USSR of July 12, 1952, increased the capacity of the reservoir from 14 to 19 cubic kilometers.[11] The reservoir was filled between 1955 and 1958, and had a surface of 2,150 square kilometers (Borodaenko and Golubytskii 2010). Wetlands, at least their lower parts, disappeared forever under the waters of the reservoir. As everywhere that the high modernist state's quixotic plans to dam rivers were put into practice (see, for instance, McCully 2001; Isaacman and Isaacman 2013; Štanzel 2017), nature and people were drastically reshuffled for economic reasons. Around 34,000 people—9,210 households from 75 settlements—were resettled and 210,600 hectares of land were flooded.[12]

Unlike the other hydropower plants built on the Dnipro, Kakhovka was not built exclusively to generate electricity but also as a reservoir to irrigate the fertile but arid lands in Kherson, Zaporizhia region, and Crimea.[13] Its construction was seen as the final phase of the locals' ancient struggle with the aridity of the steppes.[14]

After building the dam and the reservoir a large number of popular and scientific publications praised the changes made to Ukraine's steppes and consequently to the life of people from the region brought about by Soviet science and technology. One of the publications that included the reservoir in recommended excursion routes of the area compared the wealth produced by wetlands with the wealth brought by the reservoir. The author acknowledged the importance of wetlands in the life of locals since time immemorial but attributed the same value to the reservoir (Zhigalov 1962). Special attention was given to the heroic workers. Following

Ivan Michurin's motto, "We cannot wait for kindnesses from nature; our task is to wrest them from her" (Brain 2010b: 93), they were praised for subduing nature to human needs. Scientists also tried to create a positive image for the Kakhovka reservoir. The reservoir was not even finished when the first students started to write their diplomas, and scientists began to give conferences on the new hydrotechnical construction (Olenenko 2019).

Not every expert was thrilled about the disappearance of the wetlands. Responding to a call before the final vote on the construction of the dam, several experts examined the plans of the Ukrainian branch of the institute Hydroenergoproekt, the body responsible for planning the dam.[15] Some suggested that "the need to protect should be dictated by the economic importance of the industrial facilities and settlements" such as the city of Nikopol, the Kamensky Pod, and the manganese deposits near Nikopol.[16] The wetland did not qualify for such protection: "The project of protective measures against flooding of the wetlands is currently not recommended because of too large one-time capital investments, both for the arrangement of protective measures and for works on amelioration and agricultural development of the area."[17] Other experts expressed their concerns about the wetlands and suggested possible solutions. M. Vaskov, a senior engineer of the Department of Hydraulic Structures issued a report on January 27, 1952, in which he suggested reasons for protecting the Dnipro wetlands: they were valuable pastures, a source of vegetables and dairy products for the population in the industrial centers such as Zaporizhia, Dnipropetrovsk, and Krivy Rih, and the only places where valuable crops such as kenaf and southern hemp grow.[18] He continued to make new arguments in favor of protecting the wetlands: their ecological importance and their role as an oasis of the steppes. In contrast, their flooding would inevitably form large areas of rotten swamps and shoal near industrial centers, which would have major negative consequences.[19] He considered failing to protect the wetlands to be "a political and economic error."[20] Other experts such as P. Sherstnev concurred with this conclusion and pointed out that a rational organization of agriculture in the wetlands would repay the depreciation charge within ten to twelve years.[21] I. Sarafanov issued a report in January 1952 in which he showed that in arid years the wetlands functioned as a safety net for the collective farms' cattle.[22] All these experts as well as others criticized the narrow vision of the project, insisting that the wetlands were irreplaceable environments. Their assessments were nevertheless ignored and the wetlands were finally flooded.

In our reading these experts' positive assessments of the value of the wetlands shared the vision prevalent in the tsarist period. Wetland was prized not only for its own value but also for the value brought to agriculture. The projects from early Soviet times that aimed at the transformation of the wetlands continued to regard wetlands as an adjunct to agriculture. The central Soviet authorities' vision for the wetlands was in stark contrast to that of experts who advocated their protection and use for agriculture. By constructing the Kakhovka hydropower station and the reservoir, the central Soviet authorities intended to create a new landscape that differed radically from the previous one. By controlling the wetlands, making them disappear under a new man-made sea, the regime consolidated its economic and political presence in the territory. A technological environment controlled exclusively by the party replaced an environment that had been difficult to control.

TWO VISIONS FOR THE WETLANDS

Maxim Gorky's famous assertion that "man in changing nature, changes himself" (Brain 2010b: 93) expresses a false image. It is not individual humans who change nature but rather various political factions with their ideas about how certain spaces should look and function. A wetland represents many things to many people. The mercantile and instrumental vision of wetland was shared by two different political regimes. Perceived equally as a territory in which mosquitoes and malaria reign but also as a potential economic golden goose, the wetlands were doomed to change. Different projects that envisioned their transformation aimed at suppressing one feature, the permanent negotiation of water and land through yearly floods, by enforcing a stable environment. What was important for projects that aimed at transforming the wetlands was the stability of the man-made landscape, the lack of surprises from nature, and its predictability for humans and their economy.

Yet there are stark differences between the plans of tsarist and early Soviet experts for transforming wetlands and the plans of the central Soviet authorities from the 1950s. It is true that for both political regimes the transformation of wetlands was a means to an end: the control of a peripheral territory and the imposition of a vision of economic development that reflected central state interests. The difference between the two political regimes was in the magnitude of the development projects scheduled for the area. We explain the difference by considering the different resources, technological means, and experts the two regimes had at their disposal for the transformation of the wetlands. The foundation of universities that trained such experts came very late for the tsarist administration. For instance, the Moscow Agricultural Institute that trained land amelioration engineers was founded in 1894, and the Soviet regime benefited from this technical corps trained before the October Revolution. One of the authors involved in the GOELRO plan—the plan elaborated in December 1920 by Moscow to electrify the country—emphasized this fact in a report of 1925: a series of engineering organizations had worked on the construction of hydroelectric stations along the Dnipro "for the past twenty years" (Aleksandrov 1925: 3). The tsarist regime had an administrative bureaucracy and a propaganda machine that could hardly have matched those of the Soviets. Besides, the collectivization process dislodged the large landowners who might have opposed such a radical transformation of the wetlands.

In the southern steppes of Ukraine transforming wetlands was a way to express economic modernity, and the high modernist project found in the wetlands the perfect environment to be changed and adapted to the state's needs.

NOTES

We thank Marie Louise Aastrup, George Andrei, Liviu Giosan, Alexandra Ion, Hande Özkan, and Stelu Şerban for reading the chapter. Their comments, questions, and suggestions were very helpful as we elaborated the final draft. David Moon and Katja Bruisch's crystal-clear suggestions proved invaluable for the construction of the text. We thank Sally Fenn, our copyeditor, for her thorough work. Needless to say, we alone bear responsibility for any errors in the text.

1. However, for work on wetlands see Bruisch (2019, 2020). In this chapter, we do not provide a literature review of studies focusing on the Russian steppes. For a comprehensive review, see Moon (2013) and O'Neill (2017).

2. We cannot discuss the history of the Cossacks here. Initially they were a population of mixed ethnic backgrounds, runaway peasants who lived in unsettled peripheries. They became the defenders of the territories along the Dnipro and its lower tributaries against the Ottomans and the Tartars. For a detailed history of the Cossacks, see Subtelny (2009 [1988]).

3. For gardening in the Ukrainian part of the Danube Delta but in a different time, see Richardson (2015).

4. For a detailed history of Ukraine and its split between the Russian Empire in the East and the Habsburg Empire, see Kappeler (2007), Subtelny (2009), and Magocsi (2010).

5. See also Puteshestvennyia (1787); and Voenno-statisticheskoe (1850).

6. According to the 1888 law, each local FPC had to establish clearly which forest had protective status and which did not.

7. The tsarist government invited foreign colonists, including Germans, to settle in the steppe regions starting in the second half of the eighteenth century. The Russian government attempted to encourage the development of trade, industry, and agriculture on the vast unoccupied lands of the empire (Bartlett 1979: 3). The tsarist government offered the colonists land and various privileges (Moon 2013: 3).

8. We thank Katja Bruisch for this valuable observation.

9. Gosudarstvennyi arkhiv Rossiiskoi Federatsii [State Archive of the Russian Federation] (GARF), f. R 7541, op. 1, d. 161., ll. 1–34

10. For more details, see Brain (2010a) and Olšáková (2016).

11. GARF, f. R 5446, op. 86a, d. 6210, l. 56

12. GARF, f. R 5446, op. 86, d. 940, l. 151; GARF, f. R 5446, op. 86a, d. 6210, ll. 71–72.

13. Decree no. 4000, September 20, 1950, of the Council of Ministers of the Soviet Union.

14. GARF, f. R 5587, op. 27, d. 124, ll. 24, 60–66.

15. GARF, f. R 5446, op. 86a, d. 6210, ll. 74–75.

16. Quote from Rossiiskii gosudarstvennyi arkhiv ekonomiki [Russian State Archive of

Economics] (RGAE), f. 339, op. 1, d. 632, ll. 47–48; RGAE, f. 339, op. 1, d. 632, ll. 56, 74–75; and GARF, f. R 5446, op. 87, d. 526.

17. RGAE, f. 339, op. 1, d. 632, ll. 56, 74–75; and GARF, f. R 5446, op. 87, d. 526.

18. RGAE. f. 339, op. 1, d. 632, ll. 116–21.

19. RGAE, f. 339, op. 1, d. 632, l. 117.

20. RGAE, f. 339, op. 1, d. 632, l. 118.

21. RGAE, f. 339, op. 1, d. 632, ll. 47–70.

22. RGAE, f. 339, op. 1, d. 632, l. 84.

REFERENCES

Acemoglu, Daron, and James A. Robinson. 2012. *Why Nations Fail: The Origins of Power, Prosperity, and Poverty*. New York: Crown.

Adger, Neil W., and Cecilia Luttrell. 2000. "Property Rights and the Utilisation of Wetlands." *Ecological Economics* 35(1): 75–89.

Afanas'ev-Chuzhbinskii, Aleksandr Stepanovich. 1861. *Poezdka v Iuzhnuiu Rossiiu*. Ch.1: *Ocherki Dnepra* [A trip to Southern Russia. Part 1: Essays on the Dnipro]. St. Petersburg: Tip. Mor. min-va.

Aleksandrov, I. G. 1925. *Dneprovskoe stroitel'stvo i ego ekonomicheskoe znachenie* [Dnipro construction and its economic importance]. Kharkov: Gosudarstvennoe izdatel'stvo Ukrainy.

Anonymous. 1888. "Korrespondentsiia" [Correspondence]. *Ekaterinoslavskiie gubernskie vedomosti* 8: 3–4.

Anonymous. 1897. "Khronika" [Chronicle]. *Ekaterinoslavskiie gubernskie vedomosti* 202: 3.

Anonymous. 1945. "Parom cherez Konku" [Ferry across the Konka]. *Udarnik*, September 16: 2.

Antipa, Grigore. 1910. *Regiunea inundabila a Dunarii: Starea ei actuala si mijloacele de a o pune in valoare* [The Danube floodplain: its current status and the means to exploit it]. Bucharest: Institutul de Arte Grafice Carol Göbl.

Bakema, Reint Jacob, Geoffrey W. Howard, and Adrian P. Wood. 2009. "Multiple Use of Wetlands in Eastern Africa." In E. Maltby and T. Barker, eds., *The Wetlands Handbook*. Chichester: Wiley-Blackwell, 850–75.

Bark, L. 1880. *Lesorazvedenie na iuge Rossii* [Afforestation in the south of Russia]. St. Petersburg: Tip. V. F. Demakova.

Bartlett, Roger P. 1979. *Human Capital: The Settlement of Foreigners in Russia 1762–1804*. Cambridge: Cambridge University Press.

Bekasova, A. V. 2010. "Izuchenie Rossiiskoi imperii ekspeditsiiami 1760–1780-kh gg.: 'vzgliad' estestvoispytatelei i formirovanie predstavlenii o gosudarstvennykh bogatstvakh" [Studies of the Russian Empire by expeditions of the 1760s–1780s: the "view" of naturalists and the formation of concepts of state wealth]. *Istoriko-Biologicheskie Issledovaniia* 2(4): 13–34.

Bilivnenko, Sergij. 2012. "Rika Dnipro u povsjakdenni sil's'kogo pryberezhnogo naselennja: za materialamy istoryko-etnografichnyh ekspedycij Zaporiz'kogo naukovogo tovarystva im. Ja. Novyc'kogo" [The Dnipro River in the everyday life of the rural coastal population: based on materials of historical and ethnographic expeditions of the Ya. Novitsky Zaporizhia Scientific Society]. In V. Andreev et al., eds., *Dniprovs'ki locmany: narysy z istorii ta istoriografii* [The Dnipro pilots: essays on history and historiography]. Kherson: Kherson State Maritime Institute Publishing House, 11–24.

Blackbourn, David. 2006. *The Conquest of Nature: Water, Landscape, and the Making of Modern Germany*. New York: W. W. Norton.

Boiko, Anatolii. 2009. *Opisi Stepovoi Ukraini ostann'oi chverti XVIII–pochatku XIX stolittia* [Descriptions of the Ukrainian Steppe in the last quarter of the eighteenth to the beginning of the nineteenth century]. Compiled by Anatolii Boiko. (Dzherela z istorii Pivdennoi Ukraini, vol. 10). Zaporizhzhia.

Boiko, Ia. V. 1993. *Zaselenie Iuzhnoi Ukrainy: 1860–1890 gg.* [The settlement of Southern Ukraine: 1860–1890]. Cherkassy: Siiach.

Borodaenko, A. S., and A. V. Golubytskii. 2010. "Kakhovskoi GES 55 let" [Fifty-five years of the Kakhovka hydroelectric power station]. *Gidroenergetika Ukraini* 4: 51–54.

Bosa, Liubov. 2014. "Osoblivosti traditsiinogo prirodokoristuvannia v landshaftakh Pivdennoi Naddniprianshchini: transformatsiini ta deformatsiini protsesy" [Features of traditional nature management in the landscapes of the southern Dnipro region: transformation and deformation processes]. *Narodna tvorchist' ta etnologiia* 2: 46–55.

Brain, Stephen. 2010a. "The Great Stalin Plan for the Transformation of Nature." *Environmental History* 15(4): 670–700.

Brain, Stephen. 2010b. "Stalin's Environmentalism." *Russian Review* 69: 93–118.

Bruisch, Katja. 2019. "The State in the Swamps: Territorialization and Ecosystem Engineering in the Western Provinces of the Late Russian Empire." *Zeitschrift für Ostmitteleuropa-Forschung / Journal of East Central European Studies* 68(3): 345–68.

Bruisch, Katja. 2020. "Nature Mistaken: Resource-Making, Emotions and the Transformation of Peatlands in the Russian Empire and the Soviet Union." *Environment and History* 26(3): 359–82.

Bruisch, Katja, and Klaus Gestwa. 2016. "Expertise and the Quest for Rural Modernization in the Russian Empire and the Soviet Union." *Cahiers du monde russe* 57(1): 1–21.

Cioc, Mark. 2002. *The Rhine: An Eco-biography, 1851–2000*. Seattle: University of Washington Press.

Conquest, Robert. 1986. *The Harvest of Sorrow: Soviet Collectivization and the Terror-Famine*. New York: Oxford University Press.

Costlow, Jane. 2009. "Who Holds the Axe? Violence and Peasants in Nineteenth-Century Russian Depictions of the Forest." *Slavic Review* 68(1): 10–30.

Cybriwsky, Roman A. 2017. *Along Ukraine's River: A Social and Environmental History of the Dnipro*. Budapest: CEU Press.

Doklady. 1898. "Doklady N. I. Alekseeva na s'ezde lesnichikh" [N. I. Alekseev's Reports at the Congress of Foresters]. *Ekaterinoslavskie gubernskie vedomosti* 179 (supplement): 2.

Dokuchaev, V. V. 1892. *Nashi stepi prezhde i teper'* [Our steppes before and now]. St. Petersburg.

Dorondel, Stefan. 2016. *Disrupted Landscapes: State, Peasants and the Politics of Land in Postsocialist Romania*. Oxford: Berghahn Books.

Dove, Michael R., and Daniel Kammen M. 2015. *Science, Society and the Environment: Applying Anthropology and Physics to Sustainability*. London: Routledge.

Fedotova, Anastasia A., and Marina V. Loskutova. 2015. "Forests, Climate and the Rise of Scientific Forestry in Russia: From Local Knowledge and Natural History to Modern Experiments (1840s–early 1990s)." In D. Phillips and S. Kingsland, eds., *New Perspectives on the History of Life Sciencesand Agriculture*. Dordrecht: Springer, 113–37.

Gerchyk, I. 1900. *Plavnevye lesa Ekaterinoslavskoi gubernii i ikh zashchitnoe znachenie* [The wetlands of the Ekaterinoslav province and their protective value]. St. Petersburg: Tipografiia Spb. gradonachal'stva.

Iakovlev, S., and N. Chalenko. 1948. "Pidvishchiti rodiuchist' kormovikh ugid" [Increase the fertility of forage lands]. *Chervone Zaporizhzhia*, February 14: 3.

Isaacman, Allen F., and Barbara S. Isaacman. 2013. *Dams, Displacements, and the Delusion of Development*. Athens: Ohio University Press.

Josephson, Paul R. 2002. *Industrialized Nature: Brute Force Technology and the Transformation of the Natural World*. Washington, DC: Island Press.

Josephson, Paul R. 2016. "Introduction. The Stalin Plan for the Transformation of Nature, and the East European Experience." In Olšáková, *In the Name of the Great Work*, 1–41.

Kappeler, Andreas. 2007. *Mala istoria Ukraini* [A short history of Ukraine]. Kyiv: K. I. C.

Kolganov, A. V. et al. 2016. *Razvitie melioratsii zemel' sel'skokhoziaistvennogo naznacheniia v Rossii* [Development of reclamation of agricultural land in Russia]. Novocherkassk: RosNIIPM.

Lenin, V. I. 1970. "Doklad Vserossiiskogo Tsentral'nogo Ispolnitel'nogo Komiteta i Soveta Narodnykh Komissarov o vneshnei i vnutrennei politike" [Report of The Russian Central Executive Committee and Council of People's Commissars on Foreign and Internal Policy (December 22, 1920). In V. I. Lenin, *Polnoe sobranie sochinenii* [Collected works], 5th ed., vol. 42 (noiabr' 1920–mart 1921) [November 1920–March 1921]). Moscow: Izdatel'stvo politicheskoi literatury.

Lisitsin, P. 1900. "Lesa i lesorazvedenie v Ekaterinoslavskoi gubernii (s 1867 po 1899 g.)" [Forests and afforestation in Ekaterinoslav province (from 1867 to 1899)]. In *Ekaterinoslavskaia guberniia: Pamiatnaia knizhka i adresnyi kalendar' na 1900 god* [Ekaterinoslav province: commemorative book and address calendar for 1900]. Ekaterinoslav: Tipo-litografiia Gubernskogo pravleniia, 29–80.

Magocsi, Paul R. 2010. *A History of Ukraine: The Land and Its Peoples*. Toronto: University of Toronto Press.

Maltby, Edward, and Tom Barker, eds. 2009. *The Wetlands Handbook*. Chichester: Wiley-Blackwell.

McCully, Patrick. 2001. *Silenced Rivers: The Ecology and Politics of Large Dams*. London: Zed Books.

McNeill, John R., and Verena Winiwarter. 2004. "Breaking the Sod: Humankind, History and Soil." *Science* 304: 1627–29.

Mikhail, Alan. 2011. *Nature and Empire in Ottoman Egypt: An Environmental History*. Cambridge: Cambridge University Press.

Moon, David. 2010a. "The Russian Academy of Sciences Expeditions to the Steppes in the Late Eighteenth Century." *Slavonic and East European Review* 88(1–2): 204–36.

Moon, David. 2010b. "The Destruction of Woodlands in the Steppe Region, 1760–1914." *Istoriko-biologicheskie issledovaniia* 2(4): 51–62.

Moon, David. 2013. *The Plough That Broke the Steppes: Agriculture and Environment on Russia's Grasslands 1700–1914*. Oxford: Oxford University Press.

Moore, James W. 2016. "Introduction: Anthropocene or Capitalocene? Nature, History and the Crisis of Capitalism." In J. W. Moore, ed., *Anthropocene or Capitalocene? Nature, History and the Crisis of Capitalism*. Oakland, CA: PM Press, 1–11.

Nizhnii Dnepr. 1930. *Rabochaia gipoteza* [Working hypothesis]. Moscow.

Obzor Ustroistva lesnago upravleniia v Germanii, Avstrii i Frantsii: izvlechenie iz otcheta po zagranichnoi komandirovke Vitse-Direktora Lesnago Departmenta E. I. Shenroka [Overview of the Forest Management Office in Germany, Austria, and France: extract from the report of the business trip abroad of the vice-director of the Forestry Department E. I. Shenrock]. 1889. St. Petersburg: Tipo-Litografiia L. Bermana i G. Rabinovicha. http://www.book-ist.ru/wood/index.html.

Oldfield, Jonathan D. 2005. *Russian Nature: Exploring the Environmental Consequences of Societal Change.* Aldershot: Ashgate.

Olenenko, Anna. 2019. "'Novoe nashe more—novoe nashe gore': konflikt mezhdu ukrainskim i sovetskim v bor'be za konstruirovanie landshafta Nizhnego Podneprov'ia" ["Our New Sea Is Our New Sorrow": The Conflict between the Ukrainian and the Soviet in the Struggle for the Design of the Lower Dnieper's Landscape]. *Ab Imperio* 1: 125–52.

Olšáková, Doubravka. 2016. *In the Name of the Great Work: Stalin's Plan for the Transformation of Nature and Its Impact in Eastern Europe.* New York: Berghahn Books.

O'Neill, Kelly. 2017. *Claiming Crimea: A History of Catherine the Great's Southern Empire.* New Haven, CT: Yale University Press.

Pál, Viktor. 2017. *Technology and the Environment in State-Socialist Hungary: An Economic History.* Cham: Palgrave Macmillan.

Pallas, Petr Simon. 1999. *Nabliudeniia, sdelannye vo vremia puteshestviia po iuzhnym namestnichestvam Russkogo gosudarstva v 1793–1794 godakh* [Observations made during a trip to the southern governorates of the Russian State in 1793–1794]. Translation from German. Moscow: Science.

Payne, R. J. 2012. "A Longer Term Perspective on Human Exploitation and Management of Peat Wetlands: The Hula Valley, Israel." *Mires and Peat* 9: 1–9.

Peterson, Maya. 2016a. "Engineering Empire: Russian and Foreign Hydraulic Experts in Central Asia, 1887–1917." *Cahiers du monde russe* 57(1): 1–21.

Peterson, Maya. 2016b. "US to USSR: American Experts, Irrigation, and Cotton in Soviet Central Asia, 1929–32." *Environmental History* 21: 442–66.

Peterson, Maya. 2019. *Pipe Dreams: Water and Empire in Central Asia's Aral Basin.* Cambridge: Cambridge University Press.

Pravilova, Ekaterina. 2014. *A Public Empire: Property and the Quest for the Common Good in Imperial Russia.* Princeton, NJ: Princeton University Press.

Puteshestvennyia. 1787. *Puteshestvennyia zapiski Vasiliya Zueva ot S. Peterburga do Khersona v 1781 i 1782 godu* [Travel notes of Vasily Zuev from St. Petersburg to Kherson in 1781 and 1782]. St. Petersburg: Pri Imperatorskoi Akademii nauk.

Radkau, Joachim. 1996. "Wood and Forestry in German History: In Quest of an Environmental Approach." *Environment and History* 2(1): 63–76.

Riabinin, I. V. 1890. *Svinovodstvo v plavniakh* [Pig farming in the wetlands]. St. Petersburg: Tip. V. Demakova.

Richardson, Tanya. 2015. "(In)Accesible Land: The Changing Practices and Regulation of Gardening in the Reed Beds of Ukraine's Danube Delta." In K. van Assche and C. Iordachi, eds., *The Biopolitics of the Danube Delta: Nature, History, Policies.* Lanham, MD: Lexington Books, 197–222.

Savchuk, V. S. 2014. "Lisivniki Katerinoslava v istorii lisovoi spravi v Ukraini" [Ekaterinoslav foresters in the history of forestry in Ukraine]. *Materiali 15-i Vseukrains'koi naukovoi konferentsii "Aktual'ni pitannia istorii nauki i tekhniki" (Kyiv, 29 veresnia–1 zhovtnia 2014 r.)* [Materials of the fifteenth Ukrainian scientific conference "Topical issues in the history of science and technology" (Kyiv, September 29–October 1, 2014)], 222–25.

Scott, James C. 1998. *Seeing Like a State: How Certain Schemes to Improve Human Condition Have Failed.* New Haven, CT: Yale University Press.

Selezhinskii, Iosiph. 1899. *Lesa Ekaterinoslavskoi gubernii* [Forests of Ekaterinoslav province]. St. Petersburg: Spb. gradonachal'stva.

Shenrok, Ernest. 1874. *Oblesenie stepei Ekaterinoslavskoi gubernii sredstvami kazny* [Afforestation of the steppes of the Ekaterinoslav province by the treasury]. St. Petersburg: Maikova.

Skripets', I. 1948. "Za visokii urozhai kartopli" [For a high yield of potatoes]. *Chervone Zaporizhzhia*, February 14: 3.

Stahl, Johannes. 2010. *Rent from the Land: The Political Ecology of Postsocialist Rural Transformation*. London: Anthem Press.

Štanzel, Arnošt. 2017. *Wasserträume und Wasserräume im Staatssozialismus: Ein umwelthistorischer Vergleich anhand der tschechoslowakischen und rumänischen Wasserwirtschaft 1948–1989*. Göttingen: Vandenhoeck und Ruprecht.

Staples, John R. 2003. *Cross-Cultural Encounters on the Ukrainian Steppe: Settling the Molochna Basin, 1783–1861*. Toronto: University of Toronto Press.

Staples, John R. 2017. "Iogann Kornis i osnovanie Berdianskogo lesnichestva [Johann Cornies and the foundation of Berdyansk forestry]." In *Voprosy germanskoi istorii*. Dnipro: RVV DNU, 18–21.

Subtelny, Orest. 2009 [1988]. *Ukraine: A History*. Toronto: University of Toronto Press.

Titov, I. 1946. "Zimovka skota v plavniakh" [Wintering cattle in the wetlands]. *Kamensko-Dneprovskii gosudarstvennyi istoriko-arkheologicheskii muzei* [Kamenka-Dneprovskaia State Historical and Archaeological Museum]. F. 1.

Tvedt, Terje. 2016. *Water and Society: Changing Perceptions of Societal and Historical Development*. London: I. B. Tauris.

van Dam, Petra J. E. M. 2016. "An Amphibious Culture: Coping with Floods in the Netherlands." In P. Coates, D. Moon, and P. Warde, eds., *Local Places, Global Processes: Histories of Environmental Change in Britain and Beyond*. Oxford: Windgather Press, 78–93.

Vandergeest, Peter, and Nancy Lee Peluso. 2006a. "Empires of Forestry: Professional Forestry and State Power in Southeast Asia, Part 1." *Environment and History* 12(1): 31–64.

Vandergeest, Peter, and Nancy Lee Peluso. 2006b. "Empires of Forestry: Professional Forestry and State Power in Southeast Asia, Part 2." *Environment and History* 12(4): 359–93.

Voenno-statisticheskoe. 1850. *Voenno-statisticheskoe obozrenie Rossiiskoi imperii izdavaemoe po vysochaishemu poveleniiu pri 1-m otdelenii Departamenta General'nogo shtaba. Ekaterinoslavskaia guberniia* [Military Statistical Review of the Russian Empire, Published by the Highest Command at the First Branch of the Department of the General Staff. Ekaterinoslav Province], vol. 11, part 4. St. Petersburg: Tipografiia Departamenta General'nogo shtaba.

Weiner, Douglas. 1999. *A Little Corner of Freedom: Russian Nature Protection from Stalin to Gorbachev*. Berkeley: University of California Press.

Williams, Michael. 1991. "The Human Use of Wetlands." *Progress in Human Geography* 15(1): 1–22.

Worster, Donald. 1992. *Rivers of Empire: Water, Aridity, and the Growth of the American West*. Oxford: Oxford University Press.

Yaroshevsky, Aleksandr M. 1925. "Hydrologiceskie osobenosti nizovia dolini r. Dnestra i metodi gradishei melioratii Dnestroveskykh plavnei i terras" [Hydrological features of the lower Dniester River valley and methods for the forthcoming melioration of the Dniesterian wetlands and terraces]. *Trudy Iuzhnoi oblastnoi meliorativnoi organisatii* [Proceedings of the Southern Province Melioration Organization], vol. 4. Odessa: 3–43.

Zhigalov, A. 1962. *Kakhovskoe more: selo Belen'koe* [Kakhovka Sea: Belenkoe village]. Zapor-ozh'e: Knizhno-gazetnoe izdatel'stvo.

PART II

Nature, Economy, and Experts

4

From Weeds to Commodities

The Translation of Plants into Medicines in Early Twentieth-Century Transylvania

Ágota Ábrán

> Many do not realize what profit is hidden in this or that common weed, which could change into ringing profits in the hands of people. . . . Great value goes to waste year after year because the precious medicinal plants die by the side of the road or in the fields, without anyone seeing any gain from them. They wither there the same way the spring sun left them after luring them out of the soil—even though the gathering and drying of these plants would have brought a nice income for those who know how to handle them.
>
> —Béla Páter

This chapter proposes to analyze how scientists at the beginning of the twentieth century in Transylvania tried to move (medicinal) plants into a commodity state and make "idle" villagers into workers earning money. They did this in the hopes of boosting the economy of rural areas, improving the struggling economy of the country (first Hungary, then Romania), helping pay war debts, and eliminating dependency on imported medicines. It is a fascinating account of high modernism because it showcases its particularities in a peripheral context: botanists, chemists, and agronomists were aware of Transylvania's place as an exporter of cheap raw material in the global market of medicinal plants. They tried to shift this into an economic solution for rural poverty and migration. Their "peripheral solution to peripheral development" was to formalize the incorporation of Transylvanian plants into the capitalist world economy (Boatcă 2005: 7). Medicinal plants were already leaving Transylvania, but as uncontrolled, collected (not cultivated) plants, using the cheap, unorganized labor of village dwellers. To increase the monetary value of both plant lives and people's work, the scientists discussed in this chapter wanted to formalize, control, and organize the movement of medicinal plants to the global market. They tried to mobilize both the growth of plants (as either wild or cultivated species) and the work of villagers to solve Transylvania's economic problems. Anna Tsing argues that one way for goods and services to enter capitalist accumulation is for them to be "'salvaged,' that is, taking advantage of value produced without capitalist control" (2015: 63). In this sense, both the value produced by plants through growth and, according to the scientists, the unused time these villagers had on their hands had to be salvaged for economic gain.

The value produced by plants that are growing and people who are living their lives is created outside of capitalist control at what Tsing calls the "edges of capitalism" or "pericapitalist" sites as they are both inside and outside capitalist processes and are created through actions that cannot be easily defined as work (Tsing 2015: 63). These not quite work-like processes involve the growth of plants and what scientists here perceive as the unused time of those in villages who are "not doing real work"—meaning, work that has a direct monetary value. Unsurprisingly it was not the scientists who first tried to salvage plant and village labor, but what they identify as predatory foreign entrepreneurs. In this sense, scientists tried to move both plants and villagers from an informal site at the edge of capitalism to a more central one. Tsing (2005) argues that the generative friction of capitalism happens in patches across geographical landscapes. The site at the edge of capitalism is, therefore, actually many sites that encompass Transylvania as a peripheral exporter of medicinal plant ingredients on a global market, but also peripheral places within Transylvania, villages, pastures, and forests where plant and human movements are converted to commodities.[1] While attempting this movement of plant and villager work from the edge to the center, scientists at the same time also attempt to move pericapitalist sites to the center, whether this means villages, Transylvania, or even the whole country. As they criticize foreign predatory enterprises for their low and informal payments, they nonetheless seek to formalize the entry of medicinal plants and village labor (of children, the elderly, and women) into the same capitalist supply chain, reinforcing their economic value on the global market.

The story revolves around the first medicinal plant research station in Europe, established in the first decade of the twentieth century in Kolozsvár (Ro. Cluj Napoca, German Klausenburg), Transylvania, then part of Hungary of the Austro-Hungarian Empire, its founder, the botanist Béla Páter, and other scientists associated with his work. They found that while many plants had the potential to become a resource for human use or state economies, they had not (yet) entered human (state) plans, and often remained weeds (cf. Tsing 2005: 173). Scientists at the research station then started to assemble networks of translations that could transform weeds into commodities and villagers into collectors and cultivators of these "newly recognized" plants. I use Bruno Latour's (1987, 2005) concept of translation here, meaning the negotiation of meanings, claims, and interests that *transform* one thing into another. That is, through the process of translation, plants (landscapes, villages, and people) should not stay the same, as translation is transformative, they change. The translation of plants and people requires the practical work of organization of collection, cultivation, and drying facilities, packaging, transportation, and acquisition; chemical and agricultural experimentation; and not least the constant promotion of the economic and medical importance of these plants. Here, therefore, I want to show that the *medicinal* attribute of plants is not straightforward, it has to be attached to and supported by a network: an assemblage of people, ideas, research, and many other things. The scientists discussed in this chapter are trying more or less successfully to put together this network in order to make commodities, and "ringing money" out of the "common weed" (cf. Latour 1993 [1988], 1987). Moving plants and villagers toward a more central and visible position in a global commodity chain network required lots of organization and lobbying aimed at changing both the bodies and meaning of plants and villagers.

To retell this story, I use the work, accounts, and biographies of the scientists involved in the academic kinship network generating from this medicinal plant research station. As I conducted multisited ethnography on the medicinal plant industry in Romania between 2014 and 2015, botanists, chemists, and agricultural engineers repeatedly told and retold fragments of this story of early twentieth-century medicinal plant research. Many scientists from Romania in the fields of medicinal plant botany, phytopharmacology, and phytochemistry could trace their academic kinship back to Kolozsvár or had worked with or heard of the members of this network. These are precisely the scientists who provided me with many of the papers from the early to mid-twentieth century used in this chapter. The historical scope of this chapter relies only on materials aimed at promoting the medicinal plant industry to both political entities and others who would become entities in their commodity chain (villagers, small-landowners, village intelligentsia), research papers, memoirs, and biographies written by these scientists themselves. It thus obscures the possible issues and limitations that the scientists might have faced in practice, but the goal of the chapter is to show how what weeds are and who villagers are supposed to be had to be negotiated, translated, and transformed.[2]

I will retell the story of how and why (medicinal) plants came to be turned into commodities from the viewpoint of botanists, chemists, and phytopharmacologists. While this will overlook many of the potential tensions between local knowledge, people and plant agency, and top-down planning, it will give an insight into what James Scott calls high-modernist ideology: a "self-confidence about scientific and technical progress, the expansion of production, the growing satisfaction of human needs, the mastery of nature (including human nature), and, above all, the rational design of social order commensurate with the scientific understanding of natural laws" (Scott 1998: 4). All this, however, coupled with an understanding of Transylvania's place in the global market as a cheap raw material producer and a genuine, albeit often patronizing, interest in reversing this role by reinforcing and attaching economic value to plants and villagers. Acknowledging Jason W. Moore's statement that "capitalism is not an economic system; it is not a social system; it is *a way of organizing nature*," this chapter is meant to describe how "capitalism *as a project*" works in practice in twentieth-century Transylvania; how scientists are reimagining and trying to put in practice a kind of Nature ("external, controllable, reducible") and a kind of Society (or human Nature) in order to serve what they believe is their "nation's interest" (Moore 2015). I start by describing the work of the medicinal plant research station and the arguments of scientists supporting the need for a well-organized medicinal plant industry that would work together with the agricultural, chemical, and pharmacological industries to allow the respective countries to discontinue their role as cheap raw-material providers. Two underlying threads to be addressed here are the unused value of plants and village labor. I use value not just as a matter of judgment, but as it denotes practices of valuation, where "the value or values of something are established, assessed, negotiated, provoked, maintained, constructed and/or contested" (Doganova et al. 2014: 87). Value, as used by these scientists, denotes almost exclusively monetary or commodity value, even though in the case of plants this also sometimes translates to medicinal value. In regard to village labor, value is seen in the monetization of what scientists regard as the idleness and "unused workforce" of children, the poor, the elderly, and women. As for plants, the process of their dying without being turned into profit must be reversed.

According to scientists, both people's labour and plants must be changed physically into commodities through valuation practices such as collection, cultivation, drying, packaging, and the organization of travel and acquisition. Their economic value has to be continuously ascertained, proven, promoted, and performed, and even this does not always guarantee that "valuable medicinal plants" will not be left to "die by the side of the road or in the fields, without anyone seeing any gain from them" (Páter 1906: 6).[3] I therefore argue throughout this chapter that scientists had to *translate* weeds into medicinal plants or commodities, and "idle" villagers into workers for money in the eyes of the state, the global market, and even the villagers themselves. In this patronizing view, both plants and people gain value through entering commodity chains on the global market of medicinal plants.

Scientists saw this as a way of moving Transylvania from the "edge of capitalism" toward a more central position.

TRANSYLVANIA AS A CHEAP RAW-MATERIAL EXPORTER

Béla Páter, who was to found the first medicinal plant research station in Europe, was born in 1860 in the Austro-Hungarian Empire (today Slovakia). He studied botany in Budapest and after teaching there and in Kassa (today Košice, Slovakia), he arrived in Kolozsvár, Transylvania, in 1893 to work at the Agricultural Educational Institution and the National Seed Testing Station. Elemér Kopp (1944: 5), the first chemist of the research station, wrote that the station was initiated and incentivized by Páter, and the Transylvanian Agricultural Association (EGE—Erdélyi Gazdasági Egylet) took the first steps toward its establishment. The EGE was founded in 1844 by Count Domokos Teleki, who assembled Transylvanian large landowners in order to find solutions for problems in the region's struggling agricultural economy.

The EGE initiated the establishment of the Agricultural Educational Institution in the first place, and after a meeting on January 14, 1904, sent a proposal to the minister of agriculture asking that (1) medicinal plants be made the object of study, thus cultivating them in the country and making them marketable; (2) a medicinal plant research station be equipped in the same year within the Agricultural Educational Institution in Kolozsvár; (3) medicinal plant-processing facilities be provided ; and (4) the results of research on medicinal plants that had already begun two years earlier be published. They called for this move in their agricultural periodical, the *Erdélyi Gazda,* because in 1902 60–70 percent of the export of medicinal plants consisted of plants collected from the wild, while most imported plants came from cultivated sources. The argument was that by educating a larger population in recognizing and exploiting wild medicinal plants, exports of medicinal plants could be increased, and by spreading medicinal plant cultivation, imports of these could be diminished ("Egyleti Élet" 1904a). According to Elemér Kopp, Béla Tallián, then the minister of agriculture, "embraced the idea" and had the medicinal plant research station established within the year (1944: 5).

In the manner of a true high-modernist visionary, Kopp argues in 1944 that the birth of the research station was a natural consequence of the evolution of sciences, although it needed the wise insight of Páter. Kopp explains that the idea was ripe as the evolution of pharmacology *realized* that it needed botany and chemistry and further recognized the need for agriculture to be able to proceed with complete knowledge. Thus the need for a separate science of medicinal plants (1944) emerged. While this alluring interdisciplinary vision of the evolution of sciences was starkly opposed to more current visions of increasing disciplinary divides, the story was not that simple. Kopp argues that the professionalization of an arguably holistic science for medicinal plants as encompassing all these other disciplines stemmed from a natural evolutionary need of the sciences—as evidenced by the appearance of other such research stations throughout Europe not long after the one in Kolozsvár.

Károly Csedő (2010: 41), however, a later student of Páter's and Kopp's kinship network, claims that Páter's vision came from a line of voices warning against the increasing professionalization and disciplinary divide of chemistry, experimental pharmacology, and synthetic pharmaceuticals (see also Sneader 2005). These voices included Alexander Tschirch, an admirer of Páter (Farkas 2010: 15), "who at the end of the nineteenth and beginning of the twentieth century proclaimed that after humanity has ruined its health with many chemical agents, it will return to medicinal plants" (Csedő 2010: 41). He was also one of the first to introduce the chemical analysis of plants (phytochemistry) (Friedrich 2006). Others were Sebastian Kneipp, a Bavarian priest and the forefather of naturopathic medicine (regarding his remedies today, see Kneipp n.d.); Hugo Schulz, a German pharmacologist studying the application of medicinal plants in hospitals (Hegnauer 1978); and Ludwig Kroeber, a pharmacologist supporting the medical use of medicinal plants (Kroeber 1934). Such resistance against the "polluting" products of synthetic chemistry and its industry were already emerging in the nineteenth century when chemists themselves protested against the environmental and human damage of industrial chemistry (Bensaude-Vincent and Simon 2008: 14). The war between "synthetic and artificial" pharmaceuticals and "natural" medicinal plants was, therefore, already on the rise by the time the research station was established (see also Hedley Brooke 2007).

Short of clear backing from the evolution of sciences, the research stations had to work constantly to attach chemical and pharmacological knowledge to medicinal plants, as well as increase their economic potential through attracting village labor and conducting agricultural experiments, while promoting the benefits of and the need for such work. The legitimacy of medicinal plants and their research was not unambiguous. Indeed, although Kopp indicates that during the time Transylvania was part of Hungary, the research station benefited from the support of the minister of agriculture, the actual financial backing of the station was slim (Kopp 1944: 6). In the *Erdélyi Gazda*, the EGE itself was not shy about criticizing the state for its failure to provide any benefits through good legislation and institutions to support agriculture in Transylvania, even though most taxes came from this field. According to one author, the dire state of agriculture showed itself in large-scale out-migration among the Szekler population, alcoholism among the Romanian population, and the possible increase of both among Hungarian populations. They saw the "noble cause" of the EGE in supporting these populations in their agricultural work both by modernizing production (through education and "action") and by preventing others from taking away the people's profits (Virilis 1904). In fact after the 1867 Austro-Hungarian Compromise establishing the dual monarchy of Austro-Hungary, the Hungarian Kingdom put more effort into modernizing, mechanizing, and industrializing Hungary proper, whereas Transylvania was integrated as a supplier of raw materials based on legislation that undermined its small-scale manufacturing industry. Nobles were left indebted because of their heavy dependence on foreign capital and the failure of capitalist agriculture to emerge, combined with poor market conditions and the emancipation of serfs in 1854. In turn these nobles managed to keep Transylvanian peasants in a subsistence-oriented semiproletarian situation that led to mass emigration by the end of the nineteenth century and increased after 1900 (Verdery 1983: 202–22).

Páter himself saw the "noble cause" of medicinal plants in helping "poverty-stricken" village economies, providing earning capacity for the "little people," and offsetting "the fatal emigration from the country" (Páter 1908: 3). One step toward achieving this, for him, was the promotion of autochthonous medicinal plant species and their liberation "from the possible baseless defamation of foreign competitors," in order to create a national and international market for domestic medicinal plants. He criticized "racketeer" entrepreneurs for paying very little to those gathering medicinal plants from the wild, arguing that only the organization of village gathering and cultivation could change this predatory practice. In 1907 he wrote that out-migration dropped significantly in places where he had managed to organize gathering, and in 1909 he celebrated the complaint of the Chamber of Commerce regarding expensive production because of wage increases (Páter, in *Dr. Páter Béla élete és munkássága* 1959: 7). While Scott argues that "the modern state increasingly aspired to 'take in charge' the physical and human resources of the nation and make them more productive" (1998: 51), in the field of medicinal plants in Transylvania it seems that it was the scientists around the medicinal plant research station together with the EGE who tried to push for a high-modernist organization of the national medicinal plant industry.[4]

AFTER THE 1918 UNION WITH THE KINGDOM OF ROMANIA

Scientists' promotion of the medicinal plant industry to the state did not change after the 1918 transition of Transylvania to Romania. The research station was retained and, although Páter was moved from his position as rector to an associate rector position at the Agricultural Educational Institution, he still headed the research station (Szabó 2010: 37). Indeed, both Romanian- and Hungarian-speaking colleagues held him in great esteem, taking up the "Páter perception" (Csedő 2010: 44). This "perception" and the promotion of medicinal plants remained not only in Kolozsvár or Cluj, after 1918. Marin Chirițescu-Arva (1889–1935), an agronomist and social thinker, who studied and did research in the United States, Germany, and the Soviet Union after World War I, in his position as the director of the Agricultural Academy, stated that the goal of the research station was to help "get rid of the significant debt we have to pay to foreigners" (Chirițescu-Arva 1927: 4). He criticized Romanian politics vis-à-vis agriculture as a "sentimental-romantic solution of 'peasant landownership,' in the illusion of creating a nonindustrialized 'peasant state.'" Instead, he imagined a rational agricultural organization, based on "renting cooperatives" in a capitalist state society (Stahl 2001: 126). In 1927 he wrote a booklet about the Cluj medicinal plant research station, in which he criticized the cereal-intensive agriculture of Romania and emphasized the importance of the largely neglected industrial plants. He wrote that the Great War had made him recognize the importance of medicinal plants, the conditions for their cultivation being perfect in Romania, while the "natural flora" provided an abundance of ready-to-harvest plants. The cultivation and collection of medicinal plants, therefore, guided by the Cluj Research Station, should be financed by the state, especially since the country pays "tens of millions across the border" for medicines and perfumes made from these plants (Chirițescu-Arva 1927: 1).

Alexe S. Potlog voiced similar ideas. He was a young agricultural researcher who joined the research station in the 1920s (Csedő 2010: 41). Along with agricultural research papers, he wrote promotional materials on the importance of the medicinal plant industry for Romania in daily newspapers such as the *Universul* (Potlog 1937, 1934). He was concerned that the country was importing all the medicines it needed, while it cheaply exported the plants used to produce these medicines. He took issue mainly with the poor state of the processing part of the industry, which had almost no existing laboratories, factories, or equipment to "transform medicinal plants in different pharmaceutical specialities." This limited trade to the export of plant raw materials, while both internal and external trade were poorly organized (Potlog 1935: 5). In 1937 he even argued that a war would require a large number of pharmaceuticals: "In the medicinal plant domain little has been done. We should be condemned for tolerating the purchase of our goods abroad for very low prices, and our reimporting them in the form of pharmaceuticals, or only in more attractive packaging, and paying very high prices for them" (Potlog 1937: 1). Potlog then called for more and better-organized research stations as well as the organization and capitalization of medicinal plant cultivation and collection. He wanted the state to fund and organize the export of medicinal plants, and to include in the agricultural industrialization program the organization of the chemical and pharmaceutical industry and the establishment of medicinal laboratories and factories for aromatic plant distillation.

In 1945 in the second edition of his book on pharmaceutical botany, first published in 1926, the botanist G. P. Grințescu criticized the agricultural politics of the Romanian state for supporting large-scale agriculture to the detriment of small- and middle-scale village agriculture. He argued that the efforts of the Cluj Research Station were diminished by the Pharmaceutical College and School, which had little to show regarding the use of these plants and conducting of research in the pharmaceutical industry. In order to create the necessary quantity and quality of products for industrialization, agriculture should come from villages, argued Grințescu. He envisioned an industrial-scale organization of the agriculture of medicinal plants, promoting cultivation rather than mere gathering of the plants that "just grow in the fields." Plants could be cultivated in a smaller area that required fewer workers, in contrast to gathering them from the wild, thus enabling a more reliable way of exporting medicinal plants, as buyers could count on a steady supply of quality plants (Grințescu 1945). Although they seemed to disagree, Grințescu and Chirițescu-Arva both argued for the small- and middle-scale industrialization and organization of village cultivation and collection efforts, and both based their arguments on the works of the Cluj Research Station rather than a romantic notion of peasant landownership. Together with Páter and Potlog, these scientists continued arguing for better organization of the whole medicinal plant industry with processing done as much as possible inside of the country and arguing against the country becoming a cheap raw-material exporter, instead exporting good-quality expensive plants.

Similarly, Kopp critiques the Romanian state for especially emphasizing the export of medicinal plants after the 1918 union, which, although it expanded the scope of Páter's organization of wild-plant gathering to other parts of Romania, it did so to the detriment of research and cultivation (Kopp 1944: 20–21). Nevertheless, there were some improvements. Smaller collection sites were built, usually connected to agricultural schools (Chirițescu-Arva 1927), and through the land reform in 1920–1921 the Agricultural Ministry gave five to fifteen hectares of arable land to those farmers who decided to cultivate medicinal plants, increasing the number of such plants. The medicinal plant research station created model plots, conducted courses, and wrote pamphlets (in Hungarian, Romanian, and German) about cultivating medicinal plants to help the farmers gain the necessary knowledge to do so. They also supplied cultivators with quality seeds and propagation materials (Csedő 2010: 42–43). Páter encouraged the idea of producer and seller cooperatives, and at the first Romanian Medicinal Plant Conference, held in Cluj on September 20, 1931, the research station proposed the establishment of interim committees between the ministry and the producers for the sale of gathered wild medicinal plants based on the model of similar committees in Austria, France, and the Soviet Union. They suggested eight different geographical areas in the country for these committees, recommending different varieties of plants they could gather. According to Csedő, this meant that they researched the medicinal plant distribution across the country: "the country has such abundant raw-material reserves that, apart from their use inside the country, much remains for export" (Csedő 2010: 43, citing Páter).

Nevertheless, according to the volume edited by the library of the University of Agricultural Sciences in Budapest, Páter's drive to organize well-paid village gathering work and cultivation through cooperatives mostly failed (*Dr. Páter Béla élete és munkássága* 1959). As the frustrating criticisms and calls of other scientists also suggest, a more utilitarian organization of the industry was also unsuccessful—at least at the beginning of the twentieth century. However, three cultivation and marketing cooperatives were founded following Páter's suggestion: in 1926 the Adonis Cooperative in Cluj, in 1927 the Romanița Cooperative in Bucharest, and in 1929 the Digitalis Cooperative in Orăștie (Hu. Szászváros, German Broos), County Hunedoara (Stancu 2008: 70). Throughout the beginning of the twentieth century scientists promoted a utilitarian, well-organized national medicinal plant industry that would have advanced greater industrial utilization of indigenous plant raw material, with an emphasis on more controllable plant cultivation rather than collection from the wild, but supplemented by rationally organized wild gathering, both of which would have provided for the livelihood of villages. Although the state organization seems mostly to have failed to meet expectations, they still highlighted, raised, and transformed ideas about what weeds could be and how villagers should be transformed into gatherers and cultivators. They aimed their efforts at laying down a network—an assemblage of gatherers, cultivators, research, transportation, acquirers, and state support—that would help bodily translate the weeds from the roadside into productive plants.

NETWORKS OF TRANSLATIONS

According to Páter:

> Our foremost role should be to organize harvesting through villages and processuses [smaller districts within counties in Hungary]. But the people need a leader because they have neither expert knowledge, nor qualification, nor the scope to carry out the business part of the harvesting. The people thus harvest eagerly if they see the material success of their work. But then they must see money every day or every week, their harvested goods need to be exchanged. It is harder to have the people administer

> [manage] and market [sell] the harvested materials because they lack the intelligence needed for such things. Nor do the people have sufficient training for the identification of plants that need to be collected, the time and the way they need to be harvested, and the marketing—they need leadership. . . . As all of us who fight for progress and enlightenment want to illuminate the darkness of ignorance in people, in order to help the misery of backwardness with enlightening directions, we have to appeal to the humanely minded teacher to help in our foundational work. Handling medicinal plants is a very suitable source of income . . . for the weaker element, the women who are tied to the house, who cannot go far for day labor, who cannot leave the house for a longer time, who are tied by housework and care for children . . . , [for] girls and children ages eight to ten, who cannot yet do any meaningful hard labor . . . , and [for] the frail elderly, who cannot do hard day labor anymore. . . . Even if one cannot earn thousands with it, a lot of crowns can fall into the pockets of the poor. (Páter 1908: 7)

Páter's character fits well in the line of "engineers, planners, technocrats, high-level administrators, architects, scientists, and visionaries," who "envisioned a sweeping, rational engineering of all aspects of social life in order to improve the human condition" (Scott 1998: 88), assuming "responsibility for the great works of nation building and social transformation" (96). If Scott (1998: 88) and David Harvey (1992: 35) call these thinkers avant-garde, Páter sees himself and the village intelligentsia, from teachers to priests and even women of the nobility (Páter 1908: 7–8) as proponents of the enlightenment. These then, in his vision, have the "self-assigned cultural role" of educating those whom they see as the "technically backward, unschooled, subsistence-oriented population" (Scott 1998: 96). Thus, Páter and the other scientists at the research station started to work on translating villagers—small landowners, and those they saw as "idle villagers," the "unused workforce" of the elderly, the "womenfolk," and "the army of children" (Páter 1909: 3–4)—into cultivators and gatherers who recognize medicinal plants as *profitable, moneymaking* commodities. They did this through the already established pattern of the EGE's courses in cities, towns, and villages on organizing industrial vegetable production, how to deal with animal and plant illnesses, how to milk, feed, and better care for cows, and how to make honey. For example, among others, the *Erdélyi Gazda* reports on January 31, 1904, under the title "Popular (or Folk) Economic Lectures," about Páter holding lectures on the exploitation of medicinal plants growing in the wild in the villages of Ketesd, Szilágy County, (today Tetişu, Sălaj County, Romania) and Magyar Bikal, Kolozs County (today Bicălatu, Cluj County, Romania) ("Egyleti Élet: Népies Gazdasági Előadások" 1904). After 1918, this method was expanded to other parts of Romania as well (Chirițescu-Arva 1927; Kopp 1944).

Meanwhile, Páter persuaded other high-modernist intelligentsia into taking up, organizing, and promoting medicinal plant gathering and cultivation. I have already mentioned some of these people in the previous section, such as Chirițescu-Arva and Grințescu, and here I can mention Count Árvéd Teleki, who in his 1926 book titled *On the Usefulness of Our Plants,* in chapter 5 on medicinal plants, praises the work of Páter and the research station (Gróf Teleki 1926: 174). However, others actively promoted and organized village gatherers and cultivators. István Gidófalvy (1889–1984), a landowner, agronomist, and pupil of Páter at the Agricultural Academia in Cluj, remembers as the foundation for his love of "the cause" having been Páter's assistant in 1908 in a two-week course for priests and teachers. Persuaded by Páter, he started cultivating medicinal plants on his lands in Doboka, Kolozs County (today Dăbâca, Cluj County, Romania; Gidófalvy 2005: 112). He continued his passion for medicinal plants in the early 1930s, when, as the president of the medicinal plant section of the newly renamed Transylvanian Hungarian Agricultural Association (EMGE, Erdélyi Magyar Gazdasági Egylet), he organized and recruited increasing numbers of villages and villagers for gathering medicinal plants from the wild (Gidófalvy 2005: 113). The pharmacist András Faragó, following the calls of Páter for co-operativization, founded in 1929 the Digitalis Cooperative in Orăștie, Hunedoara County, together with fifteen cultivators, cultivating plants given to them by the research station and especially known for cultivating *Digitalis purpurea,* used for the extraction of digitalis, still used today in treating heart conditions (Csedő 2010: 46). Another follower and pupil of Páter was Kálmán Bodor (1886–1960), an agricultural engineer, a promoter of the gathering and cultivation of medicinal plants, lecturer at EMGE in the medicinal plant section, and steward on the Rédey estate in Sülelmed (today Ulmeni, Maramureș County, Romania) until the 1921 agrarian reform when he managed to buy a small part of the former estate from Count László Károlyi. Following Páter's suggestion, from the 1920s it was here that he cultivated and organized the collection of medicinal plants and sold them through contracts with Romanian, French, Austrian, German, and Hungarian acquirers (T. Veress 2010: 53–55).

It was thus the work of high-modernist scientists, engineers, and agronomists, working through the village intelligentsia, such as priests and teachers, that translated weeds into medicinal plants not only for the state but for the village workforce. This translation meant that they had to make the potential workforce from the villages see plants as economically profitable plants, and then materially making plants profitable through organizing the processing, transportation, and setting up of contracts with acquirers of medicinal plants to "monitor the condition of their sale" (Páter, in Kopp 1944: 6). This organizational work was an essential step in making weeds into commodities, and Páter dealt extensively with issues of processing such as properly gathering, drying, and packaging plants to prevent loss of quality, as well as with analyzing the state of the medicinal plant market, where and at what price to sell these plants ("Egyleti Élet" 1904b; Páter 1906, 1908, 1909). Through this organizational work, the role of the village intelligentsia was to attract and educate the village workforce to see the work with medicinal plants as economically viable on the one hand, and to handle them in a way that would achieve viability, on the other.

Páter did recognize that villagers might have had knowledge of some medicinal plants handed down, as he explained, from father to son or from mother to daughter. He argued, however, that such knowledge was kept secret, that only village medicine men and women knew about it, and that it was dying out due to the progress of culture and the sciences (Páter 1906: 5). Páter outright dismissed village plant lore as superstition that had no scientific value and hindered the progress of society (Páter 1896). In his view, the village workforce was thus unable to organize the network to properly market plants and it also lacked the competence to recognize and properly care for plants. Hence, the village intelligentsia intervened to turn them into workers with an economic vision and lay out the commodity chain of medicinal plants to fulfill this new vision for the villagers.

According to this vision, villagers had to be made aware of the economic benefit of their labor because of the context at the beginning of the twentieth-century era of economic struggle for Transylvania, which, as opposed to the Romanian popular imagery, did not improve during the interwar period and worsened after the 1929 Great Crisis (Murgescu 2010: 205–314). "Today they have to produce more and with more value," exclaimed Páter, about the need to create more productive ways of utilizing both land and workforce by cultivating medicinal plants, which, although more labor intensive, bring in more profit (Páter 1909: 3). Medicinal plants, he argued, would create a more productive workforce of precisely those people in villages he considered insufficiently productive, "who do not earn anything or earn very little" (Páter 1909: 3)—that is, the "weaker members of the family" (Páter 1906: 6). Páter's views hardly took into consideration villagers' active negotiation of a patchwork of occupations, different work opportunities, under different political circumstances (Narotzky 2016) and unremunerated labor (Daniels 1987). Indeed in much of his writing, villagers and more specifically children, women, the elderly, and the poor (e.g., beggars), appear as passive sufferers of economic injustices without the means and the wit to get out of poverty.

Villagers, therefore, needed to be organized and taught how to look at "common weeds" (Hu. *közönséges dudva*) as financially valuable medicinal plants, and to see them as capital (Páter 1906: 6). Those who were thus transformed could become what Gidófalvy called in 1939, "beautifully working gatherers" (Gidófalvy

2005: 11). It was through the work of these gatherers and cultivators that plants could enter the medicinal plant commodity chains, laid out through the networking work of the intelligentsia. Gatherers and cultivators had to learn to recognize, properly care for, dry, and package plants for them to enter this commodity chain. However, what was traveling through this commodity chain was not straightforward, and scientists had to work on attaching value to the domestic medicinal plants. While many values would be attached to plants, I want to emphasize what Páter calls the "true value" of medicinal plants: their composition and active components (Páter, in Kopp 1944: 6).

PLANTS AS MEDICINES

According to Páter:

> The goal of this research station is to serve the cause of the cultivation and sale [Hu. *értékesítés*, lit. valuation] of medicinal plants [Hu. *gyógynövények*, lit. healing plants]. It believes in achieving this goal through experimenting with the cultivation of different medicinal plants [Hu. *orvosinövények*, lit. medical plants], studying their breeding [Hu. *tenyésztés*], experimenting with their gathering and drying, monitoring the conditions of their sale, and finally, making its research object the composition of the cultivated medicinal plants and their active components, in order to deduce their true value. All these are important issues because only in this way can we liberate national medicinal plant cultivation from the possible baseless defamation of foreign competitors. Further goals of our station are to inform and guide the farmer audience and to supply cultivators with seeds and seedlings, of which we will give a smaller amount for experimental cultivation for free to the individual. In relation to cultivation, we will experiment with artificial fertilizers and analyze the effect of the artificial fertilizers in laboratories. (Páter, in Kopp 1944: 6)

Michael Marder argues that in the European philosophical thinking there has been a devaluing of vegetable life, plants being classified closer to "inanimate things" than "other living beings" and thus transformed "into raw materials for animal and human consumption" (Marder 2011: 86). In this "ensuing instrumental approach" (88), the difference between plants valued as treasures and those devalued as waste, in the context of this chapter, is whether or not scientists could prove that medicinal plants are legitimately useful for humans. To recognize plants as medicines, plant bodies have to be tied to human bodies. This is especially striking in the recognition of the medicinal part of plants, their specific use in healing human bodies. Weeds cannot be commodities, they have to be turned into medicinal plants, and that depends not only on recognition. The *medicinalness* of plants has to be made visible and produced in this case through the coworking of agricultural, chemical, botanical, and pharmacological sciences that had to connect human bodies to plant bodies. That is, plants had to be turned into legitimate healers of human bodies in an era when the chemical extraction and isolation of pharmacologically active substances and their subsequent synthesis into new, artificial, chemical molecules were gaining new territory.

Páter thus recognized that in order to promote Transylvanian plants, their superior chemical qualities had to be brought to the forefront. Indeed, one of the first things Páter acquired was the equipment of a chemical laboratory in 1909, and a new chemist, Elemér Kopp. The station aimed, on the one hand, to promote the (chemical) quality of wild gathered plants, and on the other, to promote and *increase* the (chemical) quality of cultivated plants, and, in a true high-modernist vision, to introduce wild medicinal plant species into more controllable cultivation, where they could work on *improving* them both in yield and chemical composition. Thus, by the time of Páter's retirement in 1931, the station had attempted to introduce 136 wild medicinal plant species (Farkas 2010: 11) and experimented with artificial fertilizers and their effect on the plants and their composition, thus calculating yield and profitability (Kopp 1944: 25). Kopp, therefore, worked first on emphasizing and then on improving and even extracting oil or other chemical molecule compositions from medicinal plants. For instance, they assessed the fatty oils in caraway seeds, extracted opium from poppies, and followed the changing alkaloids in sick plants (Csedő 2010: 45; Kopp 1944: 22). They worked not only on promoting local plants but also on acclimatizing plants already popular on the international market, for instance, several mint cultivars. They also conducted experiments comparing compound compositions between soil and weather conditions with several species to ascertain the best conditions for plants to make more compounds (Csedő 2010: 42). To see how significant this research was, in his summary of the scientific work done at the station, Csedő thought it essential to mention that samples of the extracted essential oils of the plants were sent to the pharmacist A. Iteanu in Bucharest, who classified them as being of very good quality (Csedő 2010: 43).

Chirițescu-Arva also highlighted that the value of medicinal plants lay in their chemical potential. He listed among the goals of the station the "cultivation of different medicinal and aromatic plants, for the production of drugs and seeds"; the "finding of new medicinal plant species (recruited from Romanian spontaneous flora), which could satisfyingly replace—from a medicinal point of view—exotic drugs"; the cultivation of the most profitable plants from the standpoint of drug production; and the finding of new species and varieties of plants that would turn out, in the laboratory, to be worthy of use in producing new drugs (Chirițescu-Arva 1927: 7). Similarly, Grințescu called for and criticized the failure of the chemical industry to work with the medicinal plant production industry (1945). Kopp also saw the natural evolution of chemical and pharmaceutical sciences in the emergence of a medicinal plant science, and thus, the pharmaceutical and medicinal plant industry were imagined as working together.

Turning plants into chemical materials would have another significant effect as well: controlling the quality of plants on the network of commodity chains. Potlog draws attention to the dangers of "fake" plants, especially poisonous ones, mixing with the medicinal plants. He emphasized the importance of exporting *well-conditioned* plants that were free of impurities and other plants. It was the Agronomic Research Institution that had to control this process: "The control of medicinal plants for export has been already introduced in other states and with good results. In Russia, a medicinal plant exporting country, the valuation of these products is done only through precise and compulsory norms" (Potlog 1939: 4). He then explains the two ways of identifying drugs: through their anatomic and morphologic properties such as their color, form, size, and internal structure, and through the chemical and physiological method, which analyzes the active components of the drugs. Potlog emphasizes that only the former, easier method had been in use, whereas the latter method would only be used when the trading of medicinal plants would be "genuinely founded on quality" (Potlog 1939: 4). Potlog imagined an ever-increasing mode for controlling the quality of plants through microscopic chemical research done with plants harvested in the fields of the Cluj Research Station and analyzed in the laboratory of the Pharmaceutical Faculty in Paris, which he presented in his paper.

The translation of weeds into medicinal plants in the early twentieth century, here, meant that on one hand scientists tried to legitimize increasing numbers of plant species as medically beneficial, and therefore financially valuable, and on the other, they actively practiced control of plant bodies by introducing them in cultivation, improving them, and making them into chemical matter. However, in order to transform this knowledge into financial value, plants and their quality also had to travel together (see Mol and Law 1994). That is, the scientists had to promote knowledge about the quality of local plants—for example, as Páter did by keeping in touch with German pharmacies—and they had to transport plants matching this expectation, as Potlog did. It is here, in ensuring that plants traveled well from the source, be it cultivation or the wild, to the shelf, that the need to transform villagers into "beautifully working gatherers" or cultivators, as well as people who recognize plant species, comes in, together with the need for the intelligentsia to organize the proper preprocessing (sorting, drying, and packaging) of the plants. A whole network of people, plants, knowledge, and facilities had to be put in place to make this traveling of ever-increasing amounts and species of plants possible.

Here, therefore, I have shown some of the negotiation, promotion, organization, and scientific work needed to develop these networks.

HIGH MODERNISM AT THE EDGE OF CAPITALISM

This account of the work of scientists is a description of high modernism in operation, as depicted by Harvey (1992), although in the context of the Cold War period, and Scott (1998). The scientists of this story are the "progressive elites" with a "strong (one might even say muscle-bound) version of beliefs in scientific and technical progress," with the "self-confidence about continued linear progress, the development of scientific and technical knowledge, the expansion of production, the rational design of social order, the growing satisfaction of human needs, and, not least, an increasing control over nature (including human nature)" (Scott 1998: 89). The vision of improving the rural economy thus intertwined with the visionary aim of changing the "habits, work, living patterns, moral conduct, and worldview" (89) of villagers who were seen as having to be "educated" to recognize both plants and the economic worth of these plants, thus "dragging" this "technically backward, unschooled, subsistence-oriented population into the twentieth century" (96). Meanwhile, plants themselves had to be brought under control through assessing their chemical properties, mapping their flora distribution, and most of all bringing them into cultivation where their yield and chemical properties could be better assessed, controlled, and most of all improved upon. What Scott (1998: 82) talks about as the internal colonization and "civilizing mission" is oriented toward both medicinal plants and the rural population who could bring these plants into commodity chains.

A network of people, research, organizational work, and acquirers had to be pulled together in order to transform both villagers and plants. Value then also had to be attached to local plants to make them travel through an international market of medicinal plants. Scientists worked on this network to make domestic plants travel well and domestic labor pay well, yet they saw the state as failing at translating plants into productive commodities and villagers into a productive workforce. Scientists tried to make the state, the acquirers of medicinal plants, and the villagers value "common weeds" as medicinal plant commodities, thus attempting to envision a part of nature they saw as missing from the state's productive lens. This high-modernist vision was meant to shift Transylvania's role away from being a cheap raw-material provider by drawing people and plants into the medicinal plant commodity chain. This is a small but, I believe, exciting story that emphasizes how high modernism can play out at the "edges of capitalism" and how capitalism is a project that works on reimaging and practically trying to remake what Nature and Society mean. While this chapter does not say much about how this project actually remade itself or how it was transformed by the environment (human and nonhuman), it does show the push of a group of scientists toward a kind of Nature and a kind of Society that have resonance today as well.

NOTES

Epigraph: Páter 1906: 6.

1. I have written elsewhere on the meaning of the pericapitalist site of the spontaneous flora where "wild" medicinal plants come from and how weed labor is converted to commodities (Ábrán 2018).

2. See Vladislava Vladimirova (chapter 10 in this volume) for a detailed analysis of how adaptation and resistance can play out as an answer to state and scientific transformation and control attempts with both people and animals.

3. See Stefan Dorondel and Stelu Şerban (introduction in this volume) for more on the idea of unproductive landscapes, of nature becoming wasted unless used by humans, and the need to improve upon it.

4. Compare this to Jiří Janáč (chapter 6 in this volume) on the modernization of waters in Czechia for a more involved relationship of the state, in part the Austro-Hungarian Empire, in the management of nature (in this case, water).

REFERENCES

Ábrán, Ágota. 2018. "Unwrapping the Spontaneous Flora: On the Appropriation of Weed Labour." *Studia Universitatis Babes-Bolyai* 63(1): 55–72.

Bensaude-Vincent, Bernadette, and Jonathan Simon. 2008. *Chemistry: The Impure Science*. London: Imperial College Press.

Boatcă, Manuela. 2005. "Peripheral Solutions to Peripheral Development: The Case of Early 20th Century Romania." *Journal of World-Systems Research* 11(1): 3–26.

Chiriţescu-Arva, Marin. 1927. *Staţiunea de Plante Medicinale* [The medicinal plant station]. Cluj: Cartea Românească.

Csedő, Károly. 2010. "Szemelvények Páter Béla szakmai tevékenységéből" [Excerpts from the professional activity of Béla Páter]. In Farkas and Lázár, *Páter Béla Emlékezete*, 40–48.

Daniels, Arlene Kaplan. 1987. "Invisible Work." *Social Problems* 34(5): 403–15. https://doi.org/10.2307/800538.

Doganova, Liliana, Martin Giraudeau, Claes-Fredrik Helgesson, Hans Kjellberg, Francis Lee, Alexandre Mallard, Andrea Mennicken, Fabian Muniesa, Ebba Sjögren, and Teun Zuiderent-Jerak. 2014. "Editorial Note: Valuation Studies and the Critique of Valuation." *Valuation Studies* 2(2): 87–96.

Dr. Páter Béla élete és munkássága [The life and work of Dr. Béla Páter]. 1959. Az Agrártudományi Egyetem Központi Könyvtárának Kiadványai, VI. sorozat. Egyéni életművek bibliográfiája. Budapest: Mezőgazdasági Kiadó.

"Egyleti Élet: A Gyógyfüvek Termesztése" [The life of the association: the nature of healing herbs]. 1904a. *Erdélyi Gazda* 36(5): 57.

"Egyleti Élet: Népies Gazdasági Előadások" [Social life: folk economic lectures].1904b. *Erdélyi Gazda* 36(5): 57.

Farkas, Zoltán. 2010. "Páter Béla életművének követői 100 év távlatából" [The followers of the lifework of Béla Páter from the perspective of 100 years]. In Farkas and Lázár, *Páter Béla Emlékezete*, 10–17.

Farkas, Zoltán, and László Lázár, eds. 2010. *Páter Béla Emlékezete* [The memory of Béla Páter]. Kolozsvár: Az Erdélyi Múzeum-Egyesület kiadása.

Friedrich, Christoph. 2006. "Alexander Tschirch: Koryphäe pharmazeutischer Botanik" [Alexander Tschirch: a genius of pharmaceutical botany]. *Pharmazeutische Zeitung*. http://www.pharmazeutische-zeitung.de/index.php?id=2087.

Gidófalvy, István. 2005. *Önéletrajz* [Autobiography]. Kolozsvár: Művelődés.

Grințescu, G. P. 1945. *Cultura și recolta plantelor farmaceutice* [The cultivation and harvesting of pharmaceutical plants], 2nd ed. Bucharest: Editura Universul.

Gróf Teleki, Árvéd. 1926. *Növényeink Hasznáról* [On the usefulness of our plants]. Cluj-Kolozsvár: Lapkiadó Rt.

Harvey, David. 1992. *The Condition of Postmodernity: An Enquiry into the Origins of Cultural Change*. Cambridge, MA: Blackwell.

Hedley Brooke, John. 2007. "Overtaking Nature? The Changing Scope of Organic Chemistry in the Nineteenth Century." In Bernadette Bensaude-Vincent and William R. Newman, eds., *The Artificial and the Natural: An Evolving Polarity*. Cambridge, MA: MIT Press, 275–92.

Hegnauer, R. 1978. "Arzneipflanzen--Gestern, Heute und Morgen" [Medicinal plants—yesterday, today, and tomorrow]. *Planta Medica* 34(5): 1–25.

Kneipp. n.d. "The Kneipp Story." Accessed September 4, 2017. http://www.kneippus.com/the-kneipp-story/.

Kopp, Elemér. 1944. *A kolozsvári gyógynövénykísérleti intézet története* [The history of the Kolozsvár medicinal plant station]. Kolozsvár: Lengyel Albert könyvnyomdája.

Kroeber, Ludwig. 1934. *Das neuzeitliche Kräuterbuch: Die Arzneipflanzen Deutschlands in alter und neuer Betrachtung* [The modern herbal book: the medicinal plants of Germany in old and new considerations]. Stuttgart: Hippokrates-Verlag GMBH.

Latour, Bruno. 1987. *Science in Action: How to Follow Scientists and Engineers through Society*. Cambridge, MA: Harvard University Press.

Latour, Bruno. 1993. *The Pasteurization of France*. Cambridge, MA: Harvard University Press.

Latour, Bruno. 2005. *Reassembling the Social: An Introduction to Actor-Network-Theory*. New York: Oxford University Press.

Marder, Michael. 2011. "Plant-Soul: The Elusive Meanings of Vegetative Life." *Environmental Philosophy* 8(1): 83–99.

Mol, Annemarie, and John Law. 1994. "Regions, Networks and Fluids: Anaemia and Social Topology." *Social Studies of Science* 24(4): 641–71.

Moore, Jason W. 2015. *Capitalism in the Web of Life: Ecology and the Accumulation of Capital*. E-book. London: Verso.

Murgescu, Bogdan. 2010. *România și Europa: Acumularea decalajelor economice (1500–2010)* [Romania and Europe: the accumulation of economic gaps (1500–2010)]. Iași: Polirom.

Narotzky, Susana. 2016. "Where Have All the Peasants Gone?" *Annual Review of Anthropology* 45(1): 301–18.

Páter, Béla. 1896. "A nép néhány orvosi növényéről" [About some medicinal plants of the folk]. *Természettudományi Közlöny* 28(317): 22–26.

Páter, Béla. 1906. *A vadon termő gyógynövények valamint a gazdasági termékek gyógyszertári értékesítésének rövid ismertetése* [A brief description of the sale of wild medicinal plants and farm products in pharmacies]. Budapest: Pátria irodalmi vállalat és nyomdai részvénytársaság.

Páter, Béla. 1908. *A gyógynövények gyűjtéséről és azok értékesítéséről* [About the gathering and sale of medicinal plants]. Különlenyomat a "Család és iskola" 1907. évi 18. és 19. számaiból. [Offprint from *Family and School* 1907, nos. 18 and 19]. Kolozsvár: Gombos Ferencz Könyvsajtója.

Páter, Béla. 1909. *Milyen növényeket termeszszen a kisgazda* [What medicinal plants does the small landowner cultivate]. Budapest: Az Országos Magyar Gazdasági Egyesület, Köztelek Olcsó Könyvtára.

Potlog, A. S. 1934. *Valorificarea Plantelor Medicinale* [The sale of medicinal plants]. Extras din "Pagina Agricolă"—Ziarul Universul , no. 300 [Offprint from the "Agricultural Page," *Universe Journal*, no. 300]. Cluj: Tipografia Universală.

Potlog, A. S. 1935. *Posibilităţi de cultură şi de valorificarea plantelor medicinale* [Possibilities for the cultivation and sale of medicinal plants]. Extras din revista "Viaţa Agricolă," no. 10 [Offprint from the journal *Agricultural Life*]. Cluj: Tipografia Ardealul.

Potlog, A. S. 1937. *Plantele medicinale în România*. Extras [Medicinal plants in Romania]. Din "Pagina Agricolă"—Ziarul Universul, no. 223 [Offprint from the "Agricultural Page," *Universe Journal*, no. 223]. Cluj: Tipografia Naţională S.A.

Potlog, A. S. 1939. "Contribuţiuni la identificarea drogurilor (caracterizarea anatomomorfologică a câtorva droguri vegetale)" [Contributions to the identification of drugs (anatomical and morphological characterization of some vegetal drugs)]. *Buletinul Facultăţii de Agronomie Din Cluj* 3: 3–30.

Scott, James C. 1998. *Seeing Like a State: How Certain Schemes to Improve the Human Condition Have Failed*. New Haven, CT: Yale University Press.

Sneader, Walter. 2005. *Drug Discovery: A History*. Chichester: John Wiley .

Stahl, Henri H. 2001. *Gânditori şi curente de istorie socială românească* [Thinkers and schools in Romanian social history]. E-book. Editura Universităţii din Bucureşti. http://ebooks.unibuc.ro/Sociologie/henri/index.htm.

Stancu, Emilia. 2008. "Din istoria farmaciei româneşti interbelice" [From the history of Romanian interwar pharmacy]. *Noema* 7: 61–71.

Szabó, T. Attila. 2010. "Páter Béla és a kolozsvári gyógynövénykísérleti állomás (1904–1919): Dokumentumai és emlékei (1967–1987)" [Béla Páter and the Kolozsvár medicinal plant station (1904–1919): documents and memories (1967–1987)]. In Farkas and Lázár, *Páter Béla Emlékezete*, 31–39.

T. Veress, Éva. 2010. "Páter Béla és tanyítványai, Gidófalvy István, Bodor Kálmán" [Béla Páter and his pupils: István Gidófalvy, Kálmán Bodor]. In Farkas and Lázár, *Páter Béla Emlékezete*, 49–57.

Tsing, Anna Lowenhaupt. 2005. *Friction: An Ethnography of Global Connection*. Princeton, NJ: Princeton University Press.

Tsing, Anna Lowenhaupt. 2015. *The Mushroom at the End of the World: On the Possibility of Life in Capitalist Ruins*. Princeton, NJ: Princeton University Press.

Verdery, Katherine. 1983. *Transylvanian Villagers: Three Centuries of Political, Economic, and Ethnic Change*. Berkeley: University of California Press.

Virilis. 1904. "Kolozsvári szakosztály" [The Kolozsvár department]. *Erdélyi Gazda* 36(2): 17–18.

5

The Economy of a Leashed River

State, Experts, and Politics along the Lower Danube, 1900–1940

Stelu Şerban and Stefan Dorondel

> Rodin cried exultingly: "What a marvellous thing is Science! How fortunate to be born in a century in which one can command nature!"
>
> "Where will progress stop?" threw in Camille Flammarion.
>
> —Ambroise Vollard, *Recollections of a Picture Dealer*

Perceived as "liquid highways" from ancient times (Skelton 2016), rivers have always been subjected to transformations. Yet the nineteenth century represented a turning point for alterations of the world's rivers. From the United States to the Far East, rivers were preferred dumping spots for human and industrial waste; their watercourses were regulated and constrained by levees, and their resources fiercely exploited (White 1995; Cioc 2002; Skelton 2016; Clifford 2017). As do many other natural environments, rivers of the world seem to tell a bedeviled story of ruination. The rivers of Southeast Europe have shared the same story, though the time frame has been different. For instance, the grand-scale development of the Danube began in the middle of the nineteenth century. Yet parts of the river, around the city of Vienna for example, were regulated earlier to improve navigability and to secure the steady flow of water (Hohensinner et al. 2013). The major technological "assault" began on the Lower Danube, downstream of Budapest, in the early twentieth century. The nation-states that had recently appeared on the map of Southeast Europe in the second part of nineteenth century, rising from the ruins of the Ottoman Empire, were all interested in systematizing the river—course rectification, dredging, bank consolidation, and damming—in their attempt to compete or, at least, reach the level of development of Western powers (Dorondel and Şerban, Introduction in this volume). However, they lacked the technological means, expertise, and institutions to harness the river.

This chapter aims to highlight the state-led economic transformation of the Lower Danube and its floodplain in the first decades of the twentieth century. We focus on the means the state had at its disposal—creating institutions, commissioning experts, and transferring technologies, knowledge, and experts from Western Europe—and the strategies it followed. We do this by comparing two neighboring countries' strategies for developing the Danubian zone: Romania and Bulgaria. We specifically focus on systematization of the Danube, but we look beyond flood mitigation. We focus on the state-induced transformation of the river in order to examine plans to construct a national energy system and to reclaim agricultural land from the marshes that flanked the river.

STATE, EXPERTS, AND RIVERS

An intrinsic link exists between various experts on river dynamics and the state, which was the main advocate of harnessing the Danube. We particularly focus on the ways in which knowledge, technology, and international politics constitute the ingredients of "river modernization"—that is, the scientific rationalization and abstraction of the Danube's organic complexity.

From the mid-nineteenth century, the Danube River constituted an important conduit for different technologies that were transferred and adapted in riparian Southeast European countries. The Upper and Middle Danube were subjected to technological interventions well before this date, carried out mainly in response to economic forces (Heidvogl et al. 2018). Although the river suffered multiple local-level transformations from premodern times onward (Hohensinner et al. 2013; Schmid 2013), its waters and banks only became the target of sustained engineering as part of wider governmental plans in the nineteenth century. The Lower Danube entered this game later, in the second part of the century. At first, the knowledge, technology, and experts able to induce changes in the river's geography came almost entirely from Western Europe (Jensen and Rossegger 1978). The engineers employed by the European Commission of the Danube (ECD) after 1856 brought contemporary Western ideas of what a modern river should look like and secured funds to support the transformation. However, this international institution was strictly interested in the navigability of the river—to transform it into a route for humans and merchandise (Gătejel 2016; Groparu et al. 2019). The creation of riparian national states, after 1878, provided the right conditions to establish a national corps of experts and a network of institutions charged with governing the Danube and the lands alongside it. Most of these national experts were trained in universities in Western Europe and became conduits for Western ideologies regarding new technologies, economics, change, and nature (Kostov 2009). As we pointed out in the Introduction (Dorondel and Şerban, in this volume), these institutions and experts constituted the scaffold on which the central state built its ideology. Only at this point did the vision of a Danube that was economically useful come alive, not only for navigation but for land reclamation and an extended network of electric grids. In the national political imagination the Lower Danube became what Jason Moore (2016) calls "cheap nature" and harnessing the river's power became part of nation-state building. As Filippo Menga and Erik Swyngedouw (2018) have shown, state and waterscapes mutually constitute each other through multiple mechanisms, including the material infrastructure that enables state agencies to exploit rivers (see also Menga 2015). In a similar vein, the recently constituted Romanian nation-state created state agencies to harness the Danube as part of a wider plan to demonstrate its stately character. Bulgaria attempted to take a similar path but the political atmosphere was quite different from that of its neighboring country.

As Patrick Carroll (2012) rightly shows, attempts to separate water and land involve multiple state agencies found at different levels of government and it is engineering capacity—technological knowledge and materials of construction—that makes this separation possible. What we add to this theoretical construction is the role of international and national experts directly involved in the Promethean reorganization of the natural environment. The Danube's extended role of serving the national economy was facilitated by what Mitchell Ash and Jan Surman (2012) call the "nationalization" of science and technology. They refer particularly to the institutionalization of social and natural sciences and their contribution to building a national discourse. In this chapter we look less at the experts' contribution to the national discourse and more at their contribution within national institutions that put the river, a huge natural resource, at the disposal of the newly emerged nation-state.[1] Engineers, agronomists, and natural sciences experts translated "problems of governance into problems of engineering and vice versa" (Carroll 2012: 510); they were on a "hydraulic mission"—a pledge not to let a drop of river water be wasted in its course to the sea (Molle 2009; Molle, Mollinga, and Wester 2009). Controlling the river for various economic purposes, from irrigation to electrification, required the emergence of a hydraulic bureaucracy to plan and implement the damming of the river, build levees, and generally create a new waterscape as part of the hydraulic modernization program (Molle, Mollinga, and Wester 2009; Swyngedouw 2007, 2014; Janáč, chapter 6 in this volume). These elites not only were ebullient supporters of the country's economic modernization but also sought their own social and political legitimacy (Molle, Mollinga, and Wester 2009).

This chapter focuses on a period when a corps of national experts was consolidated in Romania and Bulgaria, during the first four decades of the twentieth century. However, the ideas for transforming the Lower Danube followed an earlier European model of organizing rivers. England, Spain, and Italy, to name but a few European countries, were already engaged in a process of river "high modernism" (Scott 1998). We employ the term "river high modernism"—following Scott—to describe a combination of three elements: the modern state's obsession with imposing administrative order on nature, the power used by the state to achieve this goal, and the weak capacity of civil society to oppose these designs, when they do not directly support them. In all these European countries a corps of hydraulic experts, mostly engineers, was already contributing to controlling rivers in the name of the economy and the nation (Skelton 2016; Bocquet 2007; Parinello 2017; Swyngedouw 2014; Zeisler-Vralsted 2014). Yet, as we shall highlight in the following section, the Upper and Middle Danube represented a geographical model closer to those studied by several Romanian and Bulgarian experts.

INTERNATIONAL CONTEXT AND MODELS OF RIVER MODERNIZATION

The Danube became an international river in the context of the Treaty of Paris (1856) drawn up after the Crimean War. The defeat of the Ottoman Empire by Russia and Romania in 1877–1878 and the emergence on the map of Europe of the new nation-states, including Bulgaria and Romania, opened new political avenues for riparian states.[2] The once mighty empire lost its access to the Danube and became a mere observer in the ECD, having been replaced by the Romanian Principalities. As a main route linking the Black Sea to Western Europe the lower reaches of this unruly river were problematic for steamer owners and captains. The ECD was created to intervene technologically on the maritime section. Romania (which became independent in 1881) was selected to host the ECD headquarters in order to keep the river's mouth navigable and maintain the connection between the Black Sea and Western Europe (Ardeleanu 2020).

In 1921 a new international governing institution called the International Danube Commission (IDC) was established with its headquarters in Bratislava (Schwabach1996). The institution monitored the river from Ulm (Germany) to Brăila (Romania) and stipulated, among other things, the conditions under which some hydraulic constructions adjacent to the main Danube canal were to be built: as defense against floods, to develop irrigation networks, and as investments in economic development, land improvement, levee construction, and course rectification (Ministère 1921).

In this fluid international context, the two riparian countries had to consider models of river governance. More than anywhere else Romanian and Bulgarian experts looked to the neighboring Danubian countries upstream. Around 1900 there was a profusion of plans aimed at connecting various European rivers. One such plan was to connect the Upper and Middle Danube with the Oder, Elbe, and Vistula—the so-called Austrian canal (Janáč 2012). Another plan envisioned a canal connecting Tisza with the Adriatic Sea (Hardi 2013). Several other projects sought to connect the Danube with the Black Sea by shortening its course (Turnock 1986), the river Siret (Eastern Romania) to Bucharest and further to the Danube (Botzan et al. 1959), or the canal that was to connect Russe to the Black Sea in Bulgaria (Papazov 1981: 90–92). Such plans haunted the imagination of experts and planners at the end of the nineteenth and in the twentieth century. The avalanche of such plans, which may look utopian today, made economic sense during that period because of the boost to the cereal trade in the Great Plain of Hungary and in the Southern Romanian plain. The cereal trade and accelerated urban development, especially between Budapest and Galaţi, made the Danubian region attractive for economic expansion.

The nationalist vision of Romanian and Bulgarian technocrats and experts was fueled by these projects, which had been implemented several decades earlier in Hungary, Austria, and Germany. To accomplish the river's "modernization," around 1890 a series of governmental institutions were created in Romania. Their scope was to administer navigation on the Lower Danube between the Iron Gates and the city of Brăila and to intervene technologically on the river when needed. The creation of the Romanian Hydraulic Service (Serviciul Hidraulic Român) in 1878 and the Romanian River Navigation Service (Navigaţia Fluvială Română) in 1890 (Kostov 1995: 108; Băicoianu 1915: 148–49), both charged with dredging the river and building levees, shows the Romanian government's interest in incorporating the river into the national economy. The Bulgarian government was more sluggish in implementing similar projects—despite the 470 kilometers of the Danube bank they owned—due to political turmoil and an economic downturn (Şerban 2019).

Technocrats and science experts also started to consider the Lower Danube as a source of wealth. Dredging of the wetlands, regularization of the river Tisza, and the construction of levees on the Great Plain of Hungary, which reclaimed ten thousand square kilometers of agricultural land in the mid-nineteenth century (Pinke 2014: 100), constituted a model of river development for Romanian hydrocrats. Vasile Roşu, a governmental expert hired by the newly established Romanian Hydraulic Service, was sent to Austria and Hungary to study the way these countries successfully reclaimed agricultural fields from the former wetlands of the Danube-Tisza, Criş, and Mureş Rivers (in Transylvania, then under Hungarian rule) (Roşu 1903, 1906). The field trip was concretized in two case studies written for the Romanian government, in which he concentrated on the 1880s laws issued by the Austrian-Hungarian state regarding embankments and the responsibility shared between state and private landowners for transforming wetlands into agricultural fields. He highlighted the central role of state experts in designing, implementing, and maintaining the hydraulic works. In order to help the Romanian government understand their role in such large transformative projects Roşu detailed the organization of the Hungarian Hydraulic Service, emphasizing the central role of the Hungarian state in hydraulic improvements to the Danube. Based on these studies, in 1912 Anghel Saligny, the head of the Romanian Hydraulic Service, developed a project aiming at the embankment of the entire Romanian Danube (Botzan et al. 1991); we will elaborate on this in detail later in the chapter.

Romania followed the model of Austria and Hungary in regarding the mighty river as "white coal." This was in contrast to Poland's strategy, for instance, which had a less extensive hydrographic network and more consistent coal reserves, and was thus oriented toward basing the development of its national energy system on coal-powered plants (Lagendijk 2008; Hidvegi and Wolf 2018).

The Bulgarian state was slow in creating similar state institutions until World War I. The State Water Department, which functioned within the Ministry of Agriculture in 1906, was the only institution tasked with studying and devising technological projects to correct and dam the small rivers that run into the Danube.[3] However, only in 1908, when the department was reorganized as a water office, were they provided with financial support and experts to start work in the field. The plans to build levees along the Danube, developed in 1920–1930, aimed more at eliminating malaria, which devastated the marsh villages, than reclaiming land. The Water Service carried out a systematic study of the Danubian marshlands and proposed that the government build submersible dams. However, World War I halted any further action in levee construction.

LEVEES, POLITICS, AND DEVELOPMENT

Romania experienced a fertile period with a calm political backdrop after World War I. The significant territorial gains made after the war (Romania"s territory almost doubled in size) and the return to a two-party political system, the National Liberal Party and the National Peasants' Party, offered political stability that catalyzed economic development (Dogan 1987). The embankment of the Danube triggered an intellectual and scientific debate between two experts, Grigore Antipa and Anghel Saligny. This debate embodies the technological and scientific knowledge of the time and was carried out between a biologist, a seer of his time, and an engineer. Both educated in Germany, Antipa at the University of Jena with Ernst Haeckel, and Saligny at the Royal Technical Higher School of Charlottenburg Berlin, their visions for the Danube's economic role diverged radically. The technical discussion concerned whether the state should build submersible levees, as Antipa suggested, or nonsubmersible ones, as Saligny proposed. This was not merely a technical debate but one that touched on crucial matters of political economy. Should the state decide that the river was to be flanked by submersible levees the Danube would continue to flood the area, albeit in a controlled way, and the economy of the area would continue to rely on natural resources. The levees would maintain the richness of the river ecosystems and would preserve what today we would call biodiversity. Economic growth and river development arose from "scientific fishing" conducted with care for its sustainability (Antipa 1910, 1912).[4] Building nonsubmersible levees, as Saligny suggested, would mean a radical alteration of the entire floodplain's ecology. The Danube would be restrained in large, expensive, strong levees that would stop the flooding forever, the wetlands along the river would be dredged, and massive land improvement works would be carried out (Botzan et al. 1991: 14). The economy of the Danubian region would be entirely reliant on the production of cereals.[5]

The debate was deeply political as both participants were more than merely state experts; they were planners who occupied higher political office and had the power to make long-term plans for the country's economic development: Antipa ran the State Fisheries Department between 1893 and 1914 whereas Saligny ran the Romanian Hydraulic Service and Directorate of Land Improvement, which was an institution created especially to deal with the integration of the Danube into the national economy system.

In 1910 the liberal government issued a law drawn up by Saligny concerning building nonsubmersible levees, which met with opposition from Antipa and other hydrotechnical engineers. Very active in this group was Ion Vidraşcu, who prepared a study on the construction of levees on the Danube, drawing on the experience of levee construction on other large rivers of the world such as the Mississippi, Po, Loire, Tisza, and Rhône (Vidraşcu 1921). The polemics between Antipa and Saligny and their followers took place in the context of (but not influenced by) similar debates at the European level. For instance, in Austria a similar debate between hydraulic engineers and biologists concerned the canalization of the Danube and the reclamation of land in the Danubian floodplain (Schoder and Schmid 2017).

Antipa's technical and political economic vision triumphed in 1929, when he was invited to draft a new law on improvements to the Danube floodplain. His proposition was the construction of nonsubmersible levees for one-third of the 450,000 hectares of the floodplain while the remaining part would have submersible levees built on it or remain untouched. He suggested establishing 30 syndicates—to include the state, private investors, and landowning peasants—that would receive 350,000 hectares of floodplain for land reclamation.

Since the state's plans were constantly delayed, several local initiatives were undertaken. At the beginning of the 1930s the engineer Nicolae Greceanu organized a water syndicate composed of local landlords and private investors, such as Greceanu himself, and the state. Greceanu convinced those actors, who believed the fetid backwaters of the floodplain were economically worthless, to believe that the Promethean transformation of wetlands into productive agriculture fields was possible and desirable. He calculated the potential economic gains: whereas the leasing of fishing areas brought the state 600,000 lei per year his agricultural association would bring 40 million lei in ten years (Greceanu 1933). He built a nonsubmersible levee next to the city of Giurgiu, which reclaimed 6,400 hectares of land. Out of these, 2,100 hectares belonged to state and larger private investors, Greceanu owning 600 hectares. Other private initiatives failed in their attempt to transform the Danube's wetlands. For instance, the syndicate in Surlari, 50 kilometers east of Giurgiu was established in 1925 in order to build levees and reclaim 5,310 hectares of land. Half of this land belonged to the Royal House of Romania. Despite the fact that the Romanian state granted loans through the Rural Land Credit in order to build the levees, the works were delayed and the syndicate was declared bankrupt in 1931.[6]

The implementation of plans to transform the wetlands of the Danube into agricultural fields on a national scale faltered in Romania due to the lack of funds to implement them and their pharaonic magnitude. There were international offers of support: in 1929–1930 a French company, Société générale des industries agricoles, proposed to the Romanian government that they build levees along the Danube in order to reclaim 200,000 hectares of wetlands. In exchange, the reclaimed land would be leased at low cost to the Société générale for the next thirty years. A similar offer came from a US bank for investments (Dobre and Căzănescu 2009: 188). Both offers were turned down by the Romanian state. In the end, submersible levees were constructed in the interwar period but only to reclaim 58,000 hectares of land—15 percent of the entire floodplain (Şişeşti 1933: 17).

Post–World War I Bulgaria experienced a somber period due to significant territorial losses and the imposition of substantial war reparations. It was a period of high political instability and the emergence of radical parties such as the Agrarians and the Communist Party. This period was marked by violence and political assassinations, as in the case of the charismatic agrarian Aleksandar Stamboliyski, whose radicalism and economic reforms led to his death in 1923. The nationalist authoritarian government of Aleksandar Tsankov, which replaced the Agrarians, met a similar fate: on April 16, 1925, the explosion of a bomb in the Sveta Nedelja church of Sofia killed 130 people, including almost all the Tsankov government ministers (Crampton 2007). To the political uncertainties one should add the influx of over a quarter of a million refugees between 1912 and 1930, coming from the territories Bulgaria had lost at the end of World War I. Unable to resolve the problem of this massive population increase, in 1926 the Bulgarian government called on the League of Nations for help in drawing up plans for the resettlement of the refugees (Garcés 1995; Tooze and Ivanov 2011; Brégianni 2019). Among the measures proposed by the League of Nations was the construction of levees and drainage of 21,000 hectares of wetlands along the Danube to be transformed into settlements for refugees. The reclaimed land would have provided only a portion of the 132,000 hectares required, but the proposition was accepted by the Bulgarian government (League of Nations 1926; Batakliev 1939). For these works, the League of Nations secured an international loan of £400,000 with the possibility of extending it up to £2,250,000; in 1928 a new loan amounted to £5,000,000 (Dragostinova 2006).

To establish where the levees were to be built, several cartographic and geological surveys were carried out. With or without funds from the League of Nations the levees were scheduled to be built after the floods of 1924 devastated the Nikopol-Svishtov area (Ministère 1928b). Once the decision to build levees was made, the Bulgarian state asked for assistance from international experts provided by the League of Nations (League of Nations 1927a, 1927b). V. Fornari, the chief inspector of civil engineering, member of the Italian Central Council for Public Works, and representative of the League of Nations supervised the extensive geotechnical surveys conducted in 1926–1927 in Belene-Svishtov area. His technical plans for the levees—height, cross-section, materials—triggered a clash with Bulgarian engineers. Fornari designed nonsubmersible levees three meters higher than his Bulgarian opponents proposed and pointed out that his solution concurred with IDC prescriptions, although it was more expensive for the Bulgarian state than the local engineers' solution. Finally, Fornari's solution prevailed.[7] Also, his team proposed building two dikes in the Vidin area, twenty-three kilometers in length and one meter higher than the level reached by the 1897 catastrophic Danube flood (see figure 5.1).

The technical parameters of the levees were calculated following the methodology of the Austrian engineer Ignaz Schmied (*Die Wasserbewegung im Dammkorper*)—that is, the height, the width of the bases, the trapezoidal form, the number of steps (one or two)—and by applying the Danube's hydrological data. The project had previously been sent for consultation to experts from Hungary, Serbia, and Romania (Angelov 1933: 3). However, finances remained a key issue as international loans proved to be limited. In order to implement the technical recommendations of the 1926–1927 surveys, Boris Angelov, an undersecretary of state within the Ministry of Agriculture, suggested a financial solution.

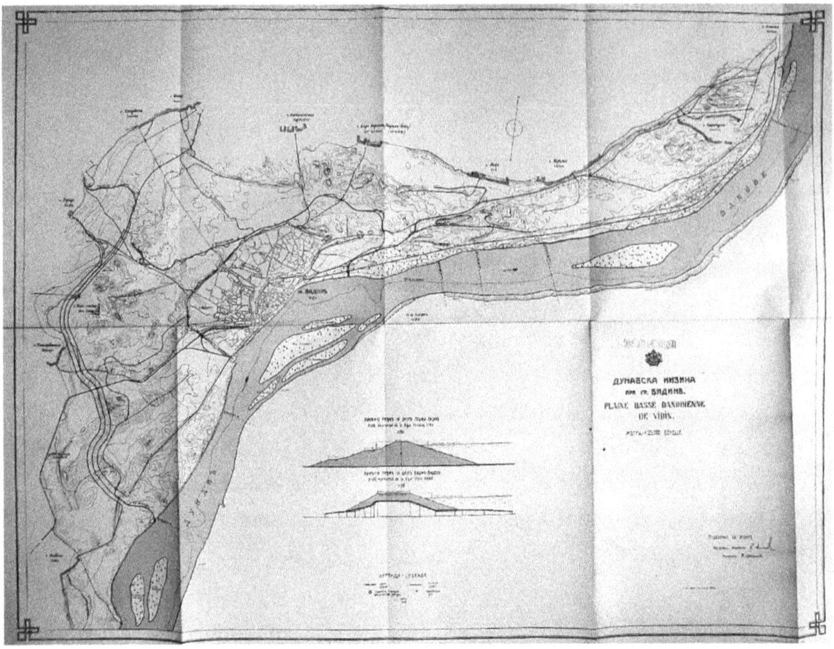

FIGURE 5.1. The map of Vidin, Bulgaria, including the metropolitan area before starting levee construction in the mid-1920s. Source: Ministère (1928a).

In the Vidin area the state was to cover 45 percent of the cost of building levees, private landowners the same amount, the rest being covered by local authorities (Ministère 1928a). As he was aware of the unbearable financial burden for landlords he suggested that the Agricultural Bank of Bulgaria could offer loans. Eventually, most of the Danube levee construction was carried out with financial support from the League of Nations.[8] The lack of funds was chronic and by the mid-1930s only 45,000 hectares of the 80,000 hectares of Bulgaria's Danube floodplain were protected by levees, most of them submersible. Plans for resettling the refugees—around 40,000 families—on reclaimed land along the Danube had failed. They were settled in areas such as Burgas, Plovdiv, and Petric, hundreds of kilometers away from the Danube (Dragostinova 2006).

Governmental experts and planners, maybe more so in Romania than in Bulgaria, made quixotic plans for transforming the Danubian landscape: huge water volumes and earth masses, as Trischler and Weinberger (2005: 58) put it, were to be moved to achieve the economic dream of a highly productive floodplain (Vlachos, chapter 2 in this volume; Dorondel and Olenenko, chapter 3 in this volume). However, there are differences between the two countries regarding the achievement of this dream. In Bulgaria the government planned a first phase (1902–1903) of building levees and dredging the backwaters mainly to eliminate malaria caused by the fetid marshes and, after World War I, to find space to relocate the surge of refugees. In Romania, a technocratic elite took their "hydraulic mission" (Molle, Mollinga, and Wester 2009) very seriously. The Romanian technocracy, who had, as every technocracy, "the tendency to give precedence to technological competence over political legitimacy, to rational administration over the hazards of public debate" (Picon 2007: 197), was better prepared in terms of technological and scientific background to develop a more coherent—if unachievable—economic plan of Danube transformation. However, in both countries the state was seen as the driver of modernity and rationalization of the Danube (Scott 1998; Dorondel and Olenenko, in this volume; Ćorović, chapter 1 in this volume). The next section will show the scale of these plans, which included an electrification system, the use of hydropower, and the different paths the two states pursued.

THE PROMISES OF IRRIGATION AND HYDROPOWER

Levee construction was intended not only to protect the riparian population against flooding but also to facilitate the emergence of national infrastructures such as electricity networks and irrigation systems (Bijker 2007; Swyngedouw 2007). The integration of these infrastructures in the national plan of development had at least two features. First, state technocrats steadily, albeit not entirely, replaced a strictly political state apparatus, as building such networks required the development of a technical corps with specific knowledge and training (as Janac shows, chapter 6 in this volume). Second, the extension of the electricity network widened access to housing improvement, which represented an incipient democratization at the grassroots level (Hughes 1983; Hausman, Hertner, and Wilkins 2008).

In 1910 governmental planners in Romania conceived a quixotic plan to connect the Siret River in Eastern Romania to Bucharest in the south and farther to the Danube (Popesco 1922). The canal, at a length of around 250 kilometers, was to be part of the national irrigation system, bringing water to 1,100,000 hectares of the savannah-like plains in Southeast Romania. In 1911, the Directorate of Land Improvement, led by Anghel Saligny, commissioned two foreign engineers, the British William Willcocks and the Italian Luigi Villoresi, to draw up technical plans (Willcocks 1914; Botzan et al. 1959: 8–17).

Wilcocks was famous for his work as an expert on the Aswan Dam construction and the Suez Canal (Molle 2009: 487), and Villoresi was head of hydraulic construction in San Benedetto on the river Po (Manfredini 2010: 102). Other canals were considered for the same purposes—irrigating large, dry plains in Dobrogea (a project dating back to the mid-nineteenth century, but then designed for navigation) and eastern parts of Romania—which would in addition have connected a large city such as Iaşi to the river Prut. In 1930, Ion Vidraşcu, an engineer who was deeply involved in the debate around the type of levee construction suitable for the Danube, prepared a study in which he argued in favor of establishing a national irrigation system using Western models, mainly the US model but also those of France, Spain, Italy, and the Soviet Union (Vidraşcu 1930). In all plans for a national irrigation system linked to the Danube and its basin, the state was the main actor in terms of finance and maintenance. Despite the abundance of plans to develop a much-needed irrigation system in the southern and eastern plains of Romania at the end of 1930s only 15,000 hectares of agricultural land was irrigated (Cismaru and Gabor 2004: 9).

The Danube was regarded by Romanian planners not only as a source of irrigation and an economic asset but also as a source of electricity. As early as 1924 the liberal government passed a law mandating the electrification of the country. A commission of Swiss, French, and Italian engineers was established to assist Romanian engineers in drafting technical provisions (Dinulescu 1981: 17; Turnock 2004: 17). In 1930 another law was assumed by the other main party, the National Peasants' Party, one in which hydropower took a more central role. At that time, Romania had a modest hydropower capacity, 7.5 percent of the country's overall power production.[9] One of the experts commissioned to expand the country's hydropower network was Dorin Pavel (1900–1979), an engineer with a doctoral degree from the Polytechnic University of Zurich, who devised a national plan for an estimated hydropower capacity of almost 6,000 megawatts (Pavel 1933).[10] He envisaged in great detail the construction of 567 hydropower plants across the country's territory by using the rich hydrographic network. Three of these hydropower plants were to be constructed on the Danube in cooperation with Yugoslavia, with an installed capacity of 870 megawatts for each part (see figure 5.2). The network was to be connected to the European grid, extending from the Pyrenees to the Soviet Union.[11] Another state-sponsored expert, Dimitrie Leonida, drew up plans for the Bicaz hydropower plant on the Bistriţa River ,which was to be connected to the Iron Gates hydropower complex.[12] The Bicaz hydropower plant was to bear King Ferdinand's name (the second king of modern Romania) "in recognition of the king's great interest in electrification" (Turnock 2008: 29).

The Bulgarian state pursued a different strategy by encouraging the formation of powerful "water syndicates" through a law passed on October 20, 1920. The syndicates were interested neither in the construction of levees nor in a national plan to integrate the river and its basin into the national economy. Instead, they favored installing irrigation and electricity systems on smaller rivers and tributaries of the Danube. By the end of the 1930s the number of private irrigation pumps on

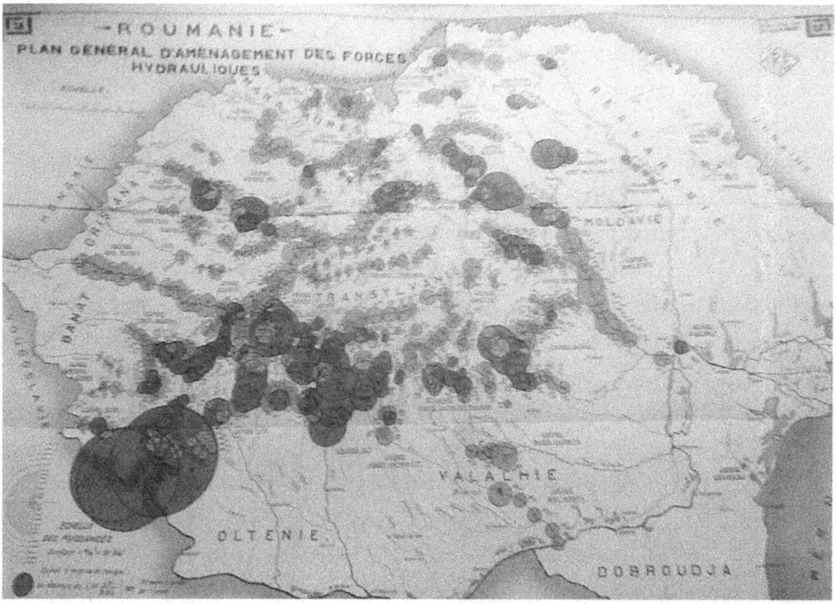

FIGURE 5.2. The plan to build a national network of hydropower plants in interwar Romania
Source: Pavel (1933).

the Danube's tributaries was relatively high. For instance, in the Lovech district, which was crossed by the Osam, Iskar, and Vit Rivers, there were thirty-six such pumps, the majority of which were installed in small towns.[13]

The Bulgarian government organized four irrigation syndicates here, providing partial funding to build canals, pumps, and power stations.[14] The syndicates organized in areas near the Danube were part of a network that in 1941 comprised 114 members, 95 irrigation syndicates, and 19 electricity-producing syndicates (Nikolov 1941: 12). Despite the modest surface irrigated, some 50,000 hectares—slightly more than 1 percent of the country's arable land (Donev 1972: 18–21)—a large number of farmers were involved and the technology available performed quite well. For example, in 1941, 36 pumping stations were capable of raising 1,000 liters per second to a height of up to 54 meters (Nikolov 1941: 30). On the Struma River in the town of Kustendil in Southwest Bulgaria, a water syndicate was organized in 1941 comprising 5,194 individuals and 17 associations; farmers and associations showed considerable interest in such integrated organizations (Nikolov 1941: 15). An example of what the hydraulic structures built by water syndicates looked like was the Zlatna Panega reservoir, which was constructed by building a dam on the Panega River, a tributary of the Iskar, close to the city of Lukovit. This facility provided water both for irrigating 900 hectares and providing electricity for the city and the surrounding villages. The reservoir held 2,250,000 cubic meters of water, ensuring an installed power capacity of 2.2 megawatts and an irrigation flow rate of 6,000 liters per second (Nikolov 1941: 23; Penchev 2009).

Despite the interest shown in the hydroelectric system by water syndicates, at the national level they proved rather ineffective; the production of electricity was low and the national grid remained underdeveloped. In the mid-1930s it was clear that the legal framework was anachronous. At the beginning of the 1920s, in parallel with the activity of water syndicates, the Association of Engineers and Architects of Bulgaria prepared a coherent national electrification plan (Penchev 2009: 34). A new commission was tasked in 1929 with the same process whose works under the auspices of the Ministry of Agriculture and Public Domain lasted until 1933 (Spirov 1999; Penchev 2009). A collection of reports on electrification in some of the more advanced European countries—Germany, Czechoslovakia, Austria, Poland, Sweden, Norway, Italy, Switzerland, France, and Belgium—was produced (Stefanov, Zlatev, and Mihailov 1931). Based on this report a new electrification project was elaborated. Yet the project never made it into a new legislative draft: the liberal government was replaced by a military government uninterested in such pedestrian projects. Thus, at the end of the interwar period Bulgaria had an installed power capacity at the national level of 130–140 megawatts, mostly held by private investors or communal and municipal investors (Spirov 1999: 272–73). One of the largest energy producers in Bulgaria was the Vucha water syndicate, which provided energy to 139 areas along the Maritsa River in south Bulgaria, including two cities, Plovdiv and Pazardjik (Penchev 2009: 35–38).

The plans for building irrigation systems and the hydropower energy system that haunted the national imagination of the two Southeast European countries must be put in the context of the development of similar enterprises in Western Europe. Examining the national imagination regarding the inclusion of the Danube in the national economy of the two countries we see that more economically advanced countries constituted the model for using nature for developmental purposes. Yet it was not always "the West" that attracted the attention of Romanian and Bulgarian experts and planners but countries such as Hungary and Austria with more advanced systems of irrigation, canalization, and development of the hydroelectric network. The "modernization" of the Danube may have started as a Western process that bore similarities to colonial politics and technologies of governing—the first decades of the ECD's existence and its works on the mouth of the river prove that. However, several decades later Western experts were replaced by local ones: it is nevertheless true that they were bearers of Western ideas about the technological transformation of nature for developmental reasons (Özkan, chapter 8 in this volume). Besides, the riparian states were as interested in the Danube as an "organic machine" as were Western powers, and in the economic benefits.

At the end of the nineteenth century the international status of the Lower Danube started to change steadily. It was still an international river, with two international commissions that supervised its economic transformation, yet national voices were able to influence the discussion. The two countries had different strategies regarding the river and its inclusion in the national economy. Bulgaria experienced political turmoil, with a surge of Bulgarian refugees from the territories lost after World War I, and encountered both political and economic instability. One may even say that Bulgaria lost two kinds of land in two types of "war": it lost land to its neighboring adversaries and potential land to the "war" on the river.[15] Romania sailed through more peaceful political waters with new territories added after World War I. Continuing the above metaphor, Romania won both types of war: it gained land from its neighboring adversaries and from the struggle against the river.

The political agendas of the two countries led to different national strategies for developing land adjacent to the Danube and to different institutional arrangements. Romania preferred first to obtain Western knowledge and experience and then to commission national experts—themselves trained in Western universities—to elaborate wider plans for including the Danube in the national economy. Bulgaria, while still looking for international expertise and funds, bet instead on national organizations, the water syndicates, which developed a more locally oriented development strategy. The Danube was less integrated into the Bulgarian economy as more attention was paid to smaller hinterland rivers.

Building levees, digging long water canals for irrigation, and establishing a national electricity system was from the beginning a high-modernist project for these two nation-states. The frantic policies of transforming the Danube for economic purposes were influenced by Western ideas about the utility of rivers, if not in a linear way then certainly through experts, both local and foreigners. Parts of the world with no direct connection were nevertheless connected through experts working in African colonies and in Eastern Europe, who made the high-modernist project truly global.

NOTES

We thank Alexandra Ion and George Andrei for their meticulous reading and comments on an earlier draft. Their comments prompted us to find new ways of restructuring the chapter and making it more readable. Stelu Serban also thanks the personnel of the State Archive of Pleven for their courtesy and assistance. This work was supported by a grant from the Romanian Ministry of National Education, CNCS—UEFISCDI, no. PN II-ID-PCE-2012-4-0587 (2013–2016).

Epigraph: Ambroise Vollard, *Recollections of a Picture Dealer*, trans. Violet M. MacDonald (Mineola, NY: Dover, 2002 [1936], 210.

1. For a more general view of these issues, see Kohlrausch and Trischler (2014).

2. See the Introduction to this volume for more details on redrawing the political map of Europe and the new geopolitical realities in the second part of the nineteenth century.

3. Tsentralen Durzhaven Arhiv–Pleven [Bulgarian State Archive, Pleven Branch] (TsDAP), fund 32 K, www.archives.government.bg/52-Pleven.

4. Antipa wrote the first modern fishing law in Romania in 1896.

5. For more about this debate, see Şerban (2019).

6. Arhivele Naţionale ale României, Depozit Central [Romanian National Archives, Central Deposit], fund 19.2.1 (Ministry of Finance, Subsection General Activity), doss. 73/1934. The Rural Land Credit was a state bank funded in 1880 with the aim of financing rural development.

7. TsDAP, fund 32K, doss. 98, ff. 5–20.

8. Arhiva Ministerului Afacerilor Externe, Guvernul României [Archive of Ministry for Foreign Affairs, Romanian Government], issue 220/1965, vol. 1, f. 4.

9. The figures regarding the entire installed power capacity of the national energy system in the interwar period vary. Rarincesco (1930) posited that the entire installed power capacity of the national system amounted to 880 megawatts in 1929 whereas Dinulescu's (1981) more cautious evaluation is 800 megawatts at the end of the interwar period.

10. Between 1949 and 1971 he was the head of the Investment Department in the Ministry of Energy. He was, obviously, recruited by communist officials who used his technical knowledge. Pavel was on various technical committees in the socialist government including the commission that decided to resume the construction of the Danube–Black Sea Canal (Tănăsescu 2010).

11. For more details on this project, see Vleuten and Kaijser (2005: 35).

12. Leonida studied at the Royal Technical Higher School of Charlottenburg Berlin (1903–1908) (Štanzel 2017).

13. TsDAP, fund 32K, part 1, doss. 271/1939, ff. 31 ff.

14. TsDAP, fund 32K, part 1, doss. 409/1938, ff. 38 ff. At the end of the 1930s the state contribution for financing the water syndicates amounted to around 17 percent (Nikolov 1941: 27).

15. We are indebted to George Andrei for this observation.

REFERENCES

Angelov, Boris. 1933. *Navodnenie po dolnata techenie na Dunav* [Floods on the lower Danube]. Sofia: n.p.

Antipa, Grigore. 1910. *Regiunea inundabilă a Dunarii: Strategia actuală şi mijloacele de a o pune în valoare* [The Danube's floodplain: its current status and the means to exploit it]. Bucharest: Institutul de Arte Grafice Carol Göbl.

Antipa, Grigore. 1912. *Memoriu relativ la organizarea comerțului și îndrumarea producțiunei pescăriilor* [Memoir concerning the organization of the commerce and production of fisheries]. Bucharest: Imprimeriile Independența.

Ardeleanu, Constantin. 2020. *The European Commission of the Danube, 1856–1948: An Experiment in International Administration*. Leiden: Brill.

Ash, Mitchell G., and Jan Surman. 2012. "The Nationalization of Scientific Knowledge in Nineteenth-Century Central Europe: An Introduction." In M. G. Ash and J. Surman, eds., *The Nationalization of Scientific Knowledge in the Habsburg Empire, 1848–1918*. Basingstoke: Palgrave Macmillan, 1–29.

Băicoianu, Constantin. 1915. *Dunărea: Privire istorică, economică și politică* [Danube: a historical, economic and political approach]. Bucharest: Tipografia Eminescu.

Batakliev, Ivan. 1939. "Dunavat i negovoto stopanstvo i politichesko znachenie za Balgarite." [The Danube and its economic and political significance for Bulgarians]. *Rodina* 1(3): 107–22.

Bijker, Wiebe E. 2007. "Dikes and Dams: Thick with Politics."*Isis* 98(1): 109–23.

Bocquet, Denis. 2007. "Engineers and the Nation in Italy (1750–1922): Local Traditions and Different Conceptions of Unity and Modernity." *History and Technology* 23(3): 227–40.

Botzan, M. et al. 1959. *Probleme de irigații și desecări ale Câmpiei Bărăganului* [Problems of irrigation systems and drainage in the Bărăgan Lowland]. Bucharest: Ed. Academiei RPR.

Botzan, M. et al. 1991. *Valorificarea hidroameliorativă a luncii Dunării românești și a deltei* [The exploitation and hydrotechnical transformation of the Danube floodplain and the Danube delta]. Craiova: EdituraTehnică.

Brégianni, Catherine. 2019. "Échange de populations, organisation d'espace et savoir-faire des institutions internationales." In C. Brégianni and R. Cussó, eds., *Politiques économiques des institutions internationales du début du XXe siècle au stade actuel de la globalisation*. Paris: Laboratoire de Démographie et d'Histoire Sociale-CRH, EHESS, 10–54.

Carroll, Patrick. 2012. "Water and Technoscientific State Formation in California." *Social Studies of Science* 42(4): 489–516.

Cioc, Marc. 2002.*The Rhine: An Eco-biography, 1851–2000*. Seattle: University of Washington Press.

Cismaru, Corneliu, and Victor Gabor. 2004. *Irigații, amenajări, reabilitări și modernizări* [Irrigations, improvements, rehabilitations, and modernizations]. Iași: Politehnium.

Clifford, Jim. 2017. *West Ham and the River Lea: A Social and Environmental History of London's Industrialized Marshland, 1839–1914*. Vancouver: University of British Columbia Press.

Crampton, R. J. 2007. *Bulgaria*. Oxford: Oxford University Press.

Dinulescu Constantin. 1981. *Electrificarea României: de la primele începuturi până la anul 1950* [The electrification of Romania: from the beginning to 1950]. Bucharest: Ed. Tehnică.

Dobre, Virgil, and Severin Cazănescu. 2009. "Probleme actuale privind realizarea 'Coridorului verde al Dunării inferioare' în strânsă corelare cu valorificarea superioară social-economică și ecologică a acestei zone" [Current issues of the "green corridor of the lower Danube" in connection with the social, economic, and ecologic use of this area]. In *Dezbatere națională: Dunărea, Lunca și Delta Dunării* [National debate: the Danube, the Danube floodplain, and Danube delta]. Bucharest: Academia de Ştiințe Agricole și Silvice Gheorge Ionescu-Şișești and Bren, 187–92.

Dogan, Mattei. 1987. "Romania: 1919–1938." In M. Weiner and E. Ozbudun, eds., *Competitive Elections in Developing Countries*. Durham, NC: Duke University Press, 369–89.

Donev, Kiril. 1972. *Irrigated Agriculture in Bulgaria*. Sofia: Sofia Press.

Dragostinova, Theodora. 2006. "Competing Priorities, Ambiguous Loyalties: Challenges of Socioeconomic Adaptation and National Inclusion of the Interwar Bulgarian Refugees." *Nationalities Papers: The Journal of Nationalism and Ethnicity* 34(5): 549–74.

Garcés, Laura. 1995. "The League of Nations' Predicament in Southeastern Europe." *World Affairs* 158(1): 3–17.

Gătejel, Luminiţa. 2016. "Overcoming the Iron Gates: Austrian Transport and River Regulation on the Lower Danube, 1830–1840." *Central European History* 49: 162–80.

Greceanu, Nicolae. 1933. *Indiguirea Malu Roşu, Băneasa, Gostinu din lunca inundabilă a Dunării* [Levee construction in Malu Roşu, Băneasa, Gostinu in the Danube floodplain]. Bucharest: Atelierele Adevarul.

Groparu, Tiberiu et al. 2019. "La place des idéologies dans la construction des paysages du delta du Danube : un regard géohistorique sur quatre siècles d'aménagement et de transformations de l'environment." In P. Vallette and J-M. Carozza, eds., *Géohistoire de l'environment et des paysages*. Paris : CNRS Edition Alpha, 205–21.

Hardi, Tamás 2013. "Non-realised Plans for the Enlargement of the Danubian Waterway." *Forum geografic: Studii şi cercetări de geografie şi protecţia mediului* 12(2): 138–46.

Hausman, William J., Peter Hertner, and Mira Wilkins. 2008. *Global Electrification: Multinational Enterprise and International Finance in the History of Light and Power, 1878–2007*. Cambridge: Cambridge University Press.

Hohensinner, Severin et al. 2013. "Changes in Water and Land: The Reconstructed Viennese Riverscape from 1500 to the Present." *Water History* 5: 145–72.

Heidvogl, Gertrud et al. 2018. "Urban Waters and the Development of Vienna between 1683 and 1910." *Environmental History* 23: 721–47.

Hidvegi, Maria, and Nikolaus Wolf. 2018. "Power Failure: The Electrification of Central-Eastern Europe, 1918–39." *European Review of History* 26(2): 219–42.

Hughes, Thomas. 1983. *Networks of Power: Electrification in Western Society 1880–1930*. Baltimore: Johns Hopkins University Press.

Janáč, Jiří. 2012. *European Coasts of Bohemia: Negotiating the Danube-Oder-Elbe Canal in a Troubled Twentieth Century*. Amsterdam: Amsterdam University Press.

Jensen, John H., and Gerhard Rossegger. 1978. "Transferring Technology to a Peripheral Economy: The Case of Lower Danube Transport Development." *Technology and Culture* 19(4): 675–702.

Kohlrausch, Martin, and Helmuth Trischler. 2014. *Building Europe on Expertise: Innovators, Organizers, Networkers*. Basingstoke: Palgrave Macmillan.

Kostov, Alexandre. 1995. "Trade and Navigation on the Lower Danube: Romania and Bulgaria, 1880–1912." In A. Kunz and J. Armstrong, eds., *Inland Navigation and Economic Development in Nineteenth-Century Europe*. Mainz: Verlag Philipp von Zabern, 105–18.

Kostov, Alexandre. 2009. "Les Ponts et Chaussées français et les pays balkaniques pendent la second moitié du XIXe et au début du XXe siècle: le cas de la Roumanie, de la Serbie et de la Bulgarie." *Quaderns d'Historia de l'Engyneria* 10: 367–88.

Lagendijk, Vincent. 2008. *Electrifying Europe: The Power of Europe in the Construction of Electricity Networks*. Amsterdam: Aksant Academic.

League of Nations 1926. *Scheme for the Settlement of Bulgarian Refugees: General Description and Principal Documents*. Lausanne: Imp. Reunies S.A.

League of Nations 1927a. *Scheme for the Settlement of Bulgarian Refugees: Third Report of the Commissioner of the League of Nations*. Geneva: Imp. Kunding.

League of Nations 1927b. *Scheme for the Settlement of Bulgarian Refugees: Fourth Report of the Commissioner of the League of Nations*. Geneva: Imp. Kunding.

Manfredini, Laerte. 2010. "Le bonifiche del primo Novecento dal progetto alla realizzazione: manufatti idraulici, sistemi idrografici ed edilizia rurale." In A.Cristina and P. M. De Marchi, eds., *Uomini e acque a San Benedetto Po. Il governo del territorio tra passato e futuro. Atti del Convegno Mantova-San Benedetto Po, 10–12 maggio 2007*. Borgo San Lorenzo: All'Insegna del Giglio, 101–8.

Menga, Filippo. 2015. "Building a Nation Through a Dam: The Case of Rogun in Tajikistan." *Nationalities Papers: The Journal of Nationalism and Ethnicity* 43(3): 479–94.

Menga, Filippo, and Erik Swyngedouw. 2018. "States of Water." In F. Menga and E. Swyngedouw, eds., *Water, Technology and the Nation-State*. London: Routledge, 1–18.

Ministère 1928a. Ministère de l'agriculture et les domaines de l'Etat. Royaume de Bulgarie, *L'amélioration de la plaine basse danubienne de Vidin: Projet élabore par le Service des Eaux après de Ministère d'agriculture et des domaines d'état*. Sofia: Imprimerie de l'Etat.

Ministère 1928b. Ministère de l'agriculture et les domaines de l'Etat. Royaume de Bulgarie, *L'amélioration des plaines basses danubiennes de Belene et Svicstov: Projet élabore par le Service des Eaux après de Ministère d'agriculture et des domaines d'état*. Sofia : Imprimerie de l'Etat.

Ministère 1921. Ministère des Affaires Etrangers, France, *Conférence internationale pour l'établissement du statut définitif du Danube*. Paris: Imprimerie Nationale.

Molle, François. 2009. "River-Basin Planning and Management: The Social Life of a Concept." *Geoforum* 40: 484–94.

Molle, François, Peter P. Mollinga, and Philippus Wester. 2009. "Hydraulic Bureaucracies and the Hydraulic Mission: Flows of Water, Flows of Power." *Water Alternatives* 2(3): 328–49.

Moore, Jason, W., ed. 2016. *Anthropocene or Capitalocene? Nature, History, and the Crisis of Capitalism*. Oakland, CA: PM Press.

Nikolov, Petar. 1941. *Vodnite sindikati u nas: sashtnost, organisatsija, postizhenija i schetovodstvo* [Water syndicates in Bulgaria: the current state, organization, achievements, and statistical data]. Sofia: T. T. Dragiev.

Papazov, Krasjo.1981. *Razvitie i izpolzavane na nationalna complex Dunav* [Development and use of the Danube National Complex]. Varna: G. Bakalov.

Parinello, Giacomo. 2017. "Charting the Flow: Water Science and State Hydrography in the Po Watershed, 1872–1917." *Environment and History* 23: 65–96.

Pavel, Dorin. 1933. *Plan général d'aménagement des forces hidrauliques en Roumanie*. Bucharest: Imprimérie Nationale.

Penchev Pencho D. 2009. *Elektrifikatsiata v stopnaskata politika na bulgarskata durzhnavnata ot osvbazhdenieto do 1944 g.* [Electrification in the political economy of the Bulgarian state from liberation until 1944]. Sofia: Universitetsko izdatelstvo Stopanstvo.

Picon, Antoine. 2007. "French Engineers and Social Thought, 18–20th Centuries: An Archaeology of Technocratic Ideals." *History and Technology* 23(3): 197–208.

Pinke, Zsolt 2014. "Modernization and Decline: An Eco-historical Perspective on Regulation of the Tisza Valley, Hungary." *Journal of Historical Geography* 45: 92–105.

Popesco, Georges. 1922. *Le relèvement économique de la Roumanie: voies de communication, augumentation et modernisation de l'outillage national, utilisation de l'énergie naturelle*. Paris: Felix Alcan.

Rarincesco, I. R. 1930. *L'électrification de la Roumanie*. Bucharest: Socec.

Roşu, Vasile. 1903. *Ameliorarea terenurilor de cultură prin drenaj şi irigaţii în Austria* [The improvement of agricultural fields through drainage and irrigation]. Bucharest: Carol Göbl.
Roşu, Vasile. 1906. *Desecările şi canalizările din Ungaria* [Drainage and canalization in Hungary]. Bucharest: Carol Göbl.
Schmid, Martin. 2013. "Towards an Environmental History of the Danube: Understanding a Great European River through its Transformation as a Socio-natural Site, c. 1500–2000." Habilitation thesis, University of Vienna.
Schoder, Angelika, and Martin Schmid. 2017. "Where Technology and Environmentalism Meet: Remaking of the Austrian Danube for Hydropower." In H. Petrić and I. Žebec Šilj, eds., *Environmentalism in Central and Southeastern Europe: Historical Perspectives*. Lanham, MD: Lexington Books, 3–20.
Schwabach, Aaron. 1996. "Diverting the Danube: The Gabcikovo-Nagymaros Dispute and International Freshwater Law." *Berkeley Journal of International Law* 14(2): 291–343.
Scott, James C. 1998. *Seeing Like a State: How Certain Schemes to Improve the Human Condition Have Failed*. New Haven, CT: Yale University Press.
Şerban, Stelu. 2019. "State, Technology and Environment on the Lower Danube: Bulgaria and Romania before the Balkan Wars." *Journal of Balkan and Near Eastern Studies* 21(2): 204–21.
Şişeşti, Gheorghe Ionescu. 1933. *Lunca Dunării şi punerea ei în valoare* [The Danube floodplain and its use]. Bucharest: Imprimeria Centrală.
Skelton, Leona J. 2016. *Tyne after Tyne: An Environmental History of a River's Battle for Protection, 1526–2015*. Cambridge: White Horse Press.
Spirov, Mire. 1999. *Electrifikatsijata na Bulgarija i neinite stroiteli* [Electrification in Bulgaria and its builders]. Vol. 1. Sofia: Heron Pres.
Štanzel, Arnošt. 2017. *Wasserträume und Wasserräume im Staatssozialismus: ein umwelthistorischer Vergleich anhand der tschechoslowakischen und rumänischen Wasserwirtschaft 1948–1989*. Göttingen: Vandenhoeck und Ruprecht.
Stefanov, Geno, Mincho Zlatev, and K. Mihailov. 1931. *Dokladi po elektrostopanstvoto v Zapadna Evropa* [Reports on energy systems in Western Europe]. Sofia: Vitosha.
Swyngedouw, Erik. 2007. "Technonatural Revolutions: The Scalar Politics of Franco's Hydrosocial Dream for Spain, 1939–1975." *Transactions of the Institute of British Geographers* 32(1): 9–28.
Swyngedouw, Erik. 2014. "'Not a Drop of Water . . .': State, Modernity and the Production of Nature in Spain, 1898–2010." *Environment and History* 20(1): 67–92.
Tănăsescu, Florin–Teodor. 2010. "Dorin Pavel şi hidrografia Dunării" [Dorin Pavel and the hydrography of the Danube]. *Ştiinţă şi inginerie* 17: 7–15.
Tooze, Adam, and Martin Ivanov. 2011. "Disciplining the 'Black Sheep of the Balkans': Financial Supervision and Sovereignty in Bulgaria, 1902–38." *Economic History Review* 64(1): 30–51.
Trischler, Helmuth, and Hans Weinberger. 2005. "Engineering Europe: Big Technologies and Military Systems in the Making of 20th Century Europe." *History and Technology* 21(1): 49–83.
Turnock, David. 1986. "The Danube–Black Sea Canal and Its Impact on Southern Romania." *GeoJournal* 12(1): 65–79.
Turnock, David. 2004. "Railway Network Development in Inter-war Romania: Economic and Strategic Motives." *Geographica pannonica* 8: 16–24.
Turnock, David. 2008. "The Drive for Modernisation in Inter-War Eastern Europe: Changes in Rurality in the Carpathian Mountains 1918–1945." *Geographica pannonica* 12(1): 12–38.

Vidraşcu, Ion. 1921. *Valorificarea regiunii inundabile a Dunării* [The use of the Danube floodplain]. Bucharest: Urbana.
Vidraşcu, Ion. 1930. *Irigaţiile în Ţara Românească* [The irrigation system in Walachia]. Bucharest: Tipografiile Române Unite.
Vleuten, Erik van der, and Arne Kaijser. 2005. "Networking Europe." *History and Technology* 21(1): 21–48.
White, Richard. 1995. *The Organic Machine: The Remaking of the Columbia River*. New York: Hill and Wang.
Willcocks, William. 1914. *Raport asupra proiectului de irigaţii şi navigaţiune în România* [Report on the irrigation projects and navigation in Romania]. Bucharest: Avântul.
Zeisler-Vralsted, Dorothy. 2014. *Rivers, Memory and Nation-Building: A History of the Volga and Mississippi Rivers*. New York: Berghahn Books.

6

From Nature to National Networks

Hydraulic Bureaucracy and the Modernization of Waters in Czechia, 1890s–1960s

Jiří Janáč

\\

This chapter focuses on the fate of the national technocratic modernist project of the Czech hydraulic mission during the twentieth century from the perspective of its key actors—members of the state water management bureaucracy, which scholars have dubbed the *hydrocracy*. The concept of hydrocracy was elaborated by the development studies specialists François Molle and Peter P. Mollinga as a modern response to Karl August Wittfogel's classic theory about the emergence of hydraulic civilizations in ancient Oriental empires.[1] They drew on the assumption that modern society is also to a certain extent dependent on the control of water, which is associated with social power (Molle and Mollinga 2009).

The emergence and activities of the hydrocracy were characterized by a specific modernizing ideology, referred to as the "hydraulic mission" (Molle and Mollinga 2009), firmly anchored in the context of the European project of "organized modernity," which was based on, among other things, politically mobilizing and organizing the masses and emphasizing collective (often national) welfare and expert (technocratic) governance (Wagner 1993). For the hydrocrats, emphasis on discipline, organization, and control was primarily related to water control. They were convinced of the moral and civilizing mission of water resource management, which through transforming the natural environment could influence and cultivate people and society. Crucial ingredients of hydrocratic thought from the beginning of the twentieth century onward included scientism, a paternalistic and utilitarian approach to nature, and a related teleological and ideological view of water (Molle and Mollinga 2009).

In the eyes of the hydrocracy, the water contained on a state's territory was primarily a crucial natural resource and at the same time an important infrastructure component, the basis of the modern infrastructural state––one produced, organized, and maintained by infrastructures (Guldi 2012; Bennet and Joyce 2010).[2] This was recognized also by political authorities, who saw expert knowledge as a crucial instrument for building the state and legitimizing its existence (Kohlrausch and Trischler 2014).As the relationship between the hydrocracy and the state grew more closely entwined, the high costs of the large hydraulic engineering projects necessary for controlling and managing water required state support and guarantees, while the state took advantage of the practical and symbolic political potential of water management projects to legitimize itself.[3] In this sense, it seems appropriate to speak of the co-construction of the state and its wet infrastructure, a process epitomized by the construction of large symbolic water structures––most famously, gigantic multipurpose dams (Pritchard 2011).

In this process, hydrocrats acted as the "system builders" of the state water management system.[4] The metaphor of the hydrologic cycle served them as a major instrument for translating the complex interactions between water, the natural environment, and society into a manageable form required by the functioning of the modern state (Linton 2010; Molle and Mollinga 2009: 329). To make water manageable, hydrocrats declared it a legible natural resource. Legibility was a basic precondition for effective central management and control, and the concept of the hydrological cycle allowed for calculations of the system's water balance based on a simple equation in which evaporation and runoff were subtracted from precipitation (Scott 1998). Furthermore, perceptions of rivers as "water flowing waste to the sea" made hydrocratic calls for the establishment of scientific, rational, and apolitical water resource management allowing for the efficient utilization of every "single drop" of water even more urgent. Such visions gave rise to nationally specific articulations of "hydraulic missions" programs for the long-term development of national water systems, which implicitly included demands for the large-scale transformation of the natural environment (Molle and Mollinga 2009).

The purpose of this chapter is to demonstrate, through a study of the Czech hydraulic mission and its exponents––hydrocrats––the substantial continuities of one modernization process in Europe, which would not come to light if this process was examined over a shorter period of time and which have remained neglected to a certain extent by contemporary Czech historians (Pullmann 2008). At the core of the chapter, beside the idea of continuity, is the issue of actually fulfilling the hydraulic mission, that is, the interaction between the Czech hydrocracy and the political authorities and the hydrocracy's ability to successfully advocate its visions in a markedly dynamic political environment, particularly after 1948, when, as a consequence of nationalization and restricted civic participation it seemed as if the main barriers limiting the transformation of these visions into reality had been completely removed.[5]

At the same time, the chapter attempts to contribute to the discussion about the relationship between water (the natural environment) and society. Water and water management as themes in contemporary Czech historical scholarship suffer from significant fragmentation––the overall picture has been split up into chronologically isolated moments (typically histories of individual projects) (see Janáč 2012; Štanzel and Durnová 2015; and Gorenčík 2001). Furthermore, two narratives that are markedly disparate in content stand in contrast to each other––one is the story of a successful engineering mission (Votruba 2003), and the other a tale of environmental degradation (see Hadrabová 1996, 1998).[6] The same can essentially be said about the (environmental) history of Eastern and Central Europe, which still wavers between a narrative that positively evaluates the holistic approach to the environment and a narrative of destruction epitomized by the "ecocide" thesis (Brain 2018).

THE BIRTH OF THE CZECH(OSLOVAK) HYDRAULIC MISSION

The Czech hydrocracy was formed in the second half of the nineteenth century, at a time when the Czech lands of Bohemia and Moravia were part of Austria-Hungary. In relatively short order, large state modernization projects were authorized and special expert institutions were established to implement them. The first was the 1884 Amelioration Fund, a framework for land improvement demanded by agricultural circles, which was followed by river improvement schemes and the 1901 Canal Act, a late attempt at modernization and infrastructural consolidation in the Austro-Hungarian Empire that sanctioned the construction of the Danube–Oder–Elbe Canal and the Vltava–Danube Canal as a part of statewide centralized waterway network (Bartovský 1946). These projects constituted the first attempt at systematically modifying the hydrological regime on the state's territory. Nonetheless, the water administration system remained significantly fragmented. A drop of rainwater, to paraphrase a popular saying used at the time by hydrocrats, fell under the purview of up to eight offices on its journey from the place where it fell to the state's borders. As a consequence, disputes over the exercising of power arose that slowed down any related construction work and resulted in problems due to insufficient coordination.[7]

The institutionally fragmented hydrocracy viewed the lack of coordination in water management in a fundamentally negative light and endeavored to promote a holistic approach to water. In the early years of World War I, inspired by the American technocratic movement, and building upon the generally accepted image of engineering as an important component of national modernization (Van Meer 2010), Czech hydrocrats articulated a bold program demanding centralized and depoliticized management of all water resources in place of fragmented construction programs. We can consider the document *A Draft Water Management Program in the Kingdom of Bohemia*, authored by then rector of Czech Technical University, Antonín Klír, to be the first comprehensive formulation of the local Czech version of the hydraulic mission (Klír 1916). Although the document did not refer to any specific foreign examples or literature, it was firmly grounded in the contemporary Western hydrocratic and technocratic discourse. Klír summed up a long-running debate and regularly referenced German water engineering projects and literature.

The objective of the program was to "complete major hydraulic engineering works in the Kingdom of Bohemia" within the span of two twenty-year plans and ensure that water (i.e., surface runoff) would be fully controlled and effectively utilized as natural and technical conditions allow (Klír 1916: 6). The ideas of centralization (management "in connection with the economic whole," i.e., management that takes all of society into account), scientification (everything was supposed to be based on "hydrological characteristics" in each drainage basin), and rationalization, or perhaps better put depoliticization (everything was supposed to be systematic, planned, and undisturbed by political influences) resulted in a proposal to establish a water management commission for administering all surface water in Bohemia and for making decisions about projects.

Klír's highly technocratic vision clearly invoked, if only implicitly, the idea of the hydrological cycle. First, hydrographically speaking, hydrocrats saw Bohemia as a unified whole, in which "everything gravitates inward" and "every intervention in hydrological conditions anywhere elicits changes in sites at lower elevations" (Klír 1916: 1). Indeed, the drainage basin of the upper Elbe River (with the Vltava as the main tributary) covers almost the entire territory of Bohemia. Similarly, the political boundaries of Moravia largely correspond to the delimitation of the Morava River drainage basin. The argument of the "naturalness" of the drainage basin as the basic organizational unit of water management administration (and by extension political administration as well), driven by modernist scientism, paved the way for the broader blending of water management with regional planning and, ipso facto, social engineering (Molle 2009).

Furthermore, Klír's aspirations concentrated nearly exclusively on surface runoff—water that is visible, controllable, and retainable in reservoirs. The ultimate goal of the proposed transformations was the "rectification" (i.e., balancing) of the disorderly natural hydrological regime. Klír proposed a typically modernist vision of "the industrialized river" as an "organic machine"--that is, a fully exploited drainage basin with hydroelectric and water-supply reservoirs on the upper course; a regulated, straightened channel with a regular stream gradient used for irrigation and drainage and antiflood protection on the middle course; and a navigable lower course.[8] In Klír's eyes, such measures would ensure balanced runoff conditions, which he believed would one day result in "the total eradication of harmful floods" and the "meeting of all energy needs (in Bohemia) through water alone" (Klír 1916: 11).

Within the emerging hydrocratic community of "water experts" (engineers) his plans were largely unanimously accepted, though some dissenters questioned how certain aspects of water management were prioritized (i.e., there was a dispute between agricultural and industrial water uses) and the chosen path of "symmetrical" development designed to harmonize all water uses. In the early twentieth century some engineers were already advocating the construction of a large number of high dams as the key to electrifying and industrializing rural regions (Radouš 1912), while others preferred navigation.

The hydraulic mission, however, faced opposition from less progressively minded circles. Sometimes, this was related to professional interests (timber floating, millers), and other times to antimodern, Romantic sentiments. Focusing mostly on aesthetics, the defenders of preservationist, Romantic ideas about untouched nature opposed the conservation-minded technocrats, who did, however, emphasize the necessity of incorporating hydraulic engineering works into the landscape in such a way as to preserve its picturesqueness and increase its "habitability" (Dostalík 2017).

Furthermore, hydrocrats had to fight the existing power hierarchies. The elites were not prepared to hand over their social and economic privileges to a technocratic authority. National independence, achieved in 1918, revealed to hydrocrats the limitations of their political influence. Calls for creating a central hydrocratic institution in the form of a separate ministry went unheard. Limited consolidation in water governance in the 1920s divided the field between two major institutional actors representing agriculture (the Ministry of Agriculture) and industry (the Ministry of Public Works), strengthened the traditional division within the hydrocracy between "hydraulic engineers" (hydrotechnicians, experts on navigation, large dams, and reservoirs), and "cultural engineers" (experts on agricultural land improvement), and constituted a fundamental barrier to pursuing a national hydraulic mission.

The agriculture lobby, which was afraid it would lose its traditional rights and that industry's needs would be prioritized (e.g., in the form of higher dams, restrictions on water available for irrigation, etc.), blocked attempts at fulfilling a fundamental precondition for effective technocratic management—the "making public" of all water resources (i.e., the elimination of private water resources outside of the state's control).[9] Despite continuous talks, a new water act was not passed in interwar Czechoslovakia (Černý 1943).

A state water management policy in Czechoslovakia was only articulated in 1931 in the form of what were known as the "fund acts." The waterways fund focused on supporting investments under the remit of the Ministry of Public Works (reservoirs, improved navigability) and the amelioration fund on financing water management projects for agriculture. These two funds once again comprised more of a construction program than anything else, one based on uncoordinated requests from varied subjects, such as government ministries, municipalities, and companies.[10] Nonetheless, the implementation of water management projects was based primarily on political priorities. Farmers demanded land drainage and river regulation, and industry pushed for improved navigability and electrification. Stakeholders in the railways, however, effectively objected to the farmers' demand, and the coal lobby, to those of industry. Thus, in essence the old Austrian programs from the turn of the century continued to be applied.

The technocratic approach embraced by hydrocrats did end up being successfully adopted in the field of research, however. The establishment of the Masaryk Academy of Labour in 1920 meant the creation of a central forum for technical research and its application in managing the "economic life" of society. The academy's Commission for Water Management, initially presided over by Klír, facilitated cooperation between experts from different fields of sciences that were connected to water issues (including hydrology, forestry, agriculture, and construction). Furthermore, the national Hydrological Research Institute (renamed the Hydrological and Hydrotechnical Research Institute in 1930) was formed in 1919 in order to meet another crucial condition for the realization of the hydraulic mission identified by Klír in his treatise. Under the leadership of Jan Smetana, an engineer with experience from French and German hydrology institutes, this institute gradually surveyed the hydrological characteristics of all the main drainage basins in the country.

FROM CONTROL TOWARD DEVELOPMENT

It was only in the 1930s that the combined effect of the Great Depression and extreme drought lent a new urgency to Klír's visions and arguments. The economic crisis confirmed technocratic arguments emphasizing technical and economic planning as a universal remedy for burning social issues, because, as one leading Czech hydrocrat put it, "no state in the world can endure the murderous chaos of unscrupulous competition."[11] The hot, dry period that lasted for several years and culminated in 1934 further reinforced the ideas of the national hydraulic mission.[12] The basic framework of the mission established earlier by Klír remained untouched. Building on the concept of the hydrological cycle, hydrocrats still identified their primary task as "balancing runoff," and to this end called for centralized and planned management on the national level.

However, the experience of the economic crisis, water shortages, and growing water pollution (the percentage of wastewater in rivers grew when water levels were low) significantly broadened the scope of the debate in two main ways. It sparked greater interest in the social context of water management and it highlighted the broader ecological dimensions of the hydrological cycle.[13]

The economic crisis stimulated a revival of rhetoric focused on water resource management's social and "civilizing" mission. Public water management projects launched in the fervent atmosphere of the second half of the 1930s were imbued with a hint of national symbolism. This was especially true of the Danube–Oder–Elbe Canal, even though the original wording of the "fund acts" sidelined this plan to a certain extent. This proposed canal fulfilled the technocratic dream of integrating waterways (and waters) throughout the territory of the state into a single navigation network (Janáč 2012).

Critical of profit-oriented, competitive liberal capitalism and the resulting particularism, technocratic circles urged the systemic, scientific settling of water disputes, so that water was rationed to users based on overall social priorities. For hydrocrats this meant emancipating water governance entirely from political influences, and managing water, as the leading hydrocrat Emil Zimmler put it, "with the exclusion of all influence from political parties, solely in the interest of the whole."[14] However, they also faced pressure from technocrats outside the hydrocracy, who emphasized the necessity of subjecting water management to a higher political authority in order to protect the systematic approach not only from competing stakeholders but also from engineers, who often supported projects without taking into account the broader legal, economic, and even water resource management context.[15]

Concerning the broadening of the hydrological cycle, drought naturally stirred up discussions. People blamed it on increased surface runoff as a result of inappropriate river regulation and channel straightening; the draining of wetlands, peatbogs, and fields; and the deforestation of mountainous areas (Votruba 1993). While hydrocrats strongly opposed this view, they readily accepted the "ecological" dimension of their endeavor. A resolution adopted by the Commission for Nature Protection and Water Management of the Masaryk Academy of Labour, drafted in cooperation with the academy's Commission for Water Management in 1935, "recognized [water resource management] as an issue in which biology and technology were joined together and indivisible"; the biological aspects here primarily meant vegetation cover and associated attempts at retaining water in the landscape through biological means, not by building reservoirs and regulating river channels.[16]

However, it was only with the outbreak of World War II, Nazi occupation, and the formation of the Protectorate of Bohemia and Moravia that the debates of the 1930s finally started to bear fruit. The position of the Czech lands within the Third Reich as a productive, industrial hinterland as well as the restricted political debate opened the doors to the realization of technocratic ideas (Renneberg and Walker 1994). Hydrocrats, who felt that the hydraulic mission's program was not carried out in interwar Czechoslovakia due to the failures of both the elites and the political system, were willing to ensure that political change would bring at least something positive in this regard.[17] The synergy between the interests of the Czech hydrocracy (leading figures remained in their posts) and the Reich administration, which viewed water primarily through the lens of spatial planning (Zeller 2005), resulted in an institutional transformation that met the main objectives of the hydrocratic program: (1) the creation of a single central water management agency that combined the water management responsibilities of the Ministry of Public Works and the Ministry of Agriculture under the roof of the newly established Ministry of Transportation and Technology; (2) work on the General Water Management Plan of the Protectorate, which was to be based on Reich legislation but would be in the remit of Czech offices; and (3) a lavishly designed construction program known as the Water Management Four-Year Plan (1940–1943), which elaborated on Klír's project but reflected technological developments and the Reich's interests (prioritizing the Danube–Oder–Elbe Canal and the creation of large reservoirs in the Vltava and Berounka drainage basins, which were supposed to supply water to the Reich's industry on the lower Elbe) (Bartovský 1940).[18]

Whereas the four-year plan comprised mainly a construction program to be implemented in the near future, the three-stage process of drafting the water management plan involved calculating water needs over the next thirty to forty years and subsequently determining the possibilities for ensuring the appropriate resources. Finally came the "hardest and most serious task of how to allocate water resources to individual [consumers] in order to achieve the greatest economic effect while perfectly controlling water management."[19] Work on the General Water Management Plan of the Protectorate was never completed.[20] It nevertheless led to the fundamental finding that "today there is no longer any doubt that there is a shortage of water."[21] Furthermore, in some drainage basins (the Morava River's), even "the building of all conceivable reservoirs will not lead to obtaining all the water needed."[22]

While the wartime economy prevented the realization of any major construction works, the redrawing of the institutional and political landscape had a profound effect on postwar developments. Following to the fullest the traditions of the hydraulic mission, immediately after liberation, the hydrocrats demanded that the state finally take full responsibility for managing water as "a national asset" (Vyskočil 1947).[23] This time around, their demands were agreed to, for the new political system with reduced political competition and nationalized industry was based on technocratic ideas. The implementation of the elements of a planned economy as part of the postwar two-year recovery plan (1946–1947) emphasized the state's need to "clearly demonstrate how much water of what quality is available in what place and at what time."[24]

Reflecting on past experiences, national authorities accepted the argument that growing tensions between water needs and water resources, intensified by visions of postwar demographic and industrial growth, required immediate attention. To this end, the Hydrological Research Institute (the successor of prewar water research institutions) was commissioned to draw up a comprehensive water management plan. Contemplating the lack of foreign models to follow, the institute concluded, "We must essentially find our own way."[25] It also pointed out the fundamental conflict between the traditional hydraulic mission ("the comprehensive balancing of water runoff") and the contemporary trend of prioritizing economically defined political goals (in terms of production) typical of Soviet five-year plans (hydropower) and American water management projects.[26] The plan proposed in April 1947 by the Hydrological Research Institute provided a clear resolution to this conflict and focused exclusively on the possibilities for allocating water to industry and for industrial growth.[27] The emphasis on the economically productive exploitation of water resources was also confirmed by the decision to prioritize the use of the middle Vltava for energy by increasing the volume of impounded water and the total length of reservoirs on it.[28]

Criticism, however, came from various sides. For example, biologists emphasized the negative impacts of hydraulic engineering works on the hydrological regime of the landscape.[29] Hydrologists argued that "without due justification, decisions could not be made to build large reservoirs just by stating that no reservoir can be big enough," a belief also shared by agricultural engineers.[30]

Although hydrocrats admitted in their writings from the second half of the 1940s that it was impossible to fully understand and control all natural conditions, they saw in such limits yet another argument in favor of central planning, which they believed would provide better opportunities for eliminating any possible negative effects caused by disturbing the water cycle. Such statements were typical of the techno-optimist modernist visions held by most hydrocrats. But the political elites remained wary of their demands even after the catastrophic drought and subsequent crop failures of the summer and autumn of 1947, when the Economic Council turned down the hydrocrats' proposal to put together a statewide plan, arguing that although creating a "statewide water balance" model (i.e., the balance between water resources and withdrawal) fell under the purview of the hydrocrats, a general plan addressing all economic consequences was the responsibility of economists.[31] The council, however, conceded the assembling of an intergovernmental commission, which would only meet for the first time in April 1948 due to the political changes.[32]

IMPLEMENTATION, 1948–1957

After the communist takeover in February 1948 the commission began to operate in a fundamentally changed political environment. Within the forming communist administration it became the supreme representative of the hydrocracy and successfully pushed forward its technocratic ideas. In a favorable environment characterized by the transition to a command economy, the complete nationalization of water (including both water rights and water management structures) and industry, and agricultural collectivization, hydrocratic elites successfully pushed forward their vision of a national hydraulic mission.

The adoption of the Stalinist project of thoroughly transforming society brought an emphasis on the role of water management as a tool for the transformation of nature as an essential "material" precondition for further changes, which through agricultural intensification and the development of industry and hydropower were to lead toward building a socialist society (Vinter 1951). Hydrocrats, critical of liberal democracy's and capitalism's incompetence with regard to ensuring the planned use of water for the benefit of society as a whole, viewed the rise of socialism as a fundamental step toward acknowledging and fulfilling the ideas of the hydraulic mission (Dub 1953) and enthusiastically accepted the role of co-creators of the communist project. Zimmler and others had been openly praising Soviet planning since 1945, and Smetana even joined the Communist Party in 1946. The drought of 1947 was blamed on the negative impacts of "capitalist" natural resource exploitation, which led to "bugs" in water management and induced climate change (drought). According to communist propaganda and most Czech hydrocrats, the comprehensive planning of water management should remediate drought.[33]

The commission's work was initially heavily influenced by Stalin's Great Plan for the Transformation of Nature, which was being formulated in the Soviet Union simultaneously with the commission's activities in autumn 1948. The plan was characterized by a highly holistic approach to nature, inspired by conservationist and ecological approaches, but compromised by Michurinian agrobiology. The plan, in the Soviet Union as well as in communist Europe, aimed at the "improvement" of the climate and soils by means of large-scale landscape planning (shelterbelts, afforestation, and amelioration) and the transformation of the hydrological regime. Josef Smrkovský, the chairman of the National Land Fund, a communist member of parliament, and a crucial exponent of Stalin's Great Plan for the Transformation of Nature in Czechoslovakia (Olšáková and Štanzel 2016: 109–11), chaired the commission, which was composed of a broad range of water experts––ecologists, biologists, hydrologists, agricultural and forest engineers, and hydraulic engineers––and representatives of affected economic sectors (energy, water supply, agriculture, aquaculture, industry, housing, etc.). Under Smrkovský's leadership, the commission prepared planning mechanisms and legislation regulating the administration and use of water resources that pointed toward nationalization and centralization.[34]

At its very first session, the commission successfully proposed the creation of the State Water Management Plan (SVP) as a critical and necessary measure for further planned development of the country. To avoid the recurrence of the critical water shortages of summer 1947 and further exacerbation by the rapid growth in water consumption that was inherent in the communists' first economic five-year plan (for the period 1949–1953), the communist government sanctioned the development of the SVP in December 1948, and work on the plan commenced in early 1949.[35] While it was primarily understood as a "preliminary plan" for state economic planning––a detailed survey of all potential water resources (including practically all surface waters)––it was also supposed to be the foundation for an operative plan for developing water management and water resources, and thus the basis for establishing water management as its own industry, a long-term goal of the hydrocrats.[36]

The SVP was drawn up based on the Bulíček project, and thus a survey of current conditions was at its core. Research was conducted in many fields: the natural sciences (geology, pedology, climatology, biology, and phytocenology), hydrology, population studies, and economics (agriculture, forestry, industry, transportation, and waste).[37] The plan was to ensure not only ample water supplies and water purity but also "biological equilibrium in nature."[38]

Beginning work on the SVP was an important step toward forming the state water management system—that is, toward transforming the territory of the Czech lands into an essentially integrated, artificially created catchment unit with a defined state water and water management balance (the ratio of available resources to water needs, calculated for the state as a whole for each year).[39] For this reason, the enduring fragmentation of water resource management needed to be eliminated.[40] In discussions held during the preparation of the Water Management Act in the early 1950s, hydrocrats advocated for concentrating water management into a central body, so that "uniformity in harmony with all of society's needs was guaranteed."[41] The intended central office was supposed to be responsible not only for all research, planning, investments, and designing in water management but also for managing the construction and operation of reservoirs. Furthermore, this was confirmed by the State Planning Office, which included water management in planning directives for 1953 as a separate industry "with a proper plan for accomplishments and needs," but still one without its own budget because it lacked a central administrative body.[42]

The exact form of water management centralization, however, became a matter of contention: on one side stood hydrotechnicians and hydraulic engineers, mainly from the former Ministry of Public Works, who continued to pursue tasks outlined in the first five-year plan and its focus on socialist industrialization and hydropower, and on the other, outspoken proponents of the Stalin plan (mostly agricultural engineers and biologists), who after the imprisonment of Josef Smrkovský (in connection with the show trial of the former general secretary of the Czechoslovak Communist Party, Rudolf Slánský, in 1952; see Kaplan 2002), were represented by the Government Committee for Improving Agriculture, Water Management, and Forestry (ZVL), which was active from 1952 until 1961.[43]

The five-year plan's unambiguously productivist wording played an important role here; fulfilling the plan required the rapid, efficient construction of multipurpose reservoirs. Ensuring water supplies for meeting agricultural and industrial plans was prioritized; it was also supposed to contribute to balancing regional development and improving workers' quality of life (Gorenčík 2001: 74). Concurrent with the preparation of the SVP in 1948, the commission discussed and prepared a long-term plan for developing hydropower, which imparted greater urgency to SVP activities. The hydropower scheme accepted in early 1949 proposed the construction of thirty large hydropower plants during three five-year-plan periods, which was presented as a necessary condition for economic and social modernization and the creation of socialist society.[44] When in late 1950 Soviet propagandists created the powerfully symbolic image of large-scale multipurpose hydropower dams as transformative technologies that lead to social and economic progress, they framed it as a part of the Stalin plan (Gestwa 2010). However, in Czechoslovakia, in the context of "great works of Communism" came quickly into conflict with other goals of the Stalin plan. This conflict took the shape of a dispute over the scope of the hydrological cycle, with a holistic, conservationist definition including broad environmental (and social) considerations on one side, and a productivist vision focused predominantly on the management of surface runoff, on the other.

The Czech hydrocrats, led by Čestmír Štoll and Smetana, held a reserved (but not dismissive) attitude toward the Soviet concept of inclusive water management.[45] They considered the proposed "forestry-engineering" and "agricultural-engineering" means for transforming nature and the hydrological regime to be secondary and questioned their effectiveness (Smetana 1954). In their eyes, the main objective of the Czech hydraulic mission remained to ensure the "complete balance of water levels," especially through the use of "radical hydraulic engineering means," that is, mainly, by impounding water in reservoirs.[46] Therefore, they rearticulated the Czech hydraulic mission in the context of symbolic Soviet construction projects and presented all the large hydraulic engineering works begun during the first five-year plan as "our great hydraulic works of socialism": besides the Váh Cascade (a system of canal-connected hydropower plants, the Orava Reservoir) and the Vltava Cascade (Lipno, Orlík, and Slapy Reservoirs), they also included a planned reservoir on the Danube; the Dobšiná, Vír, and Křižanovice Reservoirs; and the Klíčava, Fláje and Křímov, Kružberk, Žermanice, and Belanka Water Supply Reservoirs (Votruba 1954).

This conflict also carried over to the world of academia. To support the planning of great works, the Soviet-styled Czechoslovak Academy of Sciences established the Commission for Aiding the Great Works of Socialism under the leadership of Jan Smetana in 1953.[47] Unlike its predecessor, the Water Management Commission of the Masaryk Academy of Labour, this body focused in particular on hydraulic engineering issues (it was later attached to a board focused on energy) and was no longer a platform for dealing with "contentious technical-economic and ecological issues" (Votruba 1993: 152). Attempts at establishing a more inclusive scientific water management committee at the Czechoslovak Academy of Agricultural Sciences (ČSZAV), supported by proponents of the Stalin plan, were futile.[48]

In the end, the hydraulic engineering conception of the hydraulic mission celebrated success, when on January 12, 1954, the Central Water Management Administration (ÚSVH, Ústřední správa vodního hospodářství) was established. Around 27,000 employees worked there, including those at departments under regional national committees. Centralization also occurred in research; the State Hydrological Institute was transformed into the all-encompassing Hydrological Research Institute, where between 1952 and 1958 research on agricultural amelioration was also conducted (Jiroušek 1959). Conservationists associated with the Stalin Plan, however, lost the battle, and even the Ministry of Agriculture admitted that the Government Committee for Improving Agriculture, Water Management, and Forestry was largely ineffective.[49]

However, certain aspects of conservationist ideas survived. The resulting water resource management system was essentially extreme in two ways—along the vertical axis it was a centralized, technocratic system, whereas horizontally this system applied a maximally inclusive approach to water management that essentially attempted to control the entire "water cycle." The approval of the final version of the SVP in January 1954 clearly established the socialist hydraulic mission's objectives, which were bound to an optimistic vision of unlimited growth for the socialist economy (hydropower, water supply, sewers, etc.).[50] Nonetheless, and somewhat paradoxically, the plan called for biological measures in addition to engineering ones—it specifically called for ensuring the landscape's water-retention function by maintaining the right ratio of forest to farmland and the right distribution of forests with the right species compositions in drainage basins, as well as by introducing agricultural methods for land management that would guarantee optimal retention conditions.[51] Most of these measures remained on paper, however. For example, only one of the proposed systems of large protected water accumulation areas was declared in the 1950s. Work on this plan would only resume after 1970 (Plecháč 1999: 149).

Furthermore, the SVP contained many negative elements stemming from the technocratic nature of the modernist hydraulic mission. The technocratic approach and the technological optimism of its authors are illustrated by the statement that "today's irrigation [systems] do not suffer water shortages . . . because they were built at a scale suitable for the water available in a given river"; the amount of water available, however, was supposed to be increased by building new water supply reservoirs, which were intended to facilitate a fundamental increase in water consumption.[52] Technocratic tendencies were also reflected in the preference for applying "economies of scale," which was facilitated by the introduction of central management and the accumulation of water rights in the hands of the state. Instead of small-scale local water management solutions, larger, more costly, and more economically efficient projects were to be implemented. For example, regional water supply networks serving multiple villages were created instead of providing water to these municipalities from smaller local sources. Similarly, instead of a large number of small hydropower plants (as of 1950 there were 6,713 on the territory of today's Czech Republic), larger, economically more efficient hydropower plants were to be built in the future, even at the expense of already existing plants. Although this trend began in the 1930s, the SVP rapidly sped it up.[53]

The transformation of the water management system was completed with the publication of the new Water Act of 1955, which clearly defined the role of the SVP and the Central Water Management Administration, and confirmed that all water resources had been "made public." The SVP, which was to act as "an instrument of order and discipline in controlling and exploiting water resources," had two formal functions––on one hand this plan served as the basis for concrete water management measures included in the Central Water Management Administration's annual and five-year plans, and on the other (and perhaps most importantly), it was a baseline plan that "guided" the plans and investment proposals of other industries.[54] The Central Water Management Administration was authorized to check the compatibility of all planned water withdrawals and projects and evaluate them from the perspective of the emerging national water infrastructure system.[55]

Thus, in the 1950s the hydrocrats gained space for implementing their version of the hydraulic mission. The creation of an independent central office, a plan, and a budget, alongside the "making public" of water resources and the nationalization of hydraulic engineering works, created ideal conditions for the practical implementation of this mission. The objectives and ambitions of the hydrocrats to fully balance (i.e., control) runoff in Czechoslovakia within fifteen years, however, were compromised by the demands of other industries and budget cuts. Here, the reflection of political priorities was clear––when the Central Water Management Administration's budget was reduced in early 1955, "the most urgent demands of the energy industry would take up 51.6 percent" of the budget and "supplying housing estates and sources of water, 25.3 percent." Twelve large multipurpose reservoirs that were in the process of being built would "swallow" about 64 percent of all investment costs between 1956 and 1960.[56] When the budget was cut, all stakeholders agreed that it would be inappropriate "to reduce investments in projects of crucial importance" for "ensuring production."[57]

Despite the ideas contained in the SVP, the fears expressed by representatives of the Ministry of Agriculture during discussions about the establishment of the Central Water Management Administration came true in the end: from the beginning they had viewed the centralization of water management as the "gradual slashing of the Ministry of Agriculture's influence" and the "occupation of water by the construction industry."[58] These developments resulted in a fundamental, burning problem––rapid growth in river pollution, which in the late 1950s and early 1960s became "water management's main problem."[59]

Here, we can observe the initial corrosion of the hydraulic mission in the Czech lands. As a result of the reorganization of the state administration system after 1956 and in the context of the post-Stalin criticism of the overly centralized Soviet model of governance (Průcha et al. 2009: 578–91), in the second half of the 1950s water management was decentralized and the broadly inclusive approach to water management was gradually abandoned, as was the hydropower scheme (even the Vltava River Cascade remained incomplete). Transferring responsibility for the water supply and sewer networks back to the regional level and disbanding the Central Water Management Administration signified a loss of authority for the hydrocrats and a return to a fragmented river administration system that was comparable to that of the 1930s, that is, with authority split between forestry, agricultural, and water management organizations.[60] As part of the program for building an advanced socialist society that was declared at the Twentieth Congress of the Czechoslovak Communist Party in 1962, water management was reorganized and integrated into the technocratic system of managing society as the more or less standardized productive economic sector of "water production" (Šembera 1979: 91). Thus, emphasis shifted from formulating the hydraulic mission's goals to focusing on operating issues, and the vital demand for impounding water was replaced by the concept of water management systems that were focused more on distributing water than on retaining and efficiently using every single drop of water. As Juraj Mihál, one of the architects of the economization of water management, stated, reducing the hydraulic mission to just another part of the economy revealed an illusion suffered by many "enthusiastic hydraulic engineers, for whom performing water management activities had become, under the influence of professional tradition, not only the substance of life but also a life mission" (Mihál 1975: 13).

The Czech hydraulic mission was a modernization project that aspired to fully control water as a key natural resource and the foundation of a modernizing society. The chronology and basic ideas of the modernization process in water management in the Czech lands do not stand out in any way (Allan 2006). It is a typical example of water utilization in the period of industrial modernity, starting in the late nineteenth century and culminating after World War II (Beck 1992). The dominant discourse of water as a productive natural resource (or as a dangerous element) that should be subordinated and put to efficient use for society's needs, defined primarily in productivist terms (the amount of drinking water, industrial water, irrigation, and megawatts of power), held a stable position in Czechoslovakia no matter the political system, as it did in Western Europe, where it became the base of postwar water management policy. Problems with water availability were to be primarily solved by building more reservoirs and transferring water over ever greater distances.[61] The objectives of the hydraulic mission never deviated significantly from the European model, which was based on the intertwining of hydrocratic visions and the building of the (nation) state—essentially, the state's territory was formed as a sui generis artificial drainage basin unit, with ambitions of fully controlling the water cycle.[62]

The 1950s were a special period, more in terms of quantity than quality, when Czech hydrocrats had much greater opportunities to transform their visions into reality than their Western counterparts (Josephson 2002). Otherwise, the hydraulic mission developed in parallel in the Czech lands and in Western Europe––the main watershed moment came after World War II, when the hydropower paradigm, characterized by the construction of multipurpose reservoirs, came to dominate (Pritchard 2011). Continuity with interwar developments, best personified by Jan Smetana, allowed hydrocrats to somewhat avoid subordinating the hydraulic mission to the broader processes of planning socialist society, as embodied in Stalin's Great Plan for the Transformation of Nature.[63] However, even during this period, the Czech hydraulic mission faced traditional opposition from the agricultural bureaucracy, which to a certain extent was reflected in the conflict between the hydrocrats from the Ministry of the Construction Industry on one hand and the proponents of the Soviet vision of the transformation of nature and the demands and interests of other industries on the other.

State socialism provided hydrocrats with the space to implement their visions to a large extent because they were presented as purely technical solutions and because they largely resonated politically with official state ideology––similar to the situation during the protectorate. The construct of "water shortage," whether it was based on a physical lack of water (1934, 1947), on water pollution (1950s, 1960s), or on estimates of future needs (which were essentially always inaccurate; based on the protectorate plan in the 1950s), was under state socialism boosted by the demands made by the planned transformation of nature as a precondition for the transformation of society as a whole. In principle, hydrocrats endeavored to harmonize conflicting water demands through infrastructure projects and to protect and develop water resources. As a result, beginning in the 1940s creating infrastructure for the production of water was clearly prioritized over protecting water due to the often-mentioned threat of water shortages. Also, in keeping with technocratic ideas, protecting water resources came to mean "building" them––that is, increasing the volume of impounded water.

The consequences of fulfilling the hydraulic mission still have an influence today. The massive construction of multipurpose reservoirs requires the centralized management of water resources, and thus, great authority is entrusted to the hydrocracy. Continuity in building up a "national" water management system thus does not simply stem from the hydrological characteristics of the system but from the ability of the hydrocracy to take advantage of positive feedback stemming from the interplay of power and politics by taking small steps (institutionalization and

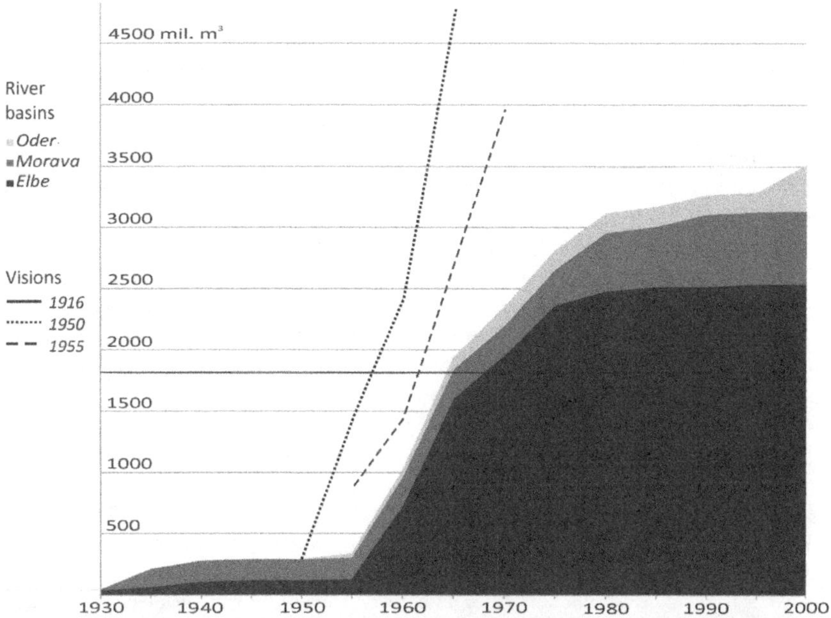

FIGURE 6.1. The chart shows the evolution of the volume of reservoir water in the Czech Republic. Each major water basin is indicated by specific shades of gray (the Elbe on the bottom, Morava in the middle, Oder on the top). The black horizontal line shows the level of water accumulation proposed by Klír for the Elbe River basin in 1916. The dotted more or less vertical curves represent the volume of accumulation expected in water management plans from the early 1950s, when the hydraulic mission climaxed in Czechoslovakia in the context of socialist industrialization. Source: Broža et al. (2005).

water infrastructure projects) to gradually accumulate and confirm its own position in maintaining and developing the entire system. By doing so, its position is largely immune to political and discursive changes (Moss 2016).

NOTES

This chapter is an English version of a text previously published in the Czech language journal *Soudobé dějiny*. The text was updated with support from the Grant Agency of the Czech Republic, project GAČR 18-05095Y.

1. This theory revolves around a causal connection between the development of irrigation systems and centralized state control. Water and access to it thus becomes an instrument for gaining and exercising power, which then contributes to the development of the bureaucratic apparatus and the authoritarian governing of society (Wittfogel 1957).

2. The formation of national water management systems has been studied in Vleuten and Disco (2004) and Disco and Vleuten (2002).

3. This aspect of water management seems to have drawn the most interest from historians (Blackbourn 2007; Gestwa 2010; Schivelbusch 2006; Josephson 2002; Swyngedouw 2015).

4. The term "system builder" comes from the field of the history of technology, mainly from the concept of "large technological systems," which have been developed since the 1970s, particularly among Anglophone historians (for details, see Janáč and Vleuten 2016).

5. James C. Scott emphasizes the positive feedback loops between high modernity and authoritarian regimes, which result in the implementation of ineffective, utopian projects (Scott 1998: 87–100).

6. Shortened versions of both texts have been published in Průcha et al. (2009: 578–91, 873–95). David Blackbourn has described a similar conflict in Germany (Blackbourn 2007).

7. Archiv Technického muzea v Brně [Archives of the Technical Museum in Brno, Czech Republic] (ATMB), fond (f.) Antonín Smrček, box (b.) 132, Pamětní spis SIA ve příčině jednotné organisace vodohospodářské činnosti v Království českém [Memorandum of the Association of Architects and Engineers of the Kingdom of Bohemia on Integrated Organization of Water Management] (1912).

8. "Industrialized river": industrialized in the sense that rivers and water resources in general were reduced to quantifiable, rationalized production units that enabled the effective exploitation of water resources (Josephson 2002; Pritchard 2011). "Organic machine": Richard White (1996) uses this metaphor to emphasize the hybrid character of rivers as parts of natural ecosystems and anthropomorphic water management systems.

9. Let us recall here groundwater's legal status as *res nulius* (property belonging to no one), whose use was not subject to, nor could be subject to, regulation, not even the obligation to report any abstraction. It came up after 1935 in discussions about water pollution and determining its sources. Archiv Akademie věd České Republiky [Archives of the Czech Academy of Sciences, Prague, Czech Republic] (AAV ČR), f. Jan Smetana, b. 31, Zápis o schůzi Ústředního výboru Poradního sboru pro čistotu vod, 28.5.1937).

10. Nonetheless, the dividing up of authority was relatively complicated and was still based on the arbitrary political logic of the Austrian construction programs, where the middle courses of rivers in agricultural areas were administered by the Ministry of Agriculture (see Act no. 49/1931 Coll. [Collection of Laws], On the State Fund for Hydrological Amelioration, and Act no. 50/1931 Coll., on the State Fund for Making Rivers Navigable, Constructing Ports, Building Valley Reservoirs, and Exploiting Hydropower).

11. AAV ČR, f. Masarykova akademie práce—presidium: pozůstalost E. Zimmlera, b. 43, rukopis článku Emila Zimmlera "O úživnosti inženýrského povolání," zaslaný redakci časopisu *Technik*, 19.8.1932.

12. Defining the term "drought" is a quite complicated matter—here we mean "hydrological drought," which is determined based on river discharge. For a comprehensive overview of drought in Czechia with an emphasis on the nineteenth and twentieth centuries, see Brázdil et al. 2015.

13. The Ministry of Agriculture established the Advisory Board for Ensuring Water Purity, whose primary goal was to define pollution (on the basis of discharge data) and to create a list of polluters. Discussions about water pollution also led to the revision of water rights in the context of redefining the common interest in water; existing regulations clearly did not suffice (AAV ČR, f. Jan Smetana, b. 31).

14. AAV ČR, f. Masarykova akademie práce (MAP), b. 43 (prezidium–pozůstalosti–Zimmler), Vyjádření Emila Zimmlera k dotazníku úsporné komise, 16.5.1937.

15. Sýkora (1935) mentioned, for example, the overly ambitious construction of weirs.

16. AAV ČR, f. MAP, b. 277, Rezoluce MAP o účelném vodním hospodářství a ochraně přírody, 1936.

17. "Failures . . . of the political system": This is illustrated well by Emil Zimmler's critical appeal for making investments in water management projects of April 1938. In it he stated that planning documents had been prepared "but have been lying more than twenty years nearly completely unutilized" (AAV ČR, f. MAP, b. 277, Návrh dr. Zimmlera ve věci řešení úpravy vodstva v ČSR, 20.4.1938). On political change and positive outcome, see, for example, Antonín Smrček who, in an interview for *Lidové noviny* welcomed the German plan for building the Danube–Oder–Elbe Canal ("Dunajsko-oderskýprůplav v popředí," *Lidové noviny*, November 17, 1938).

18. "Single management agency": see Government Regulation no. 208/42 Coll. of June 15, 1942, which reorganized the government. "General Water Management Plan of the Protectorate," Národní archivy České republiky, Prague, Czech Republic [National Archives of the Czech Republic] (NA), f. GŘVC, b. 149, Generelní vodohospodářský plán: Směrnice pro postup prací a vymezení působnosti zmocněnců u zemských úřadů, 8.3.1941.

19. NA, f. Ministerstvo veřejných prací (MVP), b. 531, Josef Bartovský, Přehled vodohospodářského a vodocestného plánování v Čechách a na Moravě: Přednáška pro schůzi vodohospodářského výboru Ústředny obchodních komor dne 15.1.1942 v Olomouci.

20. Archiv Vodohospodářské houstavu, po bočka Brno [Archive of Water Research Institute in Brno, Czech Republic], Bažant, *Vodohospodářský plán země Moravskoslezské* (1941), 3.

21. Jindřich Choleva made this claim. See Zápis o schůzi vodohospodářského výboru Ústředny obchodních a živnostenských komor, konané 15.1.1942 v Olomouci (NA, f. MVP, b. 531).

22. NA, f. Ministerstvo veřejných prací (MVP), b. 531, Josef Bartovský, Přehled vodohospodářského a vodocestného plánování v Čechách a na Moravě: Přednáška pro schůzi vodohospodářského výboru Ústředny obchodních komor dne 15.1.1942 v Olomouci.

23. Lev Ondřej Vyskočil (1890–1951) was a practicing hydrocrat associated with the Czech Technical Office and the Ministry of Agriculture. The hydrologist Jan Novotný had similar ideas (Novotný 1948).

24. AAV ČR, f. MAP, b. 277, Státní vodohospodářský 30 letý plán pro léta 1949–1979), návrh ing. dr. Jar. Bulíčka, 15.3.1947, p. 3.

25. Moravský zemský archiv [Moravian Provincial Archives] (MZA), f. Průplav Labe–Odra–Dunaj, b. 179, Generelní plan vodního hospodářství v zemi české a moravskoslezské jako základ soustavného hospodářského plánování. Státní ústav hydrologický TGM, 12.4.1947.).

26. The California Central Valley Project for irrigation and the Tennessee Valley Authority for hydropower. In Státní vodohospodářský 30 letý plán (pro léta 1949–1979), návrh ing. dr. Jar. Bulíčka, 15.3.1947, p. 2.

27. The author of this document, the director of the State Hydrological Institute, Alois Bratránek, called it "a general plan for the time being" (MZA, f. Průplav DOL, b. 179, Generelní plan vodního hospodářství, 1947).

28. NA, f. GŘVC, b. 207, Generelní plan vodního hospodářství v zemi české a moravskoslezské, Připomínky ŘVC, 19.5.1947.

29. Úlehla, for example, did not hesitate to compare the impact of farmland improvement measures implemented in South Moravia to the catastrophic dust storms that plagued the midwestern United States in the 1930s (Úlehla 1947).

30. NA, f. Státní úřad plánovací I (SÚP I), b. 105, signatura 4. odbor–suroviny, vodní zdroje, rok 1947–1949, Nejhospodárnější velikost užitných prostorů v povodí Vltavy po Štěchovice, hydrologická studie, listopad 1947.

31. NA, f. GŘVC, b. 389, Zápis o poraděkonanédne 27.4.1948 v ministerstvu techniky o postupu při vodohospodářském plánování.

32. NA, f. Ministerstvo techniky, b. 131, Přípravná komise pro generální vodohospodářský plán, leden 1948.

33. NA, f. Státní úřad plánovací II (SÚP II), b. 386, Návrhv ládního usnesení o plánu na zvelebení zemědělského a lesního hospodářství a zvýšení vodních zásob v ČSR, leden 1952.

34. NA, f. ÚPV-GSHR, b. 289, Programové prohlášení vodohospodářské komise, návrh z června 1948.

35. On the water shortages of summer 1947, see NA, f. ÚPV-GSHR, b. 289, Rezoluce vodohospodářské komisepři Ústřední plánovací komisi o vodohospodářském pětiletém investičním plánu, červen 1948. On development of the SVP, see NA, f. SÚP I, b. 105, Akce připravované komisí K21 pro sestavení vodohospodářského plánu, 5.1.1949.

36. NA, f. ÚPV-GSHR, b. 289, Návrh usnesení pro ÚPK o státním vodohospodářském plánu, 12.10.1948.

37. NA, f. ÚPV-GSHR, b. 289, Návrh usnesení pro ÚPK o státním vodohospodářském plánu, 12.10.1948, p. 2.

38. NA, f. ÚPV-GSHR, b. 289, Návrh usnesení pro ÚPK o státním vodohospodářském plánu, 12.10.1948, p. 2.

39. For more on the concept of the water management balance in Czechoslovakia, see Plainer 1977.

40. The main water management goals of the five-year plan were established within other segments of the economy. Thus, the Act on the Five-Year Plan contains thirty mentions of water management issues, from anti-erosion measures such as afforesting mountain slopes to changes in the milling and paper industry (NA, SÚP I, b. 105, Hlavní zásady pro vodohospodářskou činnost během 5LP [pětiletého plánu]).

41. It is worth noting that the Water Management Act replaced the Water Act, which implies a shift in emphasis on ownership to productive use (Kult 2010: 69–72). Quote from: NA, f. ÚPV-B (Úřad předsednictva vlády—běžná spisovna), b. 1363, Zásady zákona o hospodaření vodami—řízení dle paragrafu 12 legislativních pravidel, 30.7.1951.

42. NA, f. ÚPV-B,b. 1363, Zpráva ministra stavebního průmyslu o organizaci hospodářsko-výrobních složek vodního hospodářství, 15.7.1953.

43. Its main objective was to coordinate these three industries to limit landscape damage (which had emerged due to shortcomings in cooperation) and to draft a law on the protection of the "Czechoslovak Landscape and the Human Environment" (Růžička 1959: 137).

44. NA, f. SÚP II, b. 385, Využití vodní energie: Dlouhodobý investiční program 1949–1963. Ministerstvo techniky, březen 1949.

45. In 1948 the leading Czech hydrologist Jan Novotný was still talking about these ideas as being unproven, although both Klír and Bartovský had already to a certain extent admitted the role of vegetation cover in modifying the climate and improving the landscape's water-retention function (Novotný 1948).

46. Thus, extreme discharge levels on the Labe were to be eliminated and the river was supposed to have a constant, balanced discharge of 300 cubic meters per second (Smetana 1954: 4–5.).

47. AAV ČR, f. KVH ČSAV, b. 3, Předběžné porady o ustavení Komise pro pomoc velkým stavbám socialismu, prosinec 1952–červen 1953.

48. This is how Ladislav Votruba (1993) retrospectively assessed the work of the ČSAV's Water Management Commission.

49. NA, f. SÚP II, b. 387, Dobrozdání zemědělského odboru SÚP k pokračování ZVL, 7.10.58.

50. NA, f. ÚSVH, b. 48, *Státní vodohospodářský plán, textová část*, 172–78.

51. Afforestation, modification of vegetation species composition, erosion protection, the construction of ponds, and soil amelioration (NA, f. ÚSVH, b. 48, *Státní vodohospodářský plán, textová část*, 172–78).

52. NA, f. ÚSVH, b. 48, *Státní vodohospodářský plán, textová část*, 179.

53. NA, f. ÚSVH, b. 48, *Státní vodohospodářský plán, textová část*, 111–29.

54. NA, f. Ústřední správa vodního hospodářství (ÚSVH), b. 48, *Státní vodohospodářský plán, textová část*, 14.

55. See Act no. 11/1955 Coll., on Water Management.

56. NA, f. SÚP II, k. 385, Směrnice vodního hospodářství pro období 1956–1960, 10.2.1955, p. 4.

57. NA, f. SÚP II, k. 385, Přehled rozporů v návrhu směrnic perspektivního rozvoje vodního hospodářství pro léta 1956–1960.

58. NA, f. ÚPV-B, b. 1363, dopis ministra zemědělství Jana Nepomuckého ministru stavebního průmyslu Emanuelu Šlechtovi a náměstkovi předsedy vlády Jindřichu Uhrovi z 29.5.1953.

59. NA, f. PB ÚV KSČ 1954–1962, vol. 340, unit 431, item 5, Hlavní úkoly vodníhohospodářství při zabezpečování rozvoje národního hospodářství v ČSSR a zvyšování životní úrovně lidu, 28.2.1962, p. 2.

60. In 1958, the Ministry of Agriculture became responsible for amelioration and the administration of rivers with basins less than 150 square kilometers in area (Šembera 1979: 91).

61. See the cited works by Sara B. Pritchard, Cornelis Disco, Christoph Mauch and Thomas Zeller, Thomas Zeller, and Erik Swyngedouw.

62. In the Czech lands this approach endured until the 1970s, albeit in the context of the burgeoning environmental criticism of the hydraulic mission. See the updated version of the State Water Management plan: *Směrný vodohospodářský plán České socialistické republiky* (Prague: Státní zemědělské nakladatelství, 1976), 12.

63. Jan Smetana acted as the head hydrocrat throughout practically the entire studied period, just like the other members of the interwar Water Management Commission of the Masaryk Academy of Labour, who after the war were also involved in the Economic Council's Water Management Commission and later in the ČSAV's Water Management Commission.

REFERENCES

Allan, John Anthony. 2006. "IWRM: The New Sanctioned Discourse?" In P. P. Mollinga, A. Dixit, and K. Athukorala, eds., *Integrated Water Resources Management: Global Theory, Emerging Practice, and Local Needs*. New Delhi: Sage, 36–63.

Bartovský, Josef. 1940. *Vodohospodářský program v povodí Vltavy a Labe v Čechách: Technicko-hospodářský přehled, rozbor a náměty k revizi* [Water management program in the Vltava and Elbe river basins in Bohemia: technical and economic overview, analysis and suggestions for revision]. Prague: Ministerstvo veřejných prací.

Bartovský, Josef. 1946. Vodní cesty a vodohospodářské plánování v Čechách a na Moravě : Přehled a bilance práce dvou generací, další úkoly Dvouletka a Pětiletka [Waterways and water management planning in Bohemia and Moravia: Overview and results of work of two generations, further tasks two-year plan and five-year plan]. Prague: Společnost Dunajsko-oderského průplavu.

Beck, Ulrich. 1992. "From Industrial to Risk Society." *Theory, Culture and Society* 9(1): 97–123.

Bennett, Tony, and Joyce, Patrick, eds. 2010. *Material Powers: Cultural Studies, History and the Material Turn*. London: Routledge.

Blackbourn, David. 2007. *The Conquest of Nature: Water, Landscape and the Making of Modern Germany*. New York: Norton.

Brain, Stephen. 2018. "The Environmental History of Russia." *Oxford Research Encyclopedia of Environmental Science*. March 15. http://environmentalscience.oxfordre.com/view/10.1093/acrefore/9780199389414.001.0001/acrefore-9780199389414-e-355.

Brázdil, Rudolf et al. 2015. *Historie počasí a podnebí v Českých zemích*, sv. 11: *Sucho v Českých zemích. Minulost, současnost, budoucnost* [Drought in the Czech Lands: past, present, and future]. Brno: Centrum výzkumu globální změny AV ČR, v.v.i..

Broža, Vojtěch et al. 2005. *Přehrady Čech, Moravy a Slezska* [Dams of Bohemia, Moravia, and Silesia]. Liberec: Knihy 555.

Černý, Jaroslav. 1943. *Vodní knihy, vodní právo a vodoprávní technika* [Water books, water law, and water law techniques]. Prague: Ministerstvo zemědělství a lesnictví.

Disco, Cornelis, and Erik van derVleuten. 2002. "The Politics of Wet System Building: Balancing Interests in Dutch Water Management from the Middle Ages to the Present." *Knowledge, Technology and Policy* 14(4): 21–40.

Dostalík, Jan. 2017. "The Organicists: Planners, Planning, and the Environment in Czechoslovakia (1914–1949)." *Planning Perspectives* 32(2): 147–73.

Dub, Oto. 1953. *Vodné hospodárstvo za kapitalizmu a za socializmu* [Water management under capitalism and socialism]. Martin: Osveta.

Gestwa, Klaus. 2010. *Die Stalinschen Grossbauten des Kommunismus: Sowjetische Technik und Umweltgeschichte, 1948–1967*. Munich: R. Oldenbourg.

Gorenčík, Jan. 2001. "Slapská přehrada—asociálnost jedné velké stavby socialismu" [Slapy Dam: the antisocial character of a great work of socialism]. In Zdeněk Kárník and Jan Měchýř, eds., *K novověkým sociálním dějinám českých zemí, sv. 4: Zvraty a převraty. 1939–1992* [On the modern social history of the Czech Lands, vol. 4: Reversals and upheavals, 1939–1992]. Prague: Karolinum, 71–106.

Guldi, Jo. 2012. *Roads to Power: Britain Invents the Infrastructure State*. Cambridge, MA: Harvard University Press.

Hadrabová, Alena. 1996. "Vodní hospodářství v Československu v letech 1948–1989" [Water management in Czechoslovakia in 1948–1989]. *Acta oeconomica pragensia* 4(3): 217–34.

Hadrabová, Alena. 1998. "Nejdůležitější nástroje řízení ochrany životního prostředí v Československu v letech 1945–1989" [The most important instruments of environmental management in Czechoslovakia in 1945–1989]. *Acta oeconomica pragensia* 6(5): 147–76.

Kaplan, Karel. 2002. *StB o sobě: Výpoveď vyšetřovatele Bohumila Doubka* [StB about itself: the testimony of investigator Bohumil Doubek]. Prague: Úřad dokumentace a vyšetřování zločinů komunismu Parlamentu ČR.

Klír, Antonín. 1916. *Náčrt programu vodního hospodářství v Král. Českém* [A draft water management program in the Kingdom of Bohemia]. Prague: n.p.

Kohlrausch, Martin, and Helmuth Trischler. 2014. *Building Europe on Expertise: Innovators, Organizers, Networkers*. Basingstoke: Palgrave Macmillan.

Kult, Arnošt. 2010. *Tekoucí (povrchová) voda: Právně-filosofický pohled na rozdílné způsoby vymezování ochrany vody a vodního prostředí* [Flowing (surface) water: a legal-philosophical view of the different ways of delimitation of water conservation and water environment]. Prague: Výzkumnký ústav vodohospodářský T. G. Masaryka.

Janáč, Jiří. 2012. *European Coasts of Bohemia: Negotiating the Danube-Oder-Elbe Canal in a Troubled Twentieth Century*. Amsterdam: Amsterdam University Press.

Janáč, Jiří, and Erik van der Vleuten. 2016. "Transnational System Building across Geopolitical Shifts: The Danube–Oder–Elbe Canal, 1901–2015." *Water Alternatives* 9(2): 272–91.

Jiroušek, Josef. 1959. "Organisace a poslání vodohospodářského výzkumu" [Organization and mission of water research]. *Vodníhospodářství* 9(7): 289–92.

Josephson, Paul R. 2002. *Industrialized Nature: Brute Force Technology and the Transformation of the Natural World*. Washington, DC: Island Press.

Linton, Jamie. 2010. *What Is Water? The History of a Modern Abstraction*.Vancouver: University of British Columbia Press.

Mauch, Christof, and Thomas Zeller. 2008. *Rivers in History: Perspectives on Waterways in Europe and North America*. Pittsburgh, PA: University of Pittsburgh Press.

Mihál, Juraj. 1975. *Teoreticko-ekonomické aspekty zneškodňovania odpadných vôd* [Theoretical-economic aspects of wastewater disposal]. Bratislava: Výskumný ústav vodného hospodárstva.

Molle, François. 2009. "River-Basin Planning and Management: The Social Life of a Concept." *Geoforum* 40(3): 484–94.

Molle, François, and Peter P. Mollinga. 2009. "Hydraulic Bureaucracies and the Hydraulic Mission: Flows of Water, Flows of Power." *Water Alternatives* 2(3): 328–49.

Moss, Timothy. 2016. "Conserving Water and Preserving Infrastructures between Dictatorship and Democracy in Berlin." *Water Alternatives* 9(2): 250–71.

Novotný, Jan. 1948. *Hydrologie základem vodního hospodářství* [Hydrology as the basis of water management]. Prague: Státní ústav hydrologický T. G. Masaryka.

Olšáková, Doubravka, and Arnošt Štanzel. 2016. "Kafkaesque Paradigms: The Stalinist Plan for the Transformation of Nature in Czechoslovakia." In Doubravka Olšáková, ed., *In the Name of the Great Work: Stalin's Plan for the Transformation of Nature and Its Impact in Eastern Europe*. New York: Berghahn Books, 44–125.

Plainer, Jaroslav. 1977. *Vodohospodářská bilance* [Water management balance]. Prague: Státní zemědělské nakladatelství.

Plecháč, Václav. 1999. *Vodní hospodářství na území České Republiky, jeho vývoj a možné perspektivy* [Water management in the Czech Republic, its development and possible perspectives]. Prague: EVAN.

Pritchard, Sara B. 2011. *Confluence: The Nature of Technology and the Remaking of the Rhône*. Cambridge, MA: Harvard University Press.

Průcha, Václav et al. 2009. *Hospodářské a sociální dějiny Československa 1918–1992*, vol. 2: *Období 1945–1992* [Economic and social history of Czechoslovakia 1918–1992, vol. 2: The period 1945–1992]. Brno: Doplněk.

Pullmann, Michal. 2008. "Sociální dějiny a totalitněhistorické vyprávění" [Social history and the totalitarian history narrative]. *Soudobé dějiny* 15(3–4): 703–17.

Radouš, František. 1912. *Organizace vodních sil v Čechách* [The organization of water power in Bohemia]. Prague: Ústav ku podpoře průmyslu.

Renneberg, Monika, and Mark Walker. 1994. "Scientists, Engineers and National Socialism." In Monika Renneberg and Mark Walker, eds., *Science, Technology, and National Socialism*. Cambridge: Cambridge University Press, 1–29.

Růžička, Karel. 1959. "ZLV—první akce na ochranu krajiny na světě" [ZLV: the world's first landscape conservation action]. *Vodního spodářství* 8(3).

Schivelbusch, Wolfgang. 2006. *Three New Deals: Reflections on Roosevelt's America, Mussolini's Italy, and Hitler's Germany, 1933–1939*. New York: Metropolitan Books.

Scott, James C. 1998. *Seeing Like a State: How Certain Schemes to Improve the Human Condition Have Failed*. New Haven, CT: Yale University Press.

Šembera, Jiří. 1979. *Systém řízení vodního hospodářství* [The water management system]. Prague: Státní zemědělské nakladatelství.

Smetana, Jan. 1954. *Úprava vodstva v lidově demokratickém Československu a její vnitřní a mezinárodní význam* [Water management in the People's Democracy of Czechoslovakia and its internal and international significance]. Prague: Československá akademie věd.

Štanzel, Arnošt, and Helena Durnová. 2015. "Zkrocení přírody podle světových trendů? výstavba Vltavské Kaskády v Československu (1948–1965)" [Taming nature according to world trends? construction of the Vltava Cascade in Czechoslovakia (1948–1965)]. *Dějiny a současnost* 37(2): 14–17.

Swyngedouw, Erik. 2015. *Liquid Power: Contested Hydro-Modernities in Twentieth-Century Spain*. Cambridge, MA: MIT Press.

Sýkora, Artuš. 1947 [1935]. *O plánovitosti v našem vodním hospodářství* [On planning in our water management]. Prague: Průmyslový ústav.

Úlehla, Vladimír. 1947. *Napojme prameny* [Let's water the springs]. Prague: Život a práce.

Van Meer, Elisabeth. 2010. "'The Nation Is Technological': Technical Expertise and National Competition in the Bohemian Lands, 1800–1914." In Martin Kohlraush, Stefan Wiederkehr, and Katrin Steffen, eds., *Expert Cultures in Central Eastern Europe: The Internationalization of Knowledge and the Transformation of Nation States since World War I*. Osnabrück: Fibre Verlag, 85–104.

Vinter, Aleksandr Vasilevich. 1951. *Velikie stroiki kommunizma* [Great construction projects of communism]. Moscow: Izd-vo Akademii nauk SSSR, 1951.

Vleuten, Erik van der, and Cornelis Disco. 2004. "Water Wizards: Reshaping Wet Nature and Society." *History and Technology* 20(3): 291–309.

Votruba, Ladislav. 1954. *Naše velké vodní stavby socialism* [Our great water works of socialism]. Prague: Nakladatelství ČSAV.

Votruba, Ladislav. 1993. "Tradice ochrany přírody v práci našich techniků" [The tradition of nature conservation in the work of our technicians]. *Dějiny věd a techniky* 23(6): 147–63.

Votruba, Ladislav. 2003. "Vodohospodářské poměry, legislativa, organizace: Vodní hospodářství a nádrže" [Water management, legislation, organization: water management and reservoirs]. In Jaroslav Folta, ed., *Studie o technice v českých zemích*, sv. 9: *1945–1992* [Studies on technology in the Czech Lands, vol. 9: 1945–1992]. Prague: Encyklopedický dům, 1987–2046.

Vyskočil, Lev. 1947. *Úvahy o potřebě a zásadách nové hovodního zákona* [Reflections on the need and merits of a new water law]. Prague: Technický ústav obchodní a živnostenské komory.

Wagner, Peter. 1993. *A Sociology of Modernity: Liberty and Discipline*. London: Routledge.

White, Richard. 1996. *The Organic Machine*. New York: Hill and Wang.

Wittfogel, Karl August. 1957. *Oriental Despotism: A Comparative Study of Total Power*. New Haven, CT: Yale University Press.

Zeller, Thomas. 2005. "Molding the Landscape of Nazi Environmentalism: Alwin Seifert and the Third Reich." In Franz-Josef Brüggemeier, Mark Cioc, and Thomas Zeller, eds., *How Green Were the Nazis? Nature, Environment, and Nation in the Third Reich*. Athens: Ohio University Press, 147–70.

7

Big Dam Biographies in Central Asia

Tracing Goals, Actors, and Impacts from World War II to the Present Day

Jeanne Féaux de la Croix and Flora Roberts

During the Soviet period, social engineering was such a crucial an aspect of environmental engineering projects that perceived success in reaching social reform goals could compensate for inadequate economic outcomes and (concrete) engineering failures. The tragedy of the Belomor canal is an eloquent illustration of this phenomenon: a hugely wasteful and cruel digging project, which mobilized thousands of imprisoned laborers and cost many lives, but turned out to have almost no practical value. Maxim Gorky, however, edited a notorious anthology that celebrated the social engineering successes of Belomor in reforming "criminals" into "productive labourers" (Gorkii, Averbach, and Finn1934; Ruder 1998). Apart from the early twentieth-century focus on "reforming" workers, both Soviet and post-Soviet governments have pushed big dams as the linchpin of development, *the* strategic solution to improve human lives across the country through the placement of one great big machine.[1] In the Central Asian republics, whose populations were widely perceived by Russian bureaucrats and specialists as backward and insufficiently "enlightened" in Marxist terms, the social transformations predicted to ensue from water infrastructure projects were discussed alongside the economic gains expected from the planned environmental transformations. As in the case of the southern Macedonian irrigation projects (Vlachos, chapter 2 in this volume), an ethnicized and gendered understanding of "backwardness" shaped the way these social engineering projects were put together in the Soviet period.

Though several Soviet big dam projects have been revitalized in post-Soviet Tajikistan and Kyrgyzstan, we find a change in the meanings of "social engineering," and the political functions of these mega projects also change.

In this chapter, we use the device of dam "biographies" to conceptualize the political, social, and environmental relations through which these dams were conceived—and which they generated in turn. We track contexts of environmental and social engineering in Central Asia since the mid-twentieth century by comparing dam projects built on the Syr Darya River in the Tajik Soviet Socialist Republic with contemporary dam building on the Naryn in Kyrgyzstan, and on the Vakhsh in Tajikistan. The comparison highlights how discussions of the social and environmental transformations attendant on dam construction have gone through several metamorphoses. We find that while the social effects of dam building today differ significantly, the environmental effects have remained similar over the past seven decades. While this chapter does not discuss detailed life histories of technical experts, we track the transformation of different categories and groups of actors who experienced and participated in dam building: Soviet engineers, ditchdiggers, valley residents, and World Bank employees. The great environmental transformations generated by building dams also brought together new constellations of groups involved with dams across the momentous political transformations of the mid- and late twentieth century in Central Asia.

DIVERSE ENGINEERING GOALS, SIMILAR IMPACTS?

Historians and social scientists have long been interested in the twentieth-century dam-building binge: their intended and actual effects and the rise and fall of their appeal according to political context and era. We recognize several elements as "family traits" of the roughly 45,000 "plugs" in rivers and their power plants across the globe. Large hydraulic projects have long served economic and political ends—to control the flow of water and harness its energy, to turn "deserts into gardens," and to recruit large numbers of people as part of state projects (Worster 1985: 191–93). Big dams and other water infrastructure have also long operated as set pieces of nation building, similar in this respect to other highly visible infrastructure projects—"spectacles" such as new capital cities that consolidate government prestige and attempt to draw citizens into a common project (Klingensmith 2007: 280; Menga and Swyngedouw 2018). Dams redistribute resources across time and space, between communities and ecosystems, and thus are a means to demonstrate the techno-economic power of the state (Mitchell 2002: 21). Franklin D. Roosevelt's Tennessee Valley Authority project is an early example, and more recent examples include the Grand Ethiopian Renaissance Dam (Mains 2012) and China's Three Gorges project (Pietz 2015: 198). In contemporary Tajikistan, politicians and publicists have likened rivers to veins in the national body politic, and their water to the nation's life blood (Suyarkulova 2014: 368).

The Soviet Union drew heavily on US expertise on irrigation and dam building before World War II, but during the Cold War period began instead to "export" water engineering knowledge to projects from Southeast Asia to Egypt (Kalinovsky 2018; Zeisler-Vralsted 2014). By the early 1960s, the Soviet Union was building gigantic dams at home whose reservoirs and electricity generation capacity far outstripped the largest American dams, and replaced the United States as an international sponsor at sites such as the Egyptian Aswan High Dam (Gestwa 2010). Such gargantuan projects were (and are) always presented as fulfilling a range of needs for different sectors of the economy. In the planning phase, the potential to use dams either to store irrigation water or to generate electricity is usually prioritized, although flood control is also often promised. The political prestige of such projects frequently determines the forecasting of potential benefits for citizens, generally underplaying financial, social, and environmental costs.

These common family traits in the "social life" of big dams, however, underplay their individual biographies and "personalities," which are shaped by time, place, and different kinds of actors. The Soviet "parent" generation of dams may look similar, as material end products, to their "children" in post-Soviet Tajikistan and Kyrgyzstan. But should these dam generations meet at a family gathering, they would have serious difficulty relating to each other's ideas, outlook, and life histories. In this chapter, we therefore look in detail at the *how* of dam building in Central Asia.

TAPPING RIVERS FOR IRRIGATION AND HYDROPOWER IN CENTRAL ASIA

Kyrgyzstan and Tajikistan, the poorest of the post-Soviet republics, are sometimes called the "water towers" of Central Asia, as the glaciers and precipitation in their mountains provide 90 percent of the region's drinking and irrigation water.[2] In the 1920s and 1930s, the Kyrgyz and Tajik Soviet Socialist Republics (SSRs) were formed following several revisions in status and territory. Although named for a titular nationality, each republic was inhabited by a wide variety of ethnic and language groups: speakers of Iranian languages were concentrated in the Tajik SSR, albeit with a substantial Turkic minority, and in Kyrgyzstan speakers of Turkic languages like Kyrgyz and Uzbek predominated. The Soviet Union integrated Central Asia into the centralized planned economy, largely as a source of raw goods such as meat, wool, and cotton. Efforts in the postwar period to break away from this "colonial" pattern and diversify Central Asian economies met with limited success (Kalinovsky 2013). Today, citizens in both Kyrgyzstan and Tajikistan depend largely on a combination of subsistence agriculture and remittances from migrant workers, particularly in Russia and the Gulf states (Abashin 2014; Bahovadinova and Scarborough 2018; Reeves 2011). The two countries' political histories diverged, however: Tajikistan suffered a devastating civil war in the 1990s, while Kyrgyzstan has coped with several periods of popular unrest that have overthrown successive governments. The Kyrgyzstani government is making increasing attempts to curtail freedom of speech and opinion, but political repression and state violence are still much more severe in Tajikistan. In each case, dam histories and futures remain politically sensitive, restricting the degree to which we can include "local" voices in this overview.

Flowing several thousand kilometers from the peaks of Tajikistan and Kyrgyzstan, the Amu Darya and Syr Darya feed the intensive agriculture in the lowlands of Uzbekistan and Kazakhstan. In the pre-Soviet period, both the distribution of water and the construction of extensive irrigation systems were often organized through a combination of community-appointed water-distribution specialists (*mirob*) and khanate officials. Once Turkestan was conquered by the tsar's troops in a series of campaigns up to 1868, Russian governors had great ambitions to expand and modernize irrigation, but little appreciation for local expertise on maintaining water channels and reducing evaporation in the hot climate. Early Soviet water infrastructure such as the Great Ferghana Canal drew on mass labor coordinated by a centralized—if often haphazard—management style (Obertreis 2017; Peterson 2019), while household water use on small plots also survived. In the postwar period, the government attempted to strengthen the water sector and vastly increase agricultural land use, particularly for thirsty cotton crops, by extending irrigation networks, creating a new bureaucratic structure, and dispatching irrigation specialists from Russia.

Political ideas of decolonization and liberation were supposed to be realized through water infrastructure projects, which were expected not only to boost the economy but also to mold a new Soviet citizenry in the process (Teichmann 2007: 499). A 1960 cover of the satirical journal *Khorpushtak* (The Hedgehog) shows a Tajik man in a suit and tie, his nationality discreetly indicated by his skullcap, standing triumphantly before an industrial landscape featuring a dam. He exclaims, "Look on in envy, capitalist sirs!" to the salivating foreigners in top hats (see figure 7.1 below). Although official literature celebrated the transformation of steppe and rivers, and the fraternal collaboration between republics that this required, there was fierce competition over resources between state agencies, ministries, and republics (Cucciolla 2017; Roberts 2018).

Cascades of big dams built from the 1950s onward vastly expanded the capacity to control the rivers' seasonal flow: these "water towers" continue to supply the vast majority of electricity in Kyrgyzstan and Tajikistan. Such strong

FIGURE 7.1. "Look on in envy, Gentleman Capitalists!" Commentary: "The grandiose successes of communist construction in our country are so striking that at the sight, capitalists are filled with envy, their mouths watering." Despite the sacks of gold that identify them as capitalists—who for some reason are riding donkeys, and thus portrayed as backward—they salivate, green with envy. A smartly suited Tajik man (his ethnicity discreetly indicated by his embroidered skullcap) shows off a vista centered on an enormous dam, complete with hydroelectric station, factories, building sites, and prosperous fields of wheat and cotton. Source: Cover of *Khorpushtak* (Hedgehog) monthly magazine, issue no. 8, August 1960. Tajik language publication, with some text in Russian.

reliance on hydropower means that in low rainfall years, citizens suffer through extensive blackouts in the severe winters, and rely to a greater extent on generators, firewood, and torches for heat and light. Over the past half century, excess withdrawals from these rivers has severely dried up the Aral Sea, an outcome forecast and accepted by Soviet planners. Alongside the Chernobyl disaster, the desiccation of the Aral Sea was one of the signal catastrophes sparking public debate and environmentalist critique in the glasnost era. The centralized approach to resource management faltered after the collapse of the Soviet Union, as the differential interests in water and energy between Central Asian states became harder to reconcile. A greater focus on national interests, and mounting disengagement and distrust between neighboring states, have remodeled relationships and decision-making processes along these rivers (Dukhovny and de Schutter 2011).

It might seem surprising at first, that such small, poor countries as Tajikistan and Kyrgyzstan are again pursuing big dam projects originally designed by a much more powerful state, in the framework of a vast planned economy. Below, we will examine some of the reasons for such continuities and shifts in what these dams actually achieve socially, environmentally, and politically.

ELECTRIFICATION AND RESHAPING SPACES FOR THE COMING OF COMMUNISM

In 1920, Lenin declared communism equivalent to "Soviet Power plus the electrification of the whole country." As increasing the production of electricity became a major policy goal, a state organization for the electrification of Russia was founded, GOELRO. But, Lenin cautioned, electrification could not succeed without a concerted and ambitious campaign to increase literacy: "We must see to it that every factory and every electric power station becomes a center of enlightenment." Lenin considered this coupling of environmental and social engineering goals as necessary to achieving his ultimate goal, when communist economic development would "become a model for a future socialist Europe and Asia" (Lenin 1964: 513–18).

Social engineering goals were also closely linked to new forms of spatial and environmental planning. Many of the people tasked with designing new Soviet cities in the 1920s held that by building new physical communities they were also contributing toward fashioning new (literate) Soviet men and women. During the First Five-Year Plan alone, across the Soviet Union over sixty new towns were laid out by Soviet architects. Many were firm believers in voluntarism, holding that with careful urban planning, the new settlements' inhabitants could be guided toward socialism entirely without force (Starr 1984). Broadly defined, social engineering can encompass any set of policies designed to promote certain types of social interactions and individual behaviors, while discouraging others. In this sense, drawing new cityscapes and housing blueprints was only the beginning. Once Soviet citizens lived in socialist housing units and congregated in socialist public spaces, there was still the wider environment with which Soviet citizenry interacted, a new relationship with nature and with work to be redesigned (Brain 2010).

ENGINEERING NATURE AND SOCIETY FOR VICTORY AT FARHAD

The dams built on the Syr Darya from the 1940s onward exemplify an important trend in the way that social and environmental engineering were coupled within the Soviet Union. On both the social and environmental levels, the stated intent was to replace farm workers and farmland with industrial workers, hydropower stations, and factories. As was so often the case, the needs of agriculture were subordinated to those of industry. The farmland submerged by the dam reservoirs was written off as a loss, the population evacuated from the floodplain, and collective farms near and far were obliged to free up a significant portion of their labor force to work on dam construction.[3] Collective farm workers were drafted to distant regions to dig ditches and were rhetorically (if not always in actual fact) transformed into industrial workers by their labor. Though symbolically important, the touted transformations were at best partial and incomplete: the bulk of the labor carried out by collective farmers on construction sites was unskilled manual labor—primarily digging—carried out with basic agricultural tools (Zhinskii 1956). It may have been labor in the service of electrification and industry, but digging is digging. The planned factories did not always materialize, and when they did, their workforce was not made up mainly of local former farmworkers, who confounded Soviet social scientists by apparently preferring to remain members of kolkhozes. This did not prevent official Soviet publications from celebrating the advancing industrialization of Central Asia, albeit with less emphasis on the proportion of locals in the industrial workforce, and more on the successful outliers. In both domains, social and environmental, success could most easily be declared by moving the goalposts.

The outbreak of World War II marks a crucial turning point, as before the war, the principal concern of state-led interactions with the waters of the Syr Darya had been irrigation, not energy production. Studies of the river basin focusing on its potential to expand irrigated agriculture into the steppe had been produced regularly in the 1920s and 1930s.[4] Beginning in December 1940, the state planning commission (Gosplan) published a comprehensive plan for the waters of the Syr Darya basin, which envisioned a cascade of dams along the whole course of the river from the Kyrgyz highlands to the plains not far from the Aral Sea.[5] The first dams approved in the 1940s and 1950s prioritized industrial goals at the expense of agriculture, as both the social and environmental "footprints" of these dams—at Farhad and Kairakkum—make clear.

Even environmental engineering projects undertaken with great urgency in support of the war effort were not shorn of their concomitant social engineering objectives, which on the contrary were conceptualized as key to the projects' success. The dam and hydroplant built at Farhad in the Uzbek SSR is a good example of this. In June 1941, when Hitler's armies abruptly invaded the Soviet Union, the vast country was forced to mobilize at great speed and expended enormous efforts in dismantling and evacuating thousands of industrial enterprises, sending them eastward by train along with their workforces. Many of the evacuated factories were reassembled in the Uzbek SSR, providing a huge boost to the industrialization of that republic, while exponentially increasing local demand for electricity. It was thus because of, rather than in spite of, the outbreak of hostilities that certain dam construction projects that had been tabled before 1941 were given the green light during the war. The State Defense Committee (GOKO) ordered the construction of a dam and hydroelectric station at Farhad, a site close to the border between the Uzbek and Tajik republics at the far eastern end of the Ferghana Valley, in November 1942.[6] The Farhad hydropower plant was deemed crucial to the war effort, as it was needed to supply electricity to the Bekabad steel plant, which in turn supplied the Chkalov aircraft factory (Dukhovny and de Schutter 2011). One would imagine that social engineering goals would recede from view altogether in such extremities, but they did not.

The war did, however, temporarily reorient social engineering priorities toward fanning the flames of self-sacrificing devotion to the motherland. A peacetime dam project in Uzbekistan would not necessarily have secured the attentions of an architect, but the talented Iosif Karakis, who had designed major public buildings in Kiev before the war, was evacuated from Kiev with his family and put to work (Yunakov 2016). Karakis was charged not just with designing a dam, but with fostering and increasing the "fighting and creative spirit" of the workforce and local population by appealing to the romantic and "patriotic" (in the sense of national) local myth of Farhad and Shirin, one of the most famous love stories of Turko-Persian culture.[7] This Karakis accomplished—to the great satisfaction of the commission of Soviet architects who evaluated his work in 1943—by means of a monumental sculpture of the water-carrying hero Farhad, sculpted into the live rock face as if emerging from the rocks (Yunakov 2016: 195).

Although the Farhad project was given the green light in 1942, circumstances linked to the war would initially slow down construction. Mobilizing a workforce of sufficient size to complete the dam and power station at the accelerated tempo the war seemed to demand was a serious problem. Working conditions were harsh for the thousands of collective farm workers sent to Farhad from across the region (published figures put the total at six thousand in February 1943).[8]

The labor of Uzbek collective farmers was supplemented by deportees from farther afield, including a number of Crimean Tatars punitively resettled to the Uzbek SSR in 1944 and drafted to work on Farhadstroi.[9] Although working conditions on the site were very harsh and almost entirely reliant on human muscle, significant efforts were also expended to edify and elevate the workers. There were small but telling touches, such as the installation of a New Year's tree for local children.[10] This was the socialist version of the Christmas tree, which had been popular in urban Russian households before the Revolution, but was previously unknown in predominantly Muslim Central Asia. Theater troupes, evacuated from prestigious centers of culture to the West such as Kiev, performed for the workers; for example, members of the Gorky Drama Theater performed the "patriotic play *Russian People*."[11] Delegations of official writers, members of the local Writers' Union, visited the site in March 1943, charged with composing poems and stories celebrating the workers' achievements.[12] A radio program was broadcast from Bekabad, which began with the words "Listen Uzbekistan! Farhadstroi is talking!" The program included regular concerts and updates from engineers about the hard work being put in by collective farmers from across the republic.[13] Cumulatively, these endeavors signaled a commitment to enlighten and educate the workforce, particularly the "backward" Central Asians, among whom adoption of certain western European cultural norms and habits was taken as a sign of successful social engineering.

Another important role for construction-site newspapers, at Farhad as elsewhere, was to spur workers to greater efforts by heaping praise on those who exceeded their targets and publicizing the valuable rewards they had received. To spur socialist competition, teams of workers identified by their region of origin and occupation (collective farmers, by and large) were listed alongside their target goals and achievements.[14] A commission of Soviet architects pronounced the Farhad project a success by 1943—long before the hydropower station began producing electricity—thanks to its professed morale-boosting effects on the local community. The echoes of the lonely Farhad's mythical struggle for water, and the *collective* fulfillment of the Uzbek's people's "century-old dream" of bringing water to the Hungry Steppe had an important role to play, it was confirmed, in forging the collectivist spirit that would deliver victory to the Soviet Union. In sum, at Farhadstroi both declarations of social transformations and significant policy efforts in that direction went hand in hand with ambitions to radically alter the landscape. As we will see, social ambitions paled in subsequent dam generations, and in the case of Kairakkum town were even reversed. The power to alter environments has, however, remained a key motivation.

ENGINEERING AN INLAND SEA AT KAIRAKKUM

Before the dam at Farhad was even finished, negotiations were underway for the next in the series of dams planned for the Syr Darya. Experienced workers who had received special training and acquired new skills on the Farhad site were invited to move to the next construction site a few dozen miles upstream at Kairakkum. The project to build a dam and large reservoir at Kairakkum, in the Tajik Ferghana Valley, was approved in 1951, following an intense period of wrangling over its location (Roberts 2018). The opportunity to build another dam nearby solidified and compounded the gains of those relatively few Central Asians who had gained a new industrial (and thus enlightened) identity as technicians, cement workers, or electricians. The work of forging a native proletariat by siphoning off agricultural laborers for dam construction begun at Farhad continued at Kairakkum, and for years thereafter in the heyday of Soviet hydropower that lasted into the 1970s (Nazarov and Shozimov 2011).

The memories of outsiders and locals differ sharply with regard to the environmental impact of the dam and the reservoir of 530 square kilometers at Kairakkum. In 2017 an elderly farmer displaced in childhood by the reservoir described the area in which he had grown up as a "jangal" (a Persian word, whence the English "jungle" derives), a densely wooded area alive with foxes, wolves, and plentiful game for hunters. Part of the area flooded by the reservoir was covered by *tugai* riparian forests, described by Soviet officials objecting to the planned dam as the only source of timber in the region. A Russophone engineer, on the other hand, who surveyed the future dam site emphasized a very different topography: "In July 1951 when the first builders arrived to build a hydropower station, a desert stretched out before their eyes: not a single tree, not a single shrub. Often, smothering winds lifted the sands from place to place, and obliterated all life forms, even the prickles eaten by camels did not grow there."[15]

Of course, the reservoir flooded a wide area, encompassing diverse ecosystems, but this version of the parable of the blind encountering an elephant carries a twist: in the case of Kairakkum, only one partial version of the truth gained currency, through repeated discussion in the press and propaganda. It was surely no accident that the former toponym for the sandy area, Kairakkum (ascribed the etymology "grinding sands"), was preserved for the model Soviet town, so as to canonize the most extreme version of the transformation of nature enacted there: from desert (rather than from riparian marshland or orderly cottonfields) to inland sea.

The construction workers' town founded in proximity to the dam site would be developed into a model Soviet town—with a grid of straight, tree-lined avenues and symmetrical apartment blocks. This town was considered as important an avenue for modernization as the dam and power station itself. In a 1960 speech, the lead engineer of the Kairakkum construction project, Ya'akov Fligelman, contrasted "the beautiful, green city" of workers with the barren, windy valley dried out by the sun that had preceded it (Kalinovsky 2013).

In the course of oral history fieldwork conducted in 2017, Flora Roberts heard several versions of a (possibly apocryphal) story about Kairakkum's flagship factory. The initial plan had been for the main industrial enterprise in Kairakkum to be a bicycle factory until the chairman of the State Planning Agency intervened to prevent it, "arguing that Central Asians do not like bicycles, and that instead we should play to their strengths, and find something that would get Central Asian women out of [the] house." That is how, Roberts was told, Kairakkum got a carpet factory, still a highly cherished part of the townscape, although much diminished since Soviet times. This anecdote points to the strengths and limits of Soviet social engineering: the carpet factory provided the preconditions for multiethnic, multilingual teams to work together. But by the late 1950s, there were clearly limits to the types of work that the authorities envisioned Central Asians, and Central Asian women in particular, participating in.

Long-term residents and former residents interviewed in 2017 all made a point of emphasizing the cosmopolitan, ethnically and linguistically diverse character of Soviet-era Kairakkum, in spite of its small size. Strikingly, although the environmental transformations—intentional and less so—are ongoing and profound, the social transformations promoted in tandem have largely been reversed in subsequent decades. The settlement is still dominated by the dam and reservoir, although the progressive silting of the eastern side of the reservoir has greatly reduced the hydropower generated annually. But the town's population is very different. A strong earthquake in 1988 whose epicenter was close to Kairakkum town was soon followed by the collapse of the Soviet Union and the Tajik Civil War: in combination, these events spurred the departure of most of the nonlocal ethnic groups. In contrast to its Soviet-era identity, Kairakkum today is a predominantly Tajik town by ethnicity and language. Both the one-family wooden "Finnish homes" supplied in the 1950s to the town's elite workers, and the distinctive Khrushchev-era five-story apartment blocks, on whose stairwells a Babel of languages once echoed, are showing severe signs of age, and very little post-1990 housing stock is visible. Women have retreated from the formal labor market and occupy far fewer positions of responsibility than they did in the last Soviet decades. Thus, the major priorities in engineering the social fabric of the town of

Kairakkum—encouraging local women to enter the industrial workforce, promoting the peaceful coexistence of multiethnic communities in a modern workers' town with state of the art facilities—have largely been reversed. Soviet environmental engineering has outlived the social engineering, although the river too, continues its slow work to reverse the dam builders' efforts.

In the case of Kairakkum, the intentional environmental transformations have not only survived but also have catalyzed a chain of unintentional and profound landscape alterations. In the following sections, we follow the continuities and fractures in dam biographies, their social and environmental impacts into the post-Soviet period. Having traced the story of dams into twenty-first-century Kyrgyzstan and Tajikistan, we see changes in the available sources, the involved actors, points of friction, and social effects, despite similar physical dam designs on the very same rivers. We conclude by comparing the social and environmental impacts of these two dam "generations."

CONTEMPORARY DAM-BUILDING STRUGGLES IN TAJIKISTAN AND KYRGYZSTAN

Once the Central Asian republics, somewhat unwillingly, became independent in the early 1990s, long-suppressed tensions over the ownership and management of the dams on shared river basins emerged once more. Kyrgyzstan, Tajikistan, and Uzbekistan all suffer from chronic energy shortages and an aging distribution network. Soviet-era management of Central Asian dam cascades sought to balance the needs of upstream countries to store water for electricity production in winter, while allowing sufficient water releases for irrigation needs downstream in summer. Since 1991, water allocations and energy transfer agreements between riparian neighbors have been increasingly disputed (International Crisis Group 2014: 5–7). Built at the behest of the Uzbek SSR, but with a floodplain mainly on Tajik territory, the Farhad dam was one of several sites that became a source of tension. Having been denied a requested border change in the 1940s, the independent nation-state of Tajikistan reasserted a claim to control the area, until eventually in 2002 a squad of Tajik police seized the initiative and wrested control of the dam from the Uzbek border guards. Throughout the tenure of the Uzbek president Islam Karimov (1991–2016), control of Farhad remained one of the most serious points of contention between the two countries, and it was only under the new leadership of Shavkat Mirziyoyev that talks were held in 2018 to resolve the impasse (Mannonov 2011; Tursunov 2013; Safar 2018). The 2018 agreement, as reported by Russian news agency Interfax, resolved the border dispute by acknowledging that the territory on which the power station stands belongs to Tajikistan, but agreed that the infrastructure would continue to be managed and maintained by the Uzbek side (Interfax 2018).

Another bone of contention between Uzbekistan and Tajikistan since the 1990s has been Tajikistan's commitment to reviving a large dam-building project begun in the Soviet period. The Roghun dam on the Vakhsh River will supplement the aging Nurek dam (begun in 1961) downstream, and is intended to stake a claim as one of the world's tallest dams. Construction on Roghun began in 1982, but during the glasnost period the project became a focus of public criticism. Voices in the Tajikistani press critiqued Roghun's "gigantomania" and protested the envisioned resettlement of 42,000 people in order to flood 170 square kilometers. Construction was halted in the late 1980s, but it is no longer discernible whether this was due to public criticism or the dwindling budget (Menga 2015; Sodiqov 2013). While the new workers' town of Roghun was occupied by military forces during the Tajik Civil War (1992–1997), floods destroyed much of the work carried out on the site. Since the peace agreement in 1997, Tajikistan has been ruled by increasingly authoritarian means under President Emomali Rahmon. Relaunching the dam project in 2008, the president has made promoting Roghun his signature policy: "Roghun is not only a source of light, but a national honour and dignity" (Rahmon 2010, in UPI Energy 2010). Roghun promises to double the country's energy output (forecasted to be 3,000 megawatts; World Bank Group 2014: 11). From posters on the avenues of the capital to Tajik popular songs, Rahmon is personally associated with Roghun across the full spectrum of public media.[16]

However, the symbolism of Roghun has shifted since the late Soviet era: although still cast as a collective project, it is now associated with the "Tajik" rather than "Soviet" people, and has become the trademark of a single political figure.[17] Insistence on the most ambitious version of Roghun, rather than more modest propositions, fits with the Tajik government's drive to build the world's tallest flagpole, the largest *chaikhana* (teahouse), and the biggest national library in Central Asia. Roghun's completion will also likely benefit the state-owned TALCO aluminum plant, which consumes a large portion of the country's current hydropower output at subsidized rates, and which is effectively controlled by the president's family (Lemon and Roberts 2021). Even if the country were to double its energy output as projected, it is unclear how much the greater electricity output would benefit citizens, instead flow to industries like TALCO, or be sold to energy-hungry neighbors such as China and Pakistan. Nevertheless, in 2009–2010 the government launched a campaign encouraging citizens to buy shares in Roghun. The dam was vaunted as a "magic bullet" for Tajikistan's struggling economy, and state employees such as teachers were pressured to buy significant numbers of shares, increasing citizen debts (Eurasianet 2010). Since the modes of investment and return for Roghun are very murky, the scheme is widely considered just another way for high-ranking officials to erode citizens' salaries (Ibañez-Tirado 2015). But unlike critiques voiced in the 1980s, current skepticism about *how* the building is funded, does not necessarily translate into rejecting the idea of the project itself. The Roghun dam seems to embody, with some success, the hoped-for escape route from growing poverty.[18] And if dependence on international agencies such as the World Bank is one side of the coin, the Tajikistani government has shown itself quite adept at using World Bank approval of the Roghun project as a form of regional "soft power" (Menga and Mirumachi 2016). As the first turbine started turning on President's Day in 2018, some of the Roghun promises are being put to the test.

STOP-AND-GO DAM BUILDING IN KYRGYZSTAN

Though neighboring Kyrgyzstan has also rejuvenated late Soviet dam projects, the tone of interaction between the state, international actors, and citizens is very different: less laden with symbolism, but no less with sentiments of "hope against hope." As in Tajikistan, projects launched in the late 1980s were being dusted off by 2007. The large Kambar-Ata 1 and 2 plants (capacity foreseen at 2,200 megawatts) are intended to meet domestic demand, and to export electricity to Afghanistan and Pakistan (International Crisis Group 2014). These dams are the single prestige project to have survived the toppling of their promoter President Kurmanbek Bakiyev, in 2010. Among many other grievances, a sudden rise in energy prices was one of the bitter pills that drove people to demonstrate in that restless period (Wooden 2014). Though questions of funding and implementation are periodically debated in parliament, the basic necessity of expanding hydropower on the Naryn River is not questioned. Political turbulence has sometimes also favored these dams across successive governments. In the volatile summer of 2010, when political crises bled into ethnic violence, the opening of the smaller Kambar-Ata 2 plant on Independence Day provided a singularly positive news story and national "glue." In this sense, Kambar-Ata 1 and 2 have at times been made to play just as "national" a role as Roghun in Tajikistan, but they are not consistently made central to a national narrative, as in the case of Farhad in the 1950s, nor associated with a single figure, as with the contemporary marriage between Roghun and President Rahmon.[19]

Despite the occasional propaganda victory, all the new Naryn dam projects have caused the local population intense disappointments on several fronts. Construction has repeatedly been delayed, largely due to difficulties in securing the necessary external investment. In 2016 ongoing construction at Kambar Ata stalled and a deal with the Russian company RusHydro for the upper Naryn cascade projects was canceled amid accusations of financial mismanagement (Harris

2016). In terms of social impacts, the contrast with dam building on the same river in the 1940s and 1950s is striking. Fieldwork in the area between 2014 and 2019 revealed that no special advantages, such as preferential pricing or social and vocational training programs, were foreseen for the region affected.[20] Apart from a banner with slogans and a map of the projected dam sites in the region's capital, virtually no public information was available. The main expectation of citizens and regional administrators has been that the dam sites will provide work locally. Some local residents went so far as to invest in qualifications that would be useful in dam work (Altyn Kapalova, personal communication). Unsurprisingly, despite initial promises, it turned out that the construction company needed workers with formal qualifications, prior experience, and fluent Russian. The industry also tends to prefer retaining workers employed at previous sites to offering work to local farmers, as was done in the first generation of Soviet dams. The withdrawal of the initial promises, and the failure of the project to progress caused intense disappointment and ill-feeling (which went as far as sabotage) toward both the company and the national government. As in the Tajik town of Roghun, a sense of having received no benefits from the dam is pervasive (Féaux de la Croix and Suyarkulova 2015).

COMPARING GENERATIONS OF CENTRAL ASIAN DAMS

Who owns a river? How and with what priorities should it be administered? The post-Soviet (and as elsewhere, neoliberal) process of privatization has signaled a fundamental reworking of the role of the state, and the role of "the people" and individuals in relation to it (Alexander 2004: 256). The Soviet-era dams discussed here were envisioned as key elements in comprehensive plans to reform both people and their environs. Planners dreamed of, advocated, and at least partially realized an intricate interaction between dam labor, electricity production, urban development, agricultural production, water control, and profound transformations of environments that they considered "difficult." Soviet dams generated their own publishing industry, with biographies of dam workers, celebratory poetry, and pamphlets explaining the technology for a lay audience (Roberts 2018). National and pan-Soviet pride was yoked together under the banner of "Friendship of the Peoples." The main actors visible in the building process were Soviet state agencies, the dam workers as portrayed by the official press, and the citizens these projects were to benefit: whether through the gift of electricity, a modern flat in a new town, or a new job in a carpet factory.

The dam reservoirs drowned forests and agricultural land, with the promise of expanding irrigated fields elsewhere in the former steppe and semideserts of the lowlands. In this sense, damming the main rivers of Central Asia was a precondition for much more far-reaching environmental transformations, including the desiccation of the Aral Sea. This parent generation of dams also generated large population movements: emptying the flooded land, settling people in new towns and new agricultural settlements. The post-Soviet second generation of these dams—closely related, due to shared origins in the offices of Soviet planners—have a different range of social and political effects, some of which would have been anathema to a "Soviet" dam persona.

Since independence, dam building has become a more "insular" enterprise. Despite recent signs of a dam "détente" between upriver and downriver states, big dam sites have become nationalized in different ways in Tajikistan and Kyrgyzstan. In the Tajik case, Roghun has been intimately connected to President Rahmon, while in Kyrgyzstan a commitment to hydropower has survived successive governments and upheavals. This narrower focus is matched by a narrower range of governmental aspirations connected to dam building and an attendant paucity of public information. Despite official photographs of President Rahmon in a hard hat, dam workers are no longer celebrated as heroes: the relevant actors, often international construction companies, closely control access to their sites and documents. In Tajikistan, where Roghun is a national project, the government has actively solicited the financial support of the population, a kind of pressure unimaginable in restive Kyrgyzstan. But even considering the recordbreaking ambitions of contemporary Tajikistan, the vaunted aim is the more modest circumscribed vision of creating economic wealth, rather than the sweeping, multidimensional social transformations envisioned during the postwar Soviet period in the same areas, and for the same projects. The main "audience" for dam propaganda has shifted to potential evaluators and donors for these megaprojects.

The actual financing of these projects has a much more international dimension than in the Soviet context. Though big dams no longer enjoy the unremittingly positive reputation of the mid-twentieth century, and potential funders such as the European Bank for Reconstruction and Development have become somewhat more cautious in their backing, such reticence does not play out publicly in Central Asia. While the environmental impacts of international mining ventures in Kyrgyzstan are often publicly contested, dam projects do not provoke intense local resistance. This is surely in part because of dams' glowing official reputation as manifestations of state largesse in the later Soviet decades: another instance of the productive entanglement of "past and present for the purposes of transforming nature," as in the case of Polish forestry (Blavascunas, chapter 9 in this volume). Whether or not a single megawatt of electricity is ever produced, big dams yield significant political and financial gains for some actors. Megaprojects are known as sites not only for prestige making but also for moneymaking through contracts or controlling (and sometimes diverting) financial flows (Molle, Foran, and Kakonen 2009: 328). A further new form of internationalism is evident in that integration into the Asian or Eurasian economies: however, the type of integration is still one of exporting relatively unrefined "natural resources" rather than finished products. It is possible to view these aspects as neocolonial: being financially dependent on and judged as viable projects by countries like Russia and China but also big international institutions like the World Bank. The know-how on dam building is now an amalgam of a generation of Soviet-trained cadres, and (vastly more expensive) "international" expertise.

Tracing the biographies of Soviet and post-Soviet dams in Kyrgyzstan and Tajikistan suggests that dams are elements of Soviet environmental engineering that can be readily adapted by other kinds of governments and that invoke quite different forms of social engineering—or none at all. It is notable that other forms of ambitious environmental and social engineering, such as large-scale irrigation projects and the establishment of new industries or townscapes have not been transitioned at the scale, or with the ease, of dams. As a "family" of objects, the senior generation of Soviet dams lived on and affected a set of actors very different from today's investors: they might hardly recognize each other as relatives at a family gathering. Nevertheless, it is striking that although the place of dam building in narratives of building a (supra-)national state and sense of collectivity has fluctuated in post-Soviet iterations in Kyrgyzstan and Tajikistan, the reference point of a common "national" project remains surprisingly stable (Dorondel and Şerban, introduction to this volume). Similarly, the environmental impacts of such megaprojects, from changes in the water distribution to silting, have proved surprisingly constant.

NOTES

1. This has also been true elsewhere, particularly in the postcolonial world, as in the case of Ghana's Akosombo dam or Paraguay's Itaipú Dam.

2. Kyrgyzstan has 6.5 and Tajikistan 9.5 million inhabitants. Both are approximately the size of the United Kingdom, but the population is concentrated in the lowlands, while over 90 percent of the territory of each republic consists of sparsely populated mountains (Sultanov et al. 2018; World Bank 2021).

3. On resettlement connected to Farhad, see GARF F. P7523, op.65, d.172 (1946), and "Narod stroit Farkhadskuiu GES" [The people are building the Farhad hydropower plant], *Pravda Vostoka*, March 12, 1943, 2.

4. Rossiiskii gosudarstvennyi arkhiv ekonomiki [Russian State Archive of Economics] (RGAE), f. 8378, op. 1, d. 22 (1929–1932), meetings and resolutions on improving the management of the "land and water resources of the Syr Darya river basin."

5. RGAE, f. 4372. Gosplan, op. 38, d. 1797–1834, and Roberts 2018.

6. GARF (State Archive of the Russian Federation), f. R5446, op. 48a, d. 1578, l. 86

7. The tragic tale of Farhad, who perishes in the attempt to win the hand of Shirin by digging

an impossibly long water channel single-handedly, appears in varying versions in works by Nezami and Ferdowsi, among others.

8. "Novye otriady stroitelei" [New cohorts of builders], *Narodnaia stroika,* March 23, 1943, 2.

9. Abil' Seitbekirov, "Truba Badina zvala v Krym," *Golos Kryma,* no.7(945), February 17, 2012.

10. "Novogodniaia elka" [Christmas tree], *Narodnaia stroika,* December 29, 1943, 2.

11. "Teatr im. Gorkogo v Farhadstroe" [Gorky Theater in Farhadstroi], *Narodnaia stroika,* March 13, 1943, 2.

12. "Pisateli na stroike," [Writers at a construction site], *Narodnaia stroika,* March 23, 1943, 2.

13. "Stroiteli u mikrofona" [Builders take the microphone], *Narodnaia stroika,* March 14, 1943, 2

14. "Reshaiushchie dni" [Decisive days], *Narodnaia stroika,* December 29, 1943, 1. "Premii—otlichivshimsia," *Narodnaia stroika,* December 29, 1943, 2.

15. From an unpublished typescript of personal recollections and data on Kairakkum, authored in 1984 by G. Bernadskii, who described himself as a "veteran of the construction of Kairakkum GES."

16. The Avesta.tj news website has an entire section dedicated to Roghun-related news and opinion pieces. Popular songs include, for example, Afzalsho Shodiev's song "Roghun" (2010) or Qurbonali Rahmon's song "Roghun" (Rahmon 2010).

17. There are many other examples of dams closely associated with a single figurehead, including Roosevelt's Tennessee Valley Authority and Nasser's Aswan dam.

18. Since freedom of speech is severely restricted in contemporary Tajikistan, it is not possible to accurately gauge the balance of skepticism and hopefulness: comparison with neighboring Kyrgyzstan, which enjoys greater freedom of speech, suggests that hydropower may indeed be widely assumed to be a "Good Thing."

19. The interim president Roza Otunbayeva had been asking highly critical questions of the project under Bakiyev, but once in power, she supported the project, as did the successive government (Menga 2013: 137–38).

20. The best known is *The Report by the World Commission on Dams* (2000), and other international standards recommend that such measures be included.

REFERENCES

Abashin, Sergei. 2014. "Migration from Central Asia to Russia in the New Model of World Order." *Russian Politics and Law* 52(6): 8–23.

Alexander, Catherine. 2004. "Values, Relations and Changing Bodies: Industrial Privatization in Kazakhstan." In Caroline Humphrey and Katherine Verdery, eds., *Property in Question: Appropriation, Recognition and Value Transformation in the Global Economy*. Oxford: Berg, 251–74.

Bahovadinova, Malika, and Isaac Scarborough. 2018. "Capitalism Fulfills the Final Five-Year Plan: How Soviet Era Migration Programs Came to Fruition in Post-Soviet Eurasia." In Marlène Laruelle and Caress Schenk, eds., *Eurasia on the Move: Interdisciplinary Approaches to a Dynamic Migration Region*. Washington, DC: CAP, 1–11.

Brain, Steven. 2010. "The Great Stalin Plan for the Transformation of Nature." *Environmental History* 15(4): 670–700.

Cucciolla, Riccardo Maria. 2017. "The Crisis of Soviet Power in Central Asia: The 'Uzbek Cotton Affair' (1975–1991)." PhD diss., IMT School for Advanced Studies, Lucca.

Dukhovny, Victor A., and Joop de Schooter. 2011. *Water in Central Asia: Past, Present, Future*. London: CRC Press.

Eurasianet. 2010. "Tajikistan: Forced Rogun Payments Sowing Discontent among Impoverished Tajiks." Accessed September 13, 2018. http://www.refworld.org/docid/4b966e6ec.html.

Féaux de la Croix, Jeanne, and Mohira Suyarkulova. 2015. "The Rogun Complex: Public Roles and Historic Experiences of Dam-Building in Tajikistan and Kyrgyzstan." In Alain Cariou, ed., "Water in Central Asia: Contemporary Issues and Challenges." *Cahiers d'Asie Centrale* 25: 103–32.

Gestwa, Klaus. 2010. *Die "Stalinschen Großbauten des Kommunismus": Sowjetische Technik- und Umweltgeschichte 1948–1967*. Munich: Oldenburg Press.

Gorkii, Maxim, L. L. Averbach, and S. G. Firin, eds. 1934. *Belomorsko-Baltiiskii kanal imeni Stalina: istoriia stroitel'stva, 1931–1934 gg.* [The Stalin White Sea–Baltic Canal: construction history, 1931–1934]. Moscow: OGIZ Press.

Harris, Michael. 2016. "Kyrgyzstan Cancels Five Hydropower Plant Deal with Russia." Accessed September 13, 2018. https://www.hydroworld.com/articles/2016/01/kyrgyzstan-cancels-five-hydropower-plant-deal-with-russia.html.

Ibañez-Tirado, Diana. 2015. "Everyday Disasters, Stagnation and the Normalcy of Non-Development: Roghun Dam, a Flood, and Campaigns of Forced Taxation in Southern Tajikistan." *Central Asian* Survey 34(4): 549–63.

Interfax News Agency. 2018. "Tadzhikistan i Uzbekistan dogovorilis' o radzele Farkhadskoi GES" [Tajikistan and Uzbekistan agree on the partition of the Farhad hydropower plant]. January 11. Accessed October 9, 2021. https://www.interfax.ru/amp/594998.

International Crisis Group. 2014. *Water Pressures in Central Asia: Europe and Central Asia Report, No. 233*. Brussels.

Kalinovsky, Artemy M. 2013. "'Not Some British Colony in Africa': The Politics of Decolonization and Modernization in Soviet Central Asia, 1955–1964." *Ab Imperio* 2: 191–222.

Kalinovsky, Artemy M. 2018. *Laboratory of Development: Cold War Politics and Decolonization in Soviet Tajikistan*. Ithaca, NY: Cornell University Press.

Klingensmith, Daniel. 2007. *"One Valley and a Thousand": Dams, Nationalism, and Development*. New Delhi: Oxford University Press.

LeMentec, Katiana. 2010. "Le projet des Trois Gorges: faire parler les héros et les dieux." *Etudes chinoises* 29: 259–70.

Lemon, Edward, and Flora J. Roberts. 2021. "Tajikistan: From Reconciliation to Post-Reconciliation." In Francesc Serra Massansalvador and Ferran Izquierdo-Brichs, eds., *Political Regimes and Neopatrimonialism in Central Asia: A Sociology of Power Perspective*. Singapore: Palgrave Macmillan, 199–247.

Lenin, Vladimir Ilych. 1964. *Collected Works*. Vol. 31. Moscow: Progress.

Mains, Daniel. 2012. "Blackouts and Progress: Privatization, Infrastructure, and a Developmentalist State in Jimma, Ethiopia." *Cultural Anthropology* 27(1): 3–27.

Mannonov, Akmal. 2011. *Kak Tadzhikistan vernul "Plotinu" i Farhodskoe vodokhranilishche* [How Tadzhikistan returned the "Dam" and the Farhad reservoir]. *Azia-Plus*, August 19. http://news.tj/ru/news/kak-tadzhikistan-vernul-plotinu-i-farkhodskoe-vodokhranilishche.

Menga, Filippo. 2013. "Power and Dams in Central Asia." PhD diss., University of Cagliari.

Menga, Filippo. 2015. "Building a Nation through a Dam: The Case of Rogun in Tajikistan. *Nationalities Papers* 43(3): 479–94.

Menga, Filippo, and Naho Mirumachi. 2016. "Fostering Tajik Hydraulic Development: Examining the Role of Soft Power in the Case of the Rogun Dam." *Water Alternatives* 9(2): 373–88.

Menga, Filippo, and Eryk Swyngedouw. 2018. *Water, Technology and the Nation-State*. London: Routledge.

Mitchell, Timothy. 2002. *Rule of Experts: Egypt, Techno-Politics, Modernity*. Berkeley: University of California Press.

Molle, F., T. Foran, and M. Kakonen, eds.. 2009. *Contested Waterscapes in the Mekong Region: Hydropower, Livelihoods and Governance*. London: EarthScan.

Nazarov, Ravshan, and Pulat Shozimov. 2011. "The Ferghana Valley in the Eras of Khurshchev and Brezhnev." In Frederick Starr, ed., *Ferghana Valley: The Heart of Central Asia*. New York: M. E. Sharpe, 140–63.

Obertreis, Julia. 2017. *Imperial Desert Dreams: Cotton Growing and Irrigation in Central Asia, 1860–1991*. Göttingen: V&R Unipress.

Peterson, Maya K. 2019. *Pipe Dreams: Water and Empire in Central Asia's Aral Sea Basin*. Cambridge: Cambridge University Press.

Pietz, David A. 2015. *The Yellow River: The Problem of Water in Modern China*. Cambridge, MA: Harvard University Press.

Reeves, Madeleine. 2011. "Staying Put? Towards a Relational Politics of Mobility at a Time of Migration." *Central Asian Survey* 30(3–4): 555–76.

Roberts, Flora J. 2018. "A Controversial Dam in Stalinist Central Asia: Rivalry and 'Fraternal Cooperation' on the Syr Darya." *Ab Imperio* 2: 117–43.

Ruder, Cynthia A. 1998. *Making History for Stalin: The Story of the Belomor Canal*. Gainesville: University Press of Florida.

Safar, Tohir. 2018. "Sud'ba Farkhadskoi GES: 49 let v interesakh Tadzhikistana i Uzbekistana" [The fate of the Farhad hydroelectric power station: forty-nine years in the affairs of Tajikistan and Uzbekistan]. Radio Ozodi / Radio Free Europe Radio Liberty, April 8. Accessed October 12, 2020. https://rus.ozodi.org/a/29152892.html.

Sodiqov, Alexander. 2013. "From Resettlement to Conflict: Development-Induced Involuntary Displacement and Violence in Tajikistan." In J. Heathershaw and E. Herzog, eds., *The Transformation of Tajikistan: The Sources of Statehood*. London: Routledge, 49–65.

Starr, Frederick. 1984. "Visionary Town Planning during the Cultural Revolution." In Sheila Fitzpatrick, ed., *Cultural Revolution in Russia, 1928–1931*. Bloomington: Indiana University Press, 207–40.

Sultanov, Akylbek et al. 2018. *Sotsial'nye tendentsii Kyrgyzskoi Respubliki 2012–16* [Social trends in the Kyrgyz Republic 2021–16]. Bishkek: Natsional'nyi Statisticheskii Komitet Kyrgyzskoi Respubliki.

Suyarkulova, Mohira. 2014. "Between National Idea and International Conflict: The Roghun HPP as an Anti-colonial Endeavor, Body of the Nation, and National Wealth." In K. Bromber, J. Féaux de la Croix, and K. Lange, eds., "Big Dams: Investigating Their Temporal Politics in Africa, the Middle East and Asia." *Water History* 6: 367–83.

Teichmann, Christian. 2007. Canals, Cotton, and the Limits of Decolonization in Soviet Uzbekistan, 1924–1941." *Central Asian Survey* 26(4): 499–519.

Tursunov, Sanjar. 2013. *Zachem Tadzhikistan zanimaetsia vreditel'stvom* [Why Tajikistan is engaged in sabotage]. Mahalla, April 3.

Rahmon, Emomali. 2010. "January Address to the Nation." In *UPI Energy* 2010, April 27. Accessed May 10, 2021. https://www.upi.com/Energy-News/2010/04/27/Tajikistans-huge-dam-project-hit-by-controversy/70751272376360/.

Wooden, Amanda. 2014. "Kyrgyzstan's Dark Ages: Framing and the 2010 Hydroelectric Revolution." *Central Asian Survey* 33(4): 463–81.

World Bank Group. 2014. *Key Issues for Consideration on the Proposed Rogun Hydropower Project.* Accessed September 5, 2018. http://www.worldbank.org/en/country/tajikistan/brief/final-reports-related-to-the-proposed-rogun-hpp.

World Bank. 2021. World Development Indicators Database. Accessed October 5, 2021. https://databank.worldbank.org/data/download/POP.pdf.

World Commission on Dams. 2000. *Dams and Development: A New Framework for Decision-Making.* London: Earthscan.

Worster, Donald. 1985. *Rivers of Empire: Water, Aridity and the Growth of the American West.* Oxford: Oxford University Press.

Yunakov, Oleg. *Arkhitektor Iosif Karakis: zhizn', tvorchestvo i sud'ba* [Architect Iosif Karakis: life, work, and fate]. New York: Almaz, 2016.

Zeisler-Vralsted, Dorothy. 2014. *Rivers, Memory, and Nation-building: A History of the Volga and Mississippi.* New York: Berghahn Books.

Zhimskii, A.A. 1956. "Perekritie Sir Darii na Kairakkumgesstroe" [Blocking the Syr Darya at the Kairakkum hydroplant site]. *Gidrotekhnicheskie stroitel'stvo,* no. 7.1 (August).

PART III

Imaging New Nature

8

Goats, Axes, and Uncertain Narratives

Representations of Nature and People in Asia Minor

Hande Özkan

In 1925 the Spanish traveler and author Saturnino Ximinez published his observations on Asia Minor in his book *Asia Minor in Ruins*. In his preface to the book, the French archaeologist Bernard Haussoullier describes Asia Minor as a land of "people and conditions, upon which revolutions have little effect" (Ximinez 1926: v). This description, which strips away agency from Asia Minor's human populations, stands in stark contrast to countless descriptions in the historical record that attribute agency to trees. Observing the deserted streets in a small town, Ximinez is mesmerized by the plane trees and the gardens, which made up for the life that had escaped Didim following the war that led to Turkey's independence and the demographic transformation caused by the departure of local Greeks. "The vegetable life went on in the midst of death; the myrtles and oleanders brightened the little gardens; bunches of grapes hung from the trellises; the fig trees yielded their fruit" (Ximinez 1926: 246). Recalling Edward Said's statement that "the Orient was almost a European invention, and had been since antiquity a place of romance, exotic beings, haunting memories and landscapes" (Said 1978: 9), when descriptions like Hassoullier's that question the agency of human populations coexist with anthropomorphized representations of nature, the result is an Orientalist portrayal of Asia Minor as a whole. Narratives that emphasize humans' destruction of natural beauties reinforce western travelers' discourse that depicts the region's human populations in natural terms. These comments would not be so ironic were it not for the fact that 1923 was the culmination of a decade of wars that marked the dissolution of the Ottoman Empire and concluded with the establishment of the Republic of Turkey. It seems that "the people and conditions, upon which revolutions have little effect" had produced the last nation-state to emerge out of the Ottoman Empire (Ximinez 1926: v).

This chapter has two goals. First, I will argue that travelers' discourse that dehumanized Asia Minor's populations, while simultaneously anthropomorphizing nature—specifically forests—also reflects the modernist logic under which Republican Turkey's political and scientific elites operated. Binary categories, which lie at the core of colonial thought, are, as I will show, as prevalent in Turkish intellectuals' thinking as they are in that of westerners. Second, I will also show that neither travelers' nor Turkish elites' representations of Asia Minor's human and arboreal populations are monolithic; instead, both narratives have their inherent challenges. By investigating the uncertainties of colonial discourse and its national derivatives with special attention to the human–nature nexus, I hope to contribute to discussions on Orientalism, modernity, and nationalism.

Renewed interest in deconstructing modernity brought in its wake a new fascination with anthropological approaches to documents. This new scholarship approaches documents as processes and defines them as spaces of encounter, entanglement, and uncertainty. Erik Mueggler shows that place-making and the construction of nature were not only imperial projects in an administrative sense but also textual processes, and he argues that different ways of documenting have produced different social relations (Mueggler 2005). Likewise, David Barton and Uta Papen state that "examining written texts is essential for understanding how societies operate and are organized, how institutions communicate with the public, how work is being done, how individuals and social groups organize their lives and make sense of their experiences and how cultures in all their variations are produced and reproduced" (Barton and Papen 2011: 5). In his overview of the burgeoning literature on the anthropology of documents, Matthew Hull also emphasizes the generative capacity of documents, making their ethnographic deconstruction even more compelling (Hull 2012).

In this chapter, I use the term "archival record" to refer to a variety of documents that contributed to processes of knowledge production in the Ottoman Empire and modern Turkey. While the Ottoman Empire was never under colonial rule, Europeans and Americans, nonetheless, perceived Ottoman territories and peoples through a colonial lens, categorizing them based on binaries such as civilized versus uncivilized. Looking at how this colonial logic applies to the Balkans, Maria Todorova argues that the east–west dichotomy ingrained in scholarship has interpreted not only postcolonial parts of the world, but even Eastern Europe through a colonial lens (Todorova 2005). By presenting Balkanism not as an example of Orientalism, but as a discourse on its own, Todorova shows how essentialist binaries and their tendency to construct the "other" define grand narratives such as nationalism or modernity (Todorova 2009). Likewise, Milica Bakic-Hayden illustrates the layered multiplicity of spatial and temporal othering in her work on "nesting Orientalism" (Bakic-Hayden 1995), and Ussama Makdisi presents a similar argument regarding Ottomans' internal Orientalism (Makdisi 2002). In this chapter, I adopt a similar approach to trace modern discourses of othering that have constituted representations of Asia Minor and Turkey. I examine a variety of documents ranging from travel accounts to journalists' writings to analyze the manifestation of binary categories while paying attention to the uncertainties that accompanied the construction of nature and subjects in Asia Minor through these binaries.

LOOKING AT ASIA MINOR

Travelers' accounts are one medium in the archival record where the juxtaposition of anthropomorphized representations of Asia Minor's natural landscapes with naturalized depictions of its people prevails.[1] On the other hand, writings of scientific experts in the nineteenth and twentieth centuries challenge the singularity of the colonial gaze and draw attention to the complexity and fluidity of social categories. While travelers' accounts are emblematic of colonial discourse, experts' texts challenge the inherent Orientalism of this discourse by illustrating its contradictions and irregularities. My goal in this chapter is to draw attention to the fragmented nature of the colonial gaze in order to better challenge it and "to use humanistic critique to open up the fields of struggle, to introduce a longer sequence of thought and analysis to replace the short bursts of polemical, thought-stopping fury that so imprison us" (Said 2003). As Said eloquently stated, "humanism is sustained by a sense of community with other interpreters and other societies and periods: strictly speaking therefore, there is no such thing as an isolated humanist" (2003). Therefore, such an endeavor to understand contradictions and ambivalences is even more important when considered against a background of cultural borrowing between nationalist modernity and colonial logic.

Previous scholarship on colonial modernity has relied on an unquestioned faith in totalizing narratives. More recently, a tendency to reveal colonialism's contradictions and ambivalences has replaced this earlier approach. Homi Bhabha was one of the first to make such a call when he suggested exploring the ambivalences of colonialism's mimicry (Bhabha 1984), which was later accompanied by an emphasis on the heterogeneity of colonial discourse itself. As Frederick Cooper and Ann Laura Stoler have shown, "colonial regimes were neither monolithic nor omnipotent" (1989: 609). Likewise, Nicholas Thomas has criticized the representation of the logic of colonialism "in unitary and essentialist terms" and made a plea for "an understanding of a pluralized field of colonial narratives, which are seen less as signs than as practices, or as signifying practices rather than elements of a code" (Thomas 1994: 3). Thomas's critique is especially important for highlighting the complexities of colonialism as a process and challenging its previous interpretation as a uniform, fixed narrative. Stoler's later work contributes to these discussions through an ethnographic analysis of archives that challenges their perception as repositories of knowledge and redefines them as sites of knowledge production. She places special emphasis on the contradictions that are present in official records, thereby illustrating the uncertainty that marked colonialisms' processes of construction and classification. "Grids of intelligibility were fashioned from uncertain knowledge; disquiet and anxieties registered the uncommon sense of events and things; epistemic uncertainties repeatedly unsettled the imperial conceit that all was in order" (Stoler 2002: 91). My goal in this chapter is to analyze the archival record to defragment the modern narratives that have constituted nature and people in Turkey and Asia Minor. My use of a wide range of records ranging from official to popular texts is deliberate and aims to show the differences and conflicts that permeate these artifacts. I am as interested in *how* these records represent as in what they reveal.

NATURALIZED PEOPLE, ANTHROPOMORPHIZED TREES

This chapter opened with the preface Hassoullier wrote for Ximinez's travelogue *Asia Minor in Ruins*. In the text, Ximinez builds on this preface with descriptions of Asia Minor's populations as timeless creatures whose acts derive from habit and instinct rather than intentionality and intelligence. In Bursa, encountering a crowd that has gathered to meet Mustafa Kemal, he observes: "Everybody was in the street in response to the announcement that Mustapha Kemal was coming; but this hope was thwarted. . . . The silence of such a crowd is also striking. They watch attentively and seem to be easily satisfied. Mustapha Kemal did not appear, but what matter? The crowd amused themselves just the same without relaxing their usual impassivity" (Ximinez 1926: 31). Like a pet waiting for a treat, but contending with leftovers, the residents of Bursa perform, in this description, like a flock of animals stripped of human agency. The ruins in the travelogue's title, which refer both to the relics of Greek antiquity and the desolation, apathy, and fatalism that has become stronger in the wake of a decade of wars, provided the context for such a description. "The present is death," Ximinez writes, "Asia is merely a vast necropolis; but above the crumbled temples and buried civilizations the soul of antiquity hovers imperishable" (Ximinez 1926: 269).

One implication that characterizes narratives like the one above is the imaginary of empty lands, which has served as a justification for colonial projects of displacement as well as the actions of modernizing nation-states. The result is the textual effacement of populations in time and space. Travelers' obsession with Asia Minor's ruins had as much to do with history as with the present; their infatuation with the Hellenic past was engendered by and reinforced their disdain for the backward, uncivilized crowds they constructed in their texts.[2] A second implication in such representations is the dehumanization of populations, resulting in the erasure of their subjecthood. Recalling discussions on social death, the colonial gaze leaves the human animal intact, but obliterates the human subject.[3] In the process, a rigid nature–culture binary is reaffirmed, anchoring the West in its position as the cradle of civilization.

Mark Twain's first impressions of Constantinople reproduce this Orientalist representation via grotesque depictions of the population: "Ashore, it was—well, it was an eternal circus. People were thicker than bees, in those narrow streets, and the men were dressed in all the outrageous, outlandish, idolatrous, extravagant, thunder-and-lightning costumes that ever a tailor with the delirium tremens and seven devils could conceive of. . . . The shops here are mere coops, mere boxes, bath-rooms, closets—anything you please to call them—on the first floor. The Turks sit cross-legged in them, and work and trade and smoke long pipes, and smell like—like Turks" (Twain 1869: 169–70). Another American, Walter Hawley, reproduces this discourse by classifying Asia Minor as wilderness. Joyous to visit a hospital built by American missionaries, Hawley writes:

> One of my pleasantest recollections of Konia is associated with a visit to the hospital of the American Medical Missionaries, situated a short distance from the Konak on a street through which the tramway passes. I remember well noticing for the first time the words "American Pharmacy" in large letters above the door of a modern building with glass front, for it impressed me strangely to meet something claiming to be distinctly American in the heart of Asia Minor, and to be told that in the building, separated from it by a large gateway, I would meet resident surgeons who had been born beneath the Stars and Stripes, but for humanity were passing their lives in almost a wilderness. (Hawley 1918: 227–28)

The wilderness Hawley refers to here differs from the distinctly American nineteenth-century construct that provided the middle class with a refuge from the perceived ills of modernity (Cronon 1996). The latter was what Americans have since then wanted to flee *to*, whereas the wilderness Hawley associates with Anatolia is what he wants to flee *from*. It is a type of wilderness that identifies the unrefined qualities of the people rather than the land, making the American hospital emerge like a safe haven.

Other representations that naturalize Asia Minor's populations are less subtle. Describing Eastern Turkey and its Kurdish population, Emil Lengyel, a Hungarian American, draws the reader's attention to the similarities between Kurds, goats, and swallows: "Goats find no trouble climbing the rocks and schist in the wild country of the Turkish East. Kurds find it easy too, but their pursuers look upward aghast. Hanging precariously to the mountainsides, Kurdish villages look like swallows' nests. They are made of mud and of odds and ends, and can be deserted without much regret" (Lengyel 1941: 24). Lengyel continues with a lengthy description of the peasant, whose apathy and lethargy become apparent when he writes that the peasant "vegetates rather than lives." His erasure of the population's subjecthood is even more evident when he discusses language: "The fact, for instance, that the Turk has no words of his own for 'because,' 'if,' 'when,' and 'where' deserves notice. It shows that the Turk has not an abstract mind. . . . But few languages can match Turkish in the wealth of graphic expressions. They are beautiful and as musical when spoken as when they are sung. . . . The syllables have a cadence, acquired around the campfire on the steppes. The shepherd had ample time to polish his words until they harmonized with the mysterious whispers of nature" (Lengyel 1941: 47–48).

In a different account, J. V. C. Smith emphasizes the instinctual, emotional impulses of the people, recalling Said's points on Orientalist depictions of human nature: "The power of concealing their real feelings, either of hatred or approval, is even more adroitly managed by them than by the American Indians. When their passions are let loose, like tigers from a cage, nothing short of absolute physical exhaustion subdues their intensity. When the causes that roused them to madness, either love or revenge, have passed away, they become docile, bland and mild, as the moonbeams" (Smith 1857: 29). In *The Conquest of America*, Tzvetan Todorov discusses Columbus's failure to recognize native populations as different but equal subjects (Todorov 1984). Smith's failure to recognize and accept the difference of Anatolian peasants results in a similar discourse that places them in the realm of nature and not culture, of instinct and not cognition: "Turks eat, drink, smoke and sleep; —whether they dream is uncertain" (Smith 1857: 139). Even dreaming acquires a form of abstract rationality that Turks, according to Smith, are not subject enough to possess.

Compared with the narratives that naturalize humans, the anthropomorphizing discourse of Europeans and Americans, which imbues nature with human qualities, is more complex and subtle. Natural landscapes, more specifically trees, acquire human characteristics indirectly and often due to the failures of humans. In many accounts, trees compensate for the insufficiencies and failures of human subjects, who cannot adequately respond to the demands of a modern world. One trope that makes up this discourse is the wet–dry binary, which associates aridity with barbarism and verdure with civilization.[4] Water plays a fundamental role in this binary as an embodiment of the ethos of modernity and acquires a transformational role due to its potential to cleanse the abject (Douglas 1991 [1966]). Where water, the magic ingredient that takes phenomena from the realm of the abject into social order, is absent, dirt presides, leading Lengyel and others to associate the Orient with abject pollution while emphasizing hygiene as a feature of the Occident throughout their accounts (1941). Therefore, trees, whose survival relies on water, belong in the realm of the modern, connoting the victory of civilization over a state of nature. In turn, in this representation trees acquire a type of agency that imbues them with human qualities.

This binary approach consequently creates a paradoxical representation of trees in Asia Minor. On the one hand, the vegetation of fertile parts of Asia Minor is revered because it is believed to compensate for the backwardness and senselessness of human behavior. However, at times, travelers' fascination with Asia Minor's forests is shadowed by their distaste for the extreme wilderness of the forest landscape. An example is the Scottish traveler Charles Macfarlane's dissatisfaction with the untamed, uncultivated forests in northwestern Turkey: "Plane trees, hard oaks, evergreen ilices, firs, maples, mountain ashes, elms, beeches, chestnut-trees, were growing all together, in a way I never saw before; but all were neglected and left to themselves, and growing so thickly that none could attain to their full dimensions. With a little clearing—with a slight employment of the woodman's craft—here might be some of the finest timber in the world!" (Macfarlane1850: 442). Evidently, the wilderness, assumed to prevent trees from meeting their civilizational potential, is not an inherent feature of the trees themselves; rather, the subtext of Macfarlane's critique is human failure to manage and cultivate forests based on scientific methods. In other words, Macfarlane's commentary illustrates the meaning he constructs of trees, which he expects to see not as emblems of nature, but articulations of culture.

The paradoxical disdain for Asia Minor's trees is also due to their association with Islam, often presented as the reason for the apathy of the Orient. Trees, especially cypresses, are typical of cemeteries in Anatolia, yet for the European and American travelers, the species in cemeteries appear inferior to cultivated forms of nature. One can argue that their construct of nature as regulated by civilization is not compatible with Islam, resulting in the representation of these trees as distorted and incomplete versions of nature. Highlighting the dreariness of these trees, Smith refers to a major cemetery in Istanbul as a primitive forest (1857: 160). In an ironic twist, in Bursa, Hawley prefers a cemetery with fewer trees to the darkness of the forest-like cemeteries. "The garden before it contains large numbers of small bushes with bright flowers arranged carefully in little beds, in which there is a total absence of dense trees, so that the dismal gloom of the usual Turkish cemeteries is lacking" (1918: 152).

Trees' anthropomorphization comes across more clearly in Ximinez's distinction between Muslim and Christian olive trees. Later, observing Greek ruins in an area recently depopulated due to wars, he sees death, decay, and desolation: "Only the plane trees still offer the shade of their boughs.... The vegetable life went on in the midst of death" (Ximinez 1926: 246). In this case, trees offer life and fill in for the missing human inhabitants of the region. Hawley is even more obvious in his personification of trees when he writes that with the coming of spring, "even the tall cypresses seemed to have forgotten their sadness as they put forth fresh leaves" (Hawley 1918: 26). Later in the text, he is in awe of the graceful willow trees (117). Through a description that denies any human agency to "Oriental" women, Smith represents them like a pack of animals, who huddle and take refuge behind the generous cypresses: "A few females were scattered in groups between the tall tomb-stones of the vast cemetery—the largest, without doubt, in any country of modern times; but they invariably drew down their veils as [we] neared them, and wandered off among the trunks of the tall, sombre cypresses, quite beyond the ken of vision" (Smith 1857: 210).

AMBIVALENCES OF THE EURO-AMERICAN GAZE

Travelers were not the only foreigners who observed Asia Minor's human and natural landscapes. Professional foresters invited by the Ottoman bureaucracy provide accounts that differ from travelers' representations with their attention to nuances and complexity. The nineteenth century ushered in a series of modernizing reforms initiated by the Ottoman Empire ranging from the military to the management of natural resources. After centuries of unregulated forest use by local populations, followed by forests' commercial exploitation by foreign companies, in the late nineteenth century, Ottoman rulers embarked on creating a system of scientific forestry which served as the foundation of modern Turkey's forestry regime.[5] To facilitate these reforms, Ottomans started inviting European foresters, which generated a culture of intellectual sharing between these experts and Ottoman technocrats. Some of these European foresters occupied significant positions in the Ottoman and later Turkish bureaucracy (Özdönmez and Ekizoğlu

1993: 57–68). As a result, while travelers' accounts built on the exclusive voice of the Orientalizing other, experts' discourse was formed in such instances of contact, in cooperation with Ottoman and later Turkish elites. Travelers' essentialist representations can be explained by their detachment from what they observed, whereas experts occupied a liminal position due to their involvement in government and were, therefore, more attuned to ethnographic detail and complexity. Their affective engagement in the places and with the people they served meant that their discourse was constructed through contact, encounter, and not just observation. Unlike the travelers' partial representations, experts' constructions of landscapes and populations implied shared experiences and relied on dialogic forms of knowledge production.

In terms of the attribution of human qualities, foresters' representations of trees resemble those of travelers at first glance. Moreover, a punitive logic that criminalizes rural populations accompanies their anthropomorphized representations. For example, the French forester Louis Bricogne, who was part of a French initiative to provide assistance to forestry initiatives in the late Ottoman Empire, laments over trees mutilated at the hands of peasants who, he claims, almost enjoyed their acts of savagery: "Fires, unlimited grazing, clearings and mutilations of all kinds sadly denote the barbarism of the population in these places" (Bricogne 1877: 374). However, upon closer reading it becomes clear that Bricogne's discourse differs remarkably from the totalizing rhetoric adopted by travelers. For example, a critique of the Ottoman states' resettlement policies follows his comments that blame refugees for the destruction of forests (Bricogne 1877: 373). Likewise, in a discussion of forest degradation in eastern Asia Minor, he juxtaposes peasants' destructive practices with the state's failure to educate them on conservation. Later, almost in defiance of countless travelers' fearful accounts of Asia Minor populations' disdain for foreigners, he writes, "We need to add that even if the peasants almost always appear defiant and unimpressed, they showed no proof of any hostility towards us and we were able to travel among them fully secure (Bricogne 1876: 364)."

Bricogne offers a similarly nuanced critique of nomadic tribes' destructive effects on northwestern Turkey's forests, when he draws attention to the bigger role played by the Ottoman military: "It should be added that the arsenal and the artillery exploit the best oaks and the most beautiful pines" (Bricogne 1877: 420). Perhaps to safeguard himself in light of this accusation raised vis-à-vis the state, his employer, Bricogne then emphasizes that recent actions taken by the state have improved peasants' practices; even nomadic populations, who could not be removed from forested areas, he explains, have reduced their exploitation of forests (425). Finally, after a brief comment on the remarkable reduction in forest fires, he states how all these developments have allowed for further commercialization of these forests, implicitly alluding to the state's immense power over the use of forests.

The complexity of foresters' discourse manifests itself even more clearly in Bricogne's global approach to deforestation. While it is his job to be critical of forest degradation in the Ottoman territories and to point to rural communities' assumed complicity in this process, Bricogne is cognizant of similar processes that have plagued forests in Europe and elsewhere. This balanced view differs greatly from the contrast drawn between the West and the Ottoman Empire that prevails in travel accounts. The French forester D. de Montrichard's discussion of goats' destructive presence in forests offers a similar balanced critique. While accepting that grazing is a threat to the already degraded forests around Ankara and arguing that their number needs to be reduced, Montrichard states that goats' social significance for the rural poor as a source of income cannot be overlooked in any policy change (Montrichard 1875). Likewise Montrichard's brief reflection on the French forestry mission in Turkey provides a thoughtful comment on the need for culturally relevant work when he highlights the necessity of learning another country's traditions (Montrichard 1870).

Hermann Wenzel's article on the vegetation of Sultan Daği, a mountain in southwestern Turkey, published in *Orman ve Av* (Forestry and hunting), Turkey's leading forestry journal is another good example. After a lengthy technical discussion, Wenzel portrays the Yörük, a nomadic population, in a negative light when he blames them for the destruction of the forests in the area. Yet, when he mentions the Tahtacı, a specific ethnic nomadic group who are known for their expertise in forest labor, his account acquires a more moderate tone: "They mostly fell old trees and leave the younger ones alone" (Wenzel 1933: 6). In another article published in the same journal, a German professor of agriculture, Herr Krause, adopts an unbiased tone when he alludes to deforestation along the Black Sea coast: "It is likely that forests on the Black Sea coast spread over a wider territory in the past. This has decreased with human influence over time. Actual forests start above the 500 meter line" (Krause 1934: 29). Krause's tone is almost deliberately neutral; he does not directly reference rural communities, nor does he delineate a timeline for the decrease in forest coverage. In fact, his tone resembles countless accounts of anthropogenic deforestation in other parts of the world—including Europe—obscuring the east–west binary in the representation of Asia Minor's natural landscape.

Perhaps the most generalized example of this nuanced approach comes from a German forester, Franz Heske, who worked in the Turkish bureaucracy and highlighted that local demands for land must be considered as a variable in any forestry policy. It is with this understanding that he designed ways of altering social and economic conditions for the rural population, which would in turn decrease their demand for forest products. Heske's warning that restrictive legislation alone will not protect forests is noteworthy, especially given the comparisons he draws between Turkey and examples from western Europe (Heske 1952).

IMAGINING THE NATION WITH THE GOAT AND THE AXE

As I have discussed elsewhere, what modernist nationalism attempted to accomplish in Republican Turkey was not categorically different from the projects launched by colonial administrations in various parts of the world because there are great epistemological and ideological similarities between nationalism and colonialism as projects for imagining, understanding, and governing people and places (Özkan 2018). Therefore, theoretical discussions on the foundations and heterogeneity of colonial discourse are easy to apply when analyzing nationalist modernity, which, like colonialism, is rife with its own uncertainties and ambivalences. My argument is not that Turkish nationalists were simply imitating colonial logic in the early twentieth century. Instead, in conversation with scholars such as Maria Todorova, Arun Agrawal, and K. Sivaramakrishnan (Agrawal and Sivaramakrishnan 2003), I consider the similarities in their worldviews and social categorizations as coevally produced and shared realities.

Forestry was one of the building blocks of state making and nation building in Turkey following the founding of the republic in 1923 (Özkan 2018). Forests' significance was much more than economic; they were perceived and represented through a discourse of verdure, which provided a framework to reterritorialize the nation within the context of modernity. Given Lengyel's observation that "brown is the dominant colour of Turkey," attempts to "green" the country were dialogically constructed and in constant conversation with the Orientalist gaze at eastern landscapes (1941: 54). As Eunice Blavascunas's discussion of Polish forestry (chapter 9 in this volume) illustrates, forestry has served as a nationalist project in many countries and its identification with modernity has served as a defense against outside interventions. In fact, as Stefan Dorondel and Stelu Şerban state in the introduction, this edited volume is intended to fill a gap in the existing literature by highlighting the connections between nature making and nationalism in Eastern Europe. Likewise, relying on the binary, which positioned verdure as the categorical opposite of aridity, Republican elites presented "green" as the color of civilization. As I explained above, this Republican ethos was in conversation with Ottoman modernization efforts as well as the Euro-American colonial discourse of the previous two centuries. Consequently, Republican elites' representation of Anatolia's landscapes and peoples often echoes Orientalist narratives by stripping rural populations of human agency. The only agency attributed to peasants focuses on their destructive practices, but as I will discuss below, this is not agency proper as it dehumanizes rural populations. In line with examples of modern forestry from around the world, the redefinition of forests as national patrimony went hand in hand with legislation, which abruptly redefined everyday practices as criminal acts. Criminalization and a heavy-handed forestry regime were natural outcomes of the way elites perceived peasants; consequently, the "goat" and the "axe" were major tropes through which they represented the conflict-ridden relationship between the state and rural populations. Yulian Konstantinov offers a similar discussion (chapter 11 in this volume) by analyzing the ways in which the jackal represents the anxieties generated by postsocialism in Bulgaria. As in countless examples from around the world, in Turkey the goat serves as a metonym for peasants' assumed primitiveness and ignorance, and the axe illustrates their destructiveness. Images of goats and axes that resemble mug shots fill the pages of texts ranging from popular science books to newspapers, against a backdrop in which forests represent the nation itself. Through the goat and the axe, rural populations are represented as backward and uncivilized and their practices are deemed as an assault on the nation itself. The anthropomorphization of forests is evident in the colloquialism that forests are "the fatherland's lungs." Other times, trees are feminized and defined as victims of men's heinous crimes: "We need to instill in the spirit of our society [the idea] that a tree does not need less time than a man to grow, and that killing trees that have survived hundreds of years is not different from killing men or disfiguring a beautiful face by throwing nitric acid on it" (Karacan1935: 11). This quote illustrates that in the early Republican era deforestation and desertification were considered not only as the destruction of natural resources but also as a violation of the body politic, while afforestation was conceptualized as a therapeutic method for healing it.[6] This association between the nation and forests is evidence that the type of nature Republican foresters envisioned was an articulation; nature was culturally constructed and represented through human qualities.

Alongside technopolitical analyses of grazing and forest fires (allegedly started by peasants to generate better grass for goats' consumption), goats appear abundantly in sayings and proverbs as well as visually in print media in the early twentieth century. As Lengyel observed, "the goats are everywhere. Their droppings are a feature of the Turkish countryside and town" (Lengyel 1941: 78). Some talk about goats as pests, infecting forests; others define them as little devils (Sungar 1946: 37). The politicoscientific community's hatred toward goats was so great that in Adana, goats were banned not only in forests but in farmlands as well.[7] A 1933 report on the activities and accomplishments of the Italian Forest Militia commends Italian foresters for regulating goats' access to forests. Although Turkish intellectuals and politicians avoided direct identification with fascism and defined the Turkish political model as a third way, it was common to read their enthusiastic admiration of fascist regimes' achievements. In 1944, Falih Rıfkı Atay, a leading journalist, laments the state's struggle to enforce forest regulations and discipline rural populations in an opinion piece in which he commends fascist regimes for successfully implementing top-down forestry practices. After praising Mussolini for "greening" Italy, Atay concludes his argument with an example from Nazi Germany, quoting Hitler on the need for state involvement in forest management (Atay 1944). In another example, a veteran forester describes goats as "those infidels" when he writes, "no good comes from a forest where the goat enters" (Sungar 1946: 18). Relying on references to Islam and purity, he excludes goats and, therefore, peasants from the realm of acceptable citizenship. Another forester refers to goats as "traitors" (37), reminding readers of the association between the nation and forests, and emphasizing betrayal by undesirable subjects such as ignorant rural populations who are not fit to be good citizens. In a critique of nature reserves in Australia, Suvendrini Perera and Joseph Pugliese illustrate how the criminalization of the buffalo served as a symbol of colonial distaste for indigenous communities (1998: 88). The image of the goat served as a similar discursive strategy for Turkish elites and reflected their disdain toward peasants. As the undesirable, abject face of the peasantry, goats were literal scapegoats and reminders of the precariousness of modern subjecthood, which elites saw as their monopoly.

Alongside the discursive implications of the goat, the axe represented rural populations' greed for forest products and their disregard for conservation. As a primitive tool (itself partially procured from the forest), the axe represented a threat to the state's scientific forestry policies and thus needed to be removed from the forest: "Let us put down the axe" (Karacan 35: 11). This threat was conveyed through a clinical discourse that imbued trees with human qualities in an attempt to evoke a visceral reaction from those citizens who qualified as modern subjects. To convey the pain inflicted upon the very essence of the modern nation, elites also resorted to the affect induced by sensorial representations. Consequently, the sound of an axe cutting into the flesh of a tree emerged as another common trope. Recounting his experience of first hearing, then watching an old tree fall to the ground after being continuously struck by an axe, one author writes: "I realized once more and with immense pain how late we were to save our beautiful forests from the axe microbe and how slowly we were still acting" (Gökmen1935: 15). The clinical gaze and the depiction of felling as pathology further legitimize the disdain for traditional rural practices, while reinforcing scientific forestry's role in healing the natural landscape and the body politic. The forester who defined goats as "infidels" was also concerned about the axe when he wrote that forests' biggest enemies are "the cruel axes of peasants and the poisonous teeth of goats" (Sungar 1946: 18).

Even though the destruction of forests is attributed to Anatolia's rural population, in all these accounts their agency is discursively removed from the human realm and ascribed to an auxiliary such as the axe. As a result, rural populations emerge as primitive beings who act, but lack the agency of a modern subject. Meanwhile, trees are defined through a deliberative rationality. Traveling from Ankara to the Black Sea coast, the companion of a journalist describes the forest's voice using anthropomorphized references: "The groaning that you hear is the sound of the forest. We do not notice it in the bus, but clearly, there is a slight wind outside. But do not be fooled by this quiet groan because it is the wind that tunes the sound of the forest. Now it is like a sweet lullaby. But sometimes it almost sounds like a scream and sometimes it is horrible sounds resembling thousands of monsters gathering and howling" (Sungar 1946: 39). In a section titled "The Forest's Revenge," the journalist who is the author of this travel account narrates a criminal investigation into the death of a peasant in a small town in northwestern Turkey. Sungar writes that the local prosecutor explained to them—with a smile on his face—how a peasant who was illegally felling trees with his son-in-law died on the spot when a branch fell on him. If the smile on the prosecutor's face is not chilling enough, Sungar's reaction to the incident is enough to send shivers down the reader's spine: "Hearing about any case of death evokes sorrow. However, I have to admit that listening to this incident, I am overwhelmed by a strange and indescribable feeling that is akin to joy, and in that moment I can conjure the scary image of a victim who, even when his head is cut off succeeds, in his last moments, in punishing his murderer with a stab of the knife" (Sungar 1946: 117–18). His disregard for a man's death juxtaposed with the pain he feels for a tree is a token of the discursive mechanisms through which humanity was stripped from rural populations and attributed to trees. Even in death, the peasant was not worthy of modern subjecthood. Instead, agency in the form of revenge was attributed to the tree.

Given the association between nationalism, civilization, and forests, goats and axes were obstacles in Republican Turkey's green civilizing mission. Goats and axes could change the color of civilization from green to *brown*, reversing the process of modernization. In this nationalist imaginary, while trees were anthropomorphized and imagined as the nation's body, rural communities, imagined as extensions of the goat and axe, were represented as pests to be exterminated or tumors to be excised. This is why a prominent forester could make such a bold yet indirect claim about sacrificing peasants to protect forests: "I am of the opinion that we should not refrain from any sacrifice in the conservation of our forests, whose future is in jeopardy. If peasants' animals cannot survive unless they graze in those forests, then we should risk their death" (Ali 1931: 15). While the deaths, which he sanctions here, were of the animals, the question that begs to be asked is: Would the peasants survive without their animals?

AMBIVALENCES OF NATIONAL FORESTRY POLICIES

The discourse, which represents rural populations as nonagentive subjects, while paradoxically criminalizing their place-based practices is not uniform in the archival record. This naturalization is often accompanied by a rhetoric that is more cognizant of the complexities and nuances of forest-dependent communities' engagement with the landscape. An important dimension of early Republican foresters' ethos was a detailed emphasis on traveling, observing, and cataloging. This was of course necessary to make the nation's forests legible (Scott 1998), but their narratives reveal a passion for categorizing and cataloging that is far less utilitarian and much more affective. At times, the attention to detail presents itself as borderline obsessive, while at other times their narratives appear as if they are about a beloved rather than a profession based on technical expertise. This passion for detailed exploration whose ultimate purpose was the rational and scientific application of expertise was shared with European scientists who had left their imprint on Ottoman and Turkish forestry since the nineteenth century. Their continued influence is evident in the pages of *Orman ve Av* as well as other forestry-related publications. Scrolling through the periodical it is hard to find an issue that does not have at least one translated article written by a European forester. Moreover, references to Turkish forestry students traveling to Europe for long-term education purposes or brief research trips are not random, but regularly feature in the community news section of the periodical. While this expert gaze at national territories identifies a Foucauldian sense of power and discipline over the national territory (Özkan 2002), a deeper affective engagement with landscapes permeates their accounts, and the ambivalences of their modernist logic should be understood in the context of this affective engagement.

A report by M. Ali Salih, a forester whose writings featured frequently in *Orman ve Av*, provides a good example of this complexity. Describing the forests in his administrative district, Salih initially reproduces the Orientalist narrative, structured around the binary of the all-knowing expert and the ignorant peasants. However, following an ethnographically rich discussion of peasants' crimes against forests, Salih argues that forest communities should be in charge of managing and conserving specific sections within forests. Given the hegemonic presence of state ownership and management of forests throughout modern Turkish history , Salih's call for community forestry is a significant detour from the modernist logic of the era, and fragments its totalizing discourse (Salih 1933). Likewise, discussing the forests in the Black Sea region, Esat Muhlis provides a balanced view when he talks about the reduction in forest coverage. Muhlis's article provides a perfect example of the continued collaboration between European experts and Turkish foresters because it was based on his observations during an excursion organized on behalf of the British and European Timber Trust, which had been contracted to manage the forests in the western Black Sea region. After the usual depiction of peasants as the culprits of forest degradation over centuries, Muhlis shifts the focus from the human element to natural processes and explains that the absence of trees above a certain elevation is a result of inadequate humidity and not peasants' practices (Muhlis 1933).

Disagreements over goats' role in forest degradation also disrupt the uniformity of the discourse that dehumanizes and criminalizes rural communities. Existing criticism within the Republican People's Party, which ruled through a single-party regime in the 1920s and 1930s, escalated into a bigger political debate when the Democrat Party emerged as the main opposition in 1945, resulting in an ideological war between the military-bureaucratic elite in power and the provincially backed opposition. To counter the opposition's efforts to use goats as an advantage in their populist political agenda, the government relaxed the restrictions on goats.[8] Nevertheless, attempts to override the "goat ban" played a big role in the Democrats' election campaign as well as the new Forest Act that went into effect in 1956.[9] The journalist Ahmet Künksatan adopts a broader perspective on the goat issue and argues that the conflict, which had so far depended on a forest-or-goat logic, must be reconsidered through a forest-or-peasant logic. Recognizing that the widespread presence of goats should be understood within the context of poverty that defined the social fabric of mountainous forest villages, he writes, "The goat is the poor peasant's cow. What destroys the forest is not the goat. It is the tribes of goats, raised in excessive numbers for commercial reasons" (Künksatan 1949: 40). Half a century before James Ferguson discussed development agencies' misinterpretation of the social significance of cattle in rural Lesotho (Ferguson 1994), Künksatan addressed the "goat mystique" and illustrated that goats served as a safety net for peasants despite their representation as an irrational obsession.[10] After reiterating the scientific and economic consequences of forest clearings, he writes, "Those who clear fields in the forest should be pitied as much as the forest itself" (Künksatan 1949: 28) and discusses the destitution that is at the root of such criminal practices. The solution, he argues, is not to save the forest at the expense of the people through legal restrictions that criminalize everyday practices; instead, he demands measures that will protect both the forest and the peasantry.

UNCERTAIN NARRATIVES IN THE TWENTY-FIRST CENTURY

In this chapter, I analyzed colonial and nationalist discourses on Asia Minor's populations and natural landscapes to illustrate their inherent ambivalences and to underline the uncertainties of modern thought that is at the center of both of these discourses. In line with Walter Benjamin's motto that the goal of historical analysis is to investigate the past at a current moment of danger (2019: 198), understanding the fragmented nature of modernist thinking in the past can allow us to make sense of our contemporary struggles with social categories. The tropes and social idioms discussed in this chapter do not lie dormant in archival texts. As scholars of postcoloniality have discussed, the legacies of colonial thought resonate in contemporary social life.

For example, while of less significance today, the axe is still a metonym for criminal peasant practices in Turkish forestry circles. In 2008, in the early stages of my ethnographic research, I accompanied a forest engineer from a local forestry agency in the western Black Sea region to the forest district he was tasked with managing. As we drove in the forest, we came across six forest villagers who had been working all day.[11] As we chatted about their forest work, our conversation took a different turn when my companion started asking questions about illegal felling. Since he supervised forest production in this region, he was in regular contact with these villagers, who were used to this line of questioning, whereas I was initially startled at the display of power on the part of my companion. However, my discomfort was short-lived as I soon noticed that what I perceived as questioning was almost bantering, a ritual that both parties engaged in often to negotiate power. My presence further encouraged my companion's desire to perform his role as the good scientist and statesman. These encounters often have a theatrical element and a script, with expected questions and expected responses. However, something changed when my companion asked these men where they got the wood for their axes. This question was not part of the script; perhaps he posed it to teach me something about power relations in the forest. After initially fumbling for words, the villagers regrouped and explained that they use wood from sick, dying trees or leftovers from timber production. It was obvious to everyone, myself included, that no one would necessarily fell a tree to make an axe. What stood out to me was how the axe served as a code for traditional practices and their clash with the forestry regime of the modern state.

It would have been easy to interpret this encounter as evidence of the state's strict modernist disregard for local complexities, leading to an analysis of the strong state that completely silences voices from below. Yet during my fieldwork, I was also confronted with displays of statehood that were sensitive to local needs and cognizant of the discrepancy between the law as a text and its actual implementation. Many foresters, especially seasoned forest guards, demonstrated an understanding of the constructedness of legislation, refraining from strict applications. Likewise, forest villagers categorized state foresters based on how by the book they were, underlining the performative aspect of the encounter narrated above and serving as a warning that it is not the only narrative about modern Turkish forestry. Thus, studying the heterogeneity and ambivalences of the discourse of modernity can inform ethnographies of the modern and equip us to defragment our own narratives.

NOTES

An initial version of this chapter was presented at the 2015 International Conference of Historical Geographers in England thanks to the generosity of the Kenan/Jones funds at Transylvania University. A more recent version of this chapter was presented at the Environment and Society in the Middle East writing workshop at Middlebury College in 2017. I would like to thank Robert Greely for organizing this event and for providing feedback. I would also like to thank Bridget Guarasci and Brittany Cook as well as the students present at the event for their valuable comments. Finally, a special thanks to Christopher Zollo for his personal and academic support.

1. In a critical analysis of English travelers' attitude toward the Ottoman Empire, Gerald M. MacLean defragments the assumed western gaze at the Orient and illustrates that before the eighteenth century, the colonial logic of "othering" did not exist (MacLean 2004). The travel narratives I am interested in exploring in this chapter are not from what MacLean refers to as the "century before there was a British Empire" (xiii).

2. Infatuation with Hellenism was a major component of travelers' gaze as they encountered the Orient. The yearning for the Hellenic past imagined as a golden age influenced how travelers viewed the landscape and people. At times, this political imaginary generated different representations of Asia Minor's Muslim and non-Muslim populations, some of whom were associated with the ancient past.

3. Orlando Patterson (1982) defined social death to explain the dehumanization encountered by slaves. The concept has since acquired currency in anthropology and related disciplines to explain the alienation individuals experience based on their social location and marginalization within structures.

4. I will discuss how the wet–dry binary played a role in Turkish forestry in a forthcoming publication.

5. The practice of commercial exploitation by foreign companies continued in the first decade of the republic. For a discussion of Ottoman forestry, see Dursun (2007).

6. Consequently, afforestation became a cultural project and was considered not only the state's but citizens' duty as well.

7. See "Adana Bölgesinde Koyun ve Keçi Bulundurulmayacak."

8. Prime Ministry Republican Archives, doc. 252.1.

9. Interestingly, today foresters in research institutions are discussing goats' role in preventing forest fires by their eating waste material on the ground. Moreover, goat milk and its deriva-

tives are emerging as niche market opportunities. Even in my field site in the Black Sea region, where the number of goats is negligible compared with those in the Mediterranean regions, a goat farm had opened in 2010.

10. James Ferguson (1994) demystifies the "bovine mystique" and shows that traditional livestock practices, which were viewed by development agencies as hindrances to Lesotho's development were, in fact, deliberate mechanisms adopted by the rural populations to respond to the market for migrant labor engendered by Lesotho's containment within the larger South African economy.

11. The state is the owner and manager of all forests in Turkey. Forest villagers, who are defined by law as residents who live near or adjacent to forests, do forest work for the state's forestry agency through contract work.

REFERENCES

"Adana bölgesinde koyun ve keçi bulundurulmayacak" [No sheep or goats will be kept in the region around Adana]. 1950. *Sinop Gazetesi*, August 9.

Agrawal, Arun, and K. Sivaramakrishnan, eds. 2003. *Regional Modernities: The Cultural Politics of Development in India.* Delhi: Oxford University Press

Ali B T. 1931. "Ormancılığımızın ziraatle alakası" [The relationship between our forestry (policy) and agriculture]. *Orman ve Av* 35: 1–19.

Atay, Falih Rıfkı. 1944. "Ağaç ve devlet" [The tree and the state]. *Ormancı Postası*, February 1.

Bakic-Hayden, Milica. 1995. "Nesting Orientalisms: The Case of Former Yugoslavia." *Slavic Review* 54(4): 917–31.

Barton, David, and Uta Pappen, eds. 2011. *The Anthropology of Writing: Understanding Textually Mediated Worlds.* London: Continuum International.

Benjamin, Walter. 2019. *Illuminations: Essays and Reflections.* Boston: Mariner Books, Houghton Mifflin Harcourt.

Bhabha, Homi. 1984. "Of Mimicry and Man: The Ambivalence of Colonial Discourse." *October* 28 (*Discipleship: A Special Issue on Psychoanalysis*): 125–33.

Bricogne, Louis A. A. 1877. "Les forêts de l'empire Ottoman" [The forests of the Ottoman Empire]. *Revue des Eaux et Forêts* 16 : 273–89, 321–35, 369–84, 417–31 and 465–80.

Bricogne, Louis A. A. 1876. "La mission forestière en Turquie." [The forestry mission in Turkey]. *Revue des Eaux et Forêts* 15: 361–86.

Cooper, Frederick, and Ann Laura Stoler. 1989. "Introduction, Tensions of Empire: Colonial Control and Visions of Rule." *American Ethnologist* 16(4): 609–21.

Cronon, William. 1996. "The Trouble with Wilderness; or, Getting Back to the Wrong Nature." In W. Cronon, ed., *Uncommon Ground: Rethinking the Human Place in Nature.* New York: W. W. Norton, 69–90.

Douglas, Mary. 1991 [1966]. *Purity and Danger: An Analysis of the Concepts of Pollution and Taboo.* New York: Routledge.

Dursun, Selçuk. 2007. "Forest and the State: History of Forestry and Forest Administration in the Ottoman Empire." PhD diss., Sabancı University.

Ferguson, James. 1994. *The Anti-Politics Machine: Development, Depoliticization, and Bureaucratic Power in Lesotho.* Minneapolis: University of Minnesota Press.

Gökmen, Halil. 1935. "Ormanların derdi" [Troubles of forests]. *Verim* 7–8.

Hawley, Walter A. 1918. *Asia Minor.* London: John Lane the Bodley Head and John Lane Company.

Heske, Franze. 1952. *Stepten faydalanmanın imkanları* [Opportunities to benefit from the steppe]. Translated by Selahattin İnal. Istanbul: Kader Basımevi.
Hull, Matthew. 2012. "Documents and Bureaucracy." *Annual Review of Anthropology* 41: 251–67.
Karacan, Ali Naci. 1935. "Ağaç" [Tree] *Verim* 3.
Krause, Herr. 1934. "Anadolu mecmuai nebatiyesi" [Vegetation of Anatolia]. *Orman ve Av*: 59–62: 28-31.
Künksatan, Ahmet. 1949. *Orman davamız* [The forest question]. N.p.: Yeşilgireson Matbaası.
Lengyel, Emil. 1941. *Turkey*. New York: Random House.
Macfarlane, Charles. 1850. *Turkey and Its Destiny*. London: John Murray.
MacLean, Gerald M. 2004. *The Rise of Oriental Travel: English Visitors to the Ottoman Empire, 1580–1720*. New York: Palgrave Macmillan.
Makdisi, Ussama. 2002. "Ottoman Orientalism." *American Historical Review* 107(3): 768–96.
Montrichard, D. de. 1870. "La mission forèstiére en Turquie" [The forestry mission in Turkey]. *Revue des Eaux et Forêts* 9: 223–24.
Montrichard, D. de. 1875. "Le déboisement en Asie Mineure" [Deforestation in Asia Minor]. *Revue des Eaux et Forêts* 14: 97–105.
Mueggler, Erik. 2005. "Reading, Glaciers, and Love in the Botanical Exploration of China's Borderlands." *Michigan Quarterly Review* 44(4): 722–54.
Muhlis, Esat. 1933. "Bolu, Zonguldak, Kastamonu ve civarı ormanlarında bir cevelan" [A sojourn in the forests of Bolu, Zonguldak, Kastamonu, and environs]. *Orman ve Av* 52.
Özdönmez, Metin, and Abdi Ekizoğlu. 1933. "Tanzimat ve Meşrutiyet dönemleri ormancılığında katkıları olan yabancı uzmanlar" [Foreign experts who contributed to forestry during the Tanzimat and constitutional eras]. *Istanbul Üniversitesi Orman Fakültesi Dergisi* 3–4 (43).
Özkan, Hande. 2002. "Tek parti dönemi coğrafya ve mekan anlayışları" (Geographical and spatial perceptions in the single party era). *Toplum ve Bilim* 94: 143–74.
Özkan, Hande. 2018. "Remembering Zingal: State, Citizens, and Forests in Turkey." *International Journal of Middle East Studies* 50(3): 493–511.
Patterson, Orlando. 1982. *Slavery and Social Death: A Comparative Study*. Cambridge, MA: Harvard University Press.
Perera, Suvendrini, and Joseph Pugliese. 1998. "Parks, Mines and Tidy Towns: Enviro-Panopticism, 'Post'colonialism, and the Politics of Heritage in Australia." *Postcolonial Studies* 1(1): 69–100.
Said, Edward. 1978. *Orientalism*. London: Routledge and Kegan Paul.
Said, Edward. 2003. "A Window on the World," *The Guardian*, August 1.
Salih, M. Ali. 1933. "Kocaeli Ormanları" [Forests of Kocaeli]. *Orman ve Av* 52.
Scott, James C. 1998. *Seeing Like a State: How Certain Schemes to Improve the Human Condition Have Failed*. New Haven, CT: Yale University Press.
Smith, J. V. C. 1857. *Turkey and the Turks*. Boston: James French and Company.
Stoler, Ann Laura. 2002. "Colonial Archives and the Arts of Governance." *Archival Science* 2: 87–109.
Sungar, Şakir. 1946. *Orman işletmelerinde neler gördüm* [What I observed in (state) forestry enterprises]. Eskişehir: Sesışık Kitabevi.
Thomas, Nicholas. 1994. *Colonialism's Culture: Anthropology, Travel and Government*. Princeton, NJ: Princeton University Press.
Todorov, Tzvetan. 1984. *The Conquest of America: The Question of the Other*. Translated by Richard Howard. New York: Harper and Row.

Todorova, Maria. 2005. "The Trap of Backwardness: Modernity, Temporality, and the Study of Eastern European Nationalism." *Slavic Review* 64(1): 140–64.

Todorova, Maria. 2009. *Imagining the Balkans.* Oxford: Oxford University Press.

Twain, Mark. 1869. *Innocents Abroad.* Hartford, CT: American Publishing Co.

Wenzel, Hermann. 1933. "Sultan Dağı Mecmuai Nebatiyesi ve Hayati Iktisadiye." Translated by Ali Kemal. *Orman ve Av* 53: 3–8.

Ximinez, Saturnino. 1926. *Asia Minor in Ruins.* Translated by Arthur Chambers. New York: Brentano's.

9

Constructing Forest Expertise

Foresters in the Białowieża Forest

Eunice Blavascunas

"There are four pillars, which build our sovereign homeland," Konrad Tomaszewski, director of Polish State Forests (SF) pronounced, "the Catholic Church, State Forests, agriculture and hunting. They are integrated, work together, and any attempt at their destruction will end tragically."[1] Tomaszewski strategically made this statement to more than eight thousand Polish foresters at a 2017 Catholic mass in Toruń at a time when SF (Lasy Państwowe) would carry out an ecological agenda condemned by the European Court of Justice, UNESCO, and Polish scientists. In that year under the Law and Justice Party government (PiS), SF logged more than 190,000 cubic meters of timber from "Europe's last primeval forest," to the outrage and protest of Polish and international conservationists. Why and how the Polish forester became a "pillar" of the Polish homeland in relation to logging this ancient forest is the subject of this chapter. The transhistorical narrative that unfolds here demonstrates that the state in Eastern Europe can facilitate and extend its expertise about the constructedness of "natural" forests. It does so in an era when ultra-right-wing nationalism is conceived as an alternative modernity to liberal markets and secular European identity. In this rendering, the forester in Białowieża is more than just a bureaucratic expert rationally meeting the needs of the state. PiS equates one type of state expert, the forester, with the history of Poland itself, "martyred and misunderstood." The forester is a figure who believes that Polish forests and forestry are national constructions. In that sense, the Białowieża Forest is never primeval; forests and the Polish nation are synonymous, and Białowieża stands as the forensic for this assertion.

Eastern Europe has long been used for its natural resources by internal and external imperial powers. The Polish state today, in the form of Polish forestry, developed in tandem with and opposition to Germanic and Russian/Soviet expertise (Dorondel and Şerban, introduction in this volume). When the figure of the Polish forester is scrutinized in transhistorical time, across different occupying powers and geopolitical configurations, the contemporary narrative about Białowieża takes on a certain valence. An ideologically controlled SF combines Polish history and forestry to thwart suprastate bodies, such as the EU and UNESCO and does so across changing moral sensibilities about what should be used for the common good, what is nature, and what should be preserved. These are procedures that keep the past and present firmly entangled for the purposes of transforming nature (Sivaramakrishnan 1995: 4).

PiS's highly politicized projection of the Polish forester in Tomaszewski's rendering deserves contextualization in the history of scientific forestry as a geopolitical environmental history of the region. In this rendering, the forester is not just the agent who destroys self-sufficient village communes who rely on the forest commons, as in much of the literature about state forest experts in the colonial context (Peluso 1992; Guha 1989; Nadkarni 1989; Sivaramakrishnan 1999). Logging Białowieża against the objection of the European Commission and UNESCO occurs as PiS narrowly builds a nationalist agenda against a multicultural history produced by colonization, occupation, the exchange of international experts, and resettlements across shifting borders. The Białowieża conflict cannot be understood outside of a history of the Polish state that has long experienced nationhood and modernity as a defense against occupying powers.

Significantly, the outbreak of bark beetles (*Ips typographus*) defoliating spruce offers a pretext for politicized logging to counter the conservationists' narrative that this forest is indeed primeval. The forest complex is split between the Białowieża National Park of 10,500 hectares, (with almost no logging) and commercial forestry lands of 47,500 hectares. The strict reserve is an ecological complex where bark beetle outbreaks do little overall damage to the structure of the forest. Foresters counter that the whole forest is a human artifact. The beetle justifies increases in logging to save the forest *as* forest, as a total portrait of industrious *man* and nature, yet industrious not for the sake of industry, as in the soulless private gains of capitalism, but for the love of nation.

The term "primeval" arises in the eighteenth century with the establishment of scientific forestry and as Europe confronted deforestation at the European scale (Matteson 2015; Oosthoek and Hoelzl 2018). As a lens onto this *longue durée* history, bark beetles (in the present) may provide a new, better way of imagining the Białowieża Forest and its history in a way that also helps us consider the forester's role in creating an oppositional primeval. Foresters today claim their expertise in the name of a common good of "forest protection," which is why they "sanitize" the forest. The idea of the common good is the basis of scientific forestry itself with its roots in cameralism (Grewe and Hoelzl 2018; Rajan 2006). But the European Court of Justice ruled that Poland illegally logged the forest in 2017, referring to other types of experts and other types of common goods. They paid heed to Polish and international biologists, who argued that the common good of all Europe, or perhaps the globe, includes retaining a primeval forest where disturbance ecologies (bark beetle) retain the wildness of Europe's remaining nature (Rykowski 2012).

Given the number of political formations in Poland since the mid-1800s, it seems that the techniques of scientific forestry would offer a universal language that could easily cross each regime. In other words, the forester transcends historical time and changing political regimes precisely because his techniques are always useful to the state, whether the state is imperial, nationalist, or communist. While this rings true with much of the literature on foresters (Matthews 2011; Peluso 1992; Scott 1998; Rajan 2006), it does not explain the inflection of the forester as a pillar of contemporary Polish society, a figure who is a beleaguered and heroic patriot and not simply a bureaucrat carrying out the economic rationalization of the state.

Foresters asserted that that the only way to save the forest was to overturn a 2011 law that significantly curtailed logging. The European Commission and UNESCO claimed that the logging breached rules to which Poland was a signatory, including NATURA 2000, and UNESCO's World Heritage Status, granted to the forest for its unique *natural* heritage. The European Commission took the unusual step of asking the European Court of Justice to issue a ban on logging in the forest for failures to fulfill obligations to the EU Birds and Habitat Directives. Conservation activists chained themselves to logging equipment to prevent harvesters from cutting Białowieża's trees, and considerable social unrest erupted in Poland over the fate of Białowieża.

The forester today leads an opposition to foreign expertise, which can now mean the EU, UNESCO, and the Polish conservationists, including many biologists and activists, who look to external governing and advisory bodies to protect the forest. Yet, given the history of knowledge exchange, the science of forestry has always been transnational (Barton 2002). There are reasons to ask questions about the way PiS combines culture and political economy to imagine the past and negotiate power.

Ethnic differences and conflicts in the region share important overlaps with the establishment of Polish State Forestry in Białowieża. In 2017 the Polish forester's main goal was not to assert their modernization of nature but to claim that attacks on Polish State Foresters' expertise paralleled the historical martyrdom of Poland, a martyrdom that could not also include the histories of minorities in Poland. The Belarusian inhabitants of the Białowieża Forest in Poland had to be convinced of the notion that they had a misguided ethnic identity and that they were in fact loyal to the Polish cause, that Belarusian was a regional identity, not a national or communist one.

This ethnohistory explains the type of modernity produced through the forester. This alternative modernity, represented as resource extraction for the nation, rejects a secular and immoral Europe and embraces an internal *gemeinschaft* (Gellner 1998). Historically speaking, the morality of the Prussian bureaucrat rested in abiding by rules. But the morality of the nationalist (PiS platform) resides in his love for his people, their peculiar ways, and their politics of the specific. Much of this nationalist version of Białowieża resides in transforming any difference that might make the local look foreign, including controlling the spruce bark beetle and diminishing anyone who might not fit the portrait of a loyal Pole.

THE ROMANOV EMPIRE

Białowieża had been part of a quintessential borderland for centuries where Polish magnates ruled over ethnically ambiguous peasant populations. Polish-Lithuanian kings and aristocrats had hunted in Białowieża since 1409, appointing guards from the local population. On the brink of the Russian partitions that eroded Poland's autonomy, Poland's last king, Stanisław August Poniatowski (1764–1795), cared little for hunting, and opened the forest to calculated forms of Enlightenment rationality, including timbering. Aristocratic families accustomed to treating the woods as an extension of their estates had to prove their lease rights. In 1765, Poniatowski sent the treasurer of the Grand Duchy of Lithuania, Antoni Tyzenhaus, to the forest with clerks and scribes to verify what belonged to the crown (Samojlik 2005).

Tyzenhaus prescribed tree thinning and removed dead timber from the forest. His calculations were meant to wrest power from the hands of Poland's independent and erratic aristocrats. Poland needed a strong state with a large central treasury, not to mention a well-equipped military, to ward off threats from Russia. Despite Poniatowski's efforts to rationalize forest management and streamline the state, Russia, Prussia, and Austria partitioned Poland (1772–1795).

Polish magnates clung dearly to semi-independence under an initial agreement with Russia's tsarina, Catherine, and continued exploiting forest resources. In the core of the forest, forest guards prevented logging and poaching. While Białowieża's royal trees produced potash and charcoal, they did not contribute to the growing timber trade for shipbuilding and construction. Before the partitions, each landowner would have hired his own resident forester to oversee operations (Schama 1995). Catherine dispensed aristocratic land in Białowieża to Russian aristocrats, and military personnel, some of them from her own German homeland, and further lessened the influence of Polish gentry. In 1795 she turned over management of the entire Białowieża region to Prussian foresters. They regulated the Narewka River to float pine, oak, and spruce as far as the Vistula River, just short of the Prussian border. From the Vistula, timber was hauled overland to the Nieman River and out to Russian ports on the Baltic Sea (Michaluk 1997). Whole units of the forest were cleared for timber and agriculture (Samojlik, Fedotova, and Kuijper 2014). Polish rulers had long maintained a hereditary forest protection unit of beaters and guards, peasants who were free from feudal obligations, but under Catherine's rule all beaters lost their rights and fell under the protectorate of Russian and German estate owners (Daszkiewicz, Samojlik, and Jędrzejewska 2012). Russian soldiers and military officers, often with little training in forestry, were appointed foresters under specialists chosen by Prussian foresters (Hedemann 1939: 142–50; Majchrowska 2018: 326). Despite the onset of logging, the chief forester of the Congress Kingdom of Poland, Juliusz Brincken, described the forest at this time as a "true primeval forest" and published his assessment in French to reach the wider European natural history community (Daszkiewicz, Samojlik, and Jędrzejewska 2012: 179–82).

To counteract the insufficient education of Polish foresters, Polish gentry formed their own forestry school at the newly founded University of Warsaw in 1816. Polish aristocrats, weary of the new Russian landowners, took up arms and formed an underground resistance movement. In the insurrections against Russia in 1830–1831, Białowieża became a battlefield and base for the Polish rebels. Hundreds of Polish gentry lost their lives for their insurrectionary roles, their bodies strung up in the trees. The forestry school in Warsaw was a cradle of patriotism, so Russia closed it down in 1832 and all Polish wardens lost their positions, though Polish aristocrats continued their forestry education at agronomy schools (Majchrowska 2018: 325–31). After the insurrections, tsarist officials closed and destroyed Polish Catholic Churches and converted peasants from the Uniate to the Orthodox rite (Kossak 2001: 292).[2]

Russian rulers knew that Białowieża was a prime hunting site and that it contained the last wild bison populations, but it was not until Tsar Alexander II led an imperial hunting party in 1860 that the forest received protection again as a royal hunting ground. Białowieża became the tsar's private property in 1888. While Alexander II did not log Białowieża's trees, he imported wild and exotic game. High densities of ungulates destroyed much of the forest undergrowth. In 1889 Tsar Alexander III ordered the construction of a 120-room palace with sunken Roman baths and English gardens (Bajko 2001). Although mismanaged at various times, forest cover never dropped below 70 percent in the nineteenth century (Mikusińska et al 2013). In this period, the Polish forester is an underground figure with a Messianic vision that Poles will return to the region to restore a more just rule.

WORLD WARS AND INDUSTRIAL FORESTRY

During the World War I German occupation (1915), German foresters subjected the ancient trees to calculated exploitation. The Kaiser's Military Forest Administration cut and removed 5 million cubic meters of timber from 6,500 hectares of forest. They built networks of railroad lines to transport timber out, and they set up 24 sawmills and a chemical wood-processing factory, using forced labor, in what became the city of Hajnówka (Blood 2010; Sunseri 2012). German foresters used arguments that they had applied in their former west African colony, Cameroon (Sunseri 2012: 322). The custodial model, under the forester George Escherich, was premised on the idea that old forests needed to be protected from peasants who would misuse the forest's economic potential, but at the same time a German nature conservationist, Hugo Cowentz, an initiator of a European national parks movement, called for a national park over the whole forest (Okołów 2009: 10). Thus, heavily logging the forest also resulted in a variant of national and European patrimony, the conservation movement to protect Białowieża from industry.

Although for most of Europe World War I may have ended in 1919, a brutal Polish–Soviet War started in 1919. Interwar Polish territory was not recognized internationally until 1923. Much of the Polish–Soviet War was fought over formerly Russian territory. In the 1920s Bolshevik agitators within Białowieża tried to expand the Western Belarusian Communist Party (KPZB), founded in 1922 with local KPZB cells active until 1939 (Bajko 2017). Poland rejected the aspirations of Belarusian nationalists, by closing all Belarusian schools in 1924 (Nikitiuk 1958: 55) even as Poland was one of the first of eight newly independent states to sign the Minorities Treaty, which constituted part of the Treaty of Versailles (Rudling 2014). Most Belarusians at that time were illiterate. Five years after the signing of the Treaty of Riga (1921), half of the Belarusians within Poland were considered stateless and could not vote. Poland viewed the "eastern" populations around the Białowieża Forest as "uncivilized" and in need of culture.[3] With the KPZB as an illegal organization, those wanting to organize for minority rights formed Hromada in 1927, a legal Belarusian cultural organization. Collectively, these two organizations agitated for better workers' rights in the wood factories and against unemployment (of the majority Belarusian population). The timber economy constituted the main form of employment in this era. The KPZB's tactical strikes created treacherous social relations between the newly arrived Poles (laborers and clerical workers) and the Belarusian peasants. In a 1931 May Day march, striking workers and the unemployed distributed fliers that encouraged the creation of "self-defense forces against the police and the forest guards" (Nikitiuk 1958: 50). Workers burned down one of the key forester's estates in 1934 at Zwierzyniec.

Of the population in Białowieża (then 14,000 inhabitants), 85 percent identified as Belarusian and most of these as Orthodox (Statistical Yearbook 1921). The arrival of Polish foresters at Białowieża, and the construction of the Catholic Church in 1920, initiated a civilizing relationship of foresters toward Belarusians, who would need to realize they were indeed *Polish*, that Belarusian was an old language and regional identity rather than a nationality (Woolhiser 2003), and that Poland, rather than the Soviet Union offered a better route to their aspirations (Fleming 2010).

As for the forest and its protection in this tumultuous period, Polish botanists, trained in forestry schools, created the first strict protective measures for the newly formed republic (Krasinski 1994). In April 1919 the botanist Władysław Szafer and a team of international scientists surveyed what the Białowieża Forest had suffered. Szafer was one of the founding members of the International Union for Conservation of Nature. Moved by the unprecedented damage, he gathered support for protecting the area and returned the following year with members of the Polish Nature Conservation Council to designate an appropriate spot for Poland's first national park. In 1921 the first protective reserve was formed, and in 1932 it was officially declared a National Park (Szafer 1920).

During the same period, 1921–1928, the British Century European Timber Corporation signed an agreement with the new Polish government. Century used and owned the wood, and sent the milled lumber back to the United Kingdom. State Forestry's first director, Adam Loret, abruptly terminated the contract in 1928 once the Polish state and State Forestry were firmly established. The break with the British contractor came as Polish foresters (and nature conservationists) sought to end what they saw as another exploitative foreign relationship. Although foresters and the scientific community could not agree on how the whole forest should be managed, Białowieża's first scientific research institute, the Institute for Forestry Research, was founded in 1930. It was run by the Polish foresters Jan Jerzy Karpiński and Józef Paczoski. Paczoski was Polish by ethnicity, but Russian by upbringing and spoke little Polish.[4] He spent years organizing protection for the Ukrainian steppes and committed himself to nature conservation in Białowieża.

Karpiński studied forestry in Petersburg in 1915 before the war broke out, and during his time as park director (1929–1934), he led scientific studies aimed at conservation-oriented forest management, including studies of "virgin" forest, and at reconstituting European bison illegally hunted to extinction during World War I. Thus, the Białowieża National Park started as a special unit of SF and both its directors were trained as foresters in Russian-administered territories.

Saddled again with the vagaries of war, in 1939 Białowieża was first taken under the rule of the Soviets, who used Belarusian locals to identify Polish foresters. They sent 1,404 Białowieża inhabitants (all of them Polish foresters) to work camps in Siberia, and did heavy logging in the National Park (Kossak 2001, 459–84).[5] Soviets executed the State Forest director Adam Loret in that same year (Broda 2006; Grzywacz 2014). The Soviets continued logging operations until 1941, taking more than two thousand cubic meters of wood and enabling poaching (Okołów 2009: 13).

Five days after Hitler launched Operation Barbarossa against the Soviet army on June 22, 1941, the Nazi flag flew above the Romanov palace. Białowieża fell under the protectorate of the Nazis (1941–1944) with Hermann Goering, the prime minister of the Third Reich, directly supervising plans for the forest. German soldiers removed all villagers within the forest hamlets to prevent them from abetting the partisans. Logging stopped altogether during the occupation, and Goering contemplated a huge forest protection zone, six times the size of the national park (Faliński 1968; Więcko 1984). By 1941, the Germans had eliminated all Białowieża's Jews (10 percent of the population at that time), most of whom had facilitated the timber trade. Because they were educated, Polish foresters served as officers in the war. Many foresters fought as anti-Soviet partisans, and both Germans and Soviets murdered foresters in campaigns to eliminate Polish intellectuals.

COMMUNIST POLAND

After World War II, with Białowieża split in two under the Yalta agreements (40 percent of the forest was left in Poland), the People's Republic of Poland denied national groups their histories. Poles and Belarusians obtained work in the forestry sector (although ethnic Poles continued to dominate administration). Per oral accounts with contemporary foresters and their families, some of the former Polish foresters went back to work at their prewar positions, but it is unknown what kind of ties, if any, they may have had to the resistance (Home Army) and how they survived the war. Foresters arrived from western Poland and hid their biographical details from the Party when obtaining positions. State forestry advanced Belarusians through forestry without explicitly acknowledging ethnic identity.

Initially, the postwar People's Republic of Poland (PRL) adopted wholesale the prewar organizational scheme. The General Directorate of SF administered the Białowieża Forest as with all other forested areas with the Ministry of Forestry responsible for hunting, forest policy, and wood-processing industries. The need for revenue after the war again meant the exploitation of forest timber, with a high volume of timber extracted in 1946 and widespread reports of poaching in the chaotic period at the end of the war (Okołów 2009: 13). After Poland moved to a command economy and central planning (1949), the government viewed State Forestry in Białowieża as they did any other state-owned enterprise. The forest was there to serve the needs of industry (Majchrowska 2018). In 1946–1975 the timber harvest often exceeded 50 percent of the forest cover (Gutowski 2000). Where primary forest had been cut, foresters hired local labor to reseed areas with monocultural softwood plantations. The Ministry of Forestry developed more protective guidelines for the Białowieża Forest in 1975. Shortly thereafter, Białowieża's scientists, including the head of the Forestry Research Institute, Andrzej Sokołowski, a botanist trained at the University of Warsaw, called for the expansion of the Białowieża National Park and for an end to timber exploitation (Gutowski 2000).

While the communist and postwar periods are hard to summarize with regard to their overall impact on the forest, the period's policies created social changes that resulted in blurred ethnic identity, but without subsuming the ethnic question altogether. Many inhabitants sent their children to the Technical Forestry High School in Białowieża (founded in 1929), expecting that their children would receive jobs in forestry afterward. As wood provided the main heating source, people reminisced about the forester in this era as an amiable and respected figure. In this sense, over the course of the communist period, the forester solidified his legitimacy as an expert because he gained trust among the local population, unlike during the interwar period. The local population came to see themselves through the lens of forestry as their primary identity. The forester, as a figure belonging to a multiethnic locale, is rooted in this era.

POSTSOCIALIST FOREST POLITICS

In the early 1990s, the Polish SF downsized its labor force by 75 percent. Polish forestry adopted a corporate model, including self-financing. It also ensured a monopoly on timber sales. Only since 2014 has SF been required to pay 2 percent of its income from wood sales into the national treasury. Despite this corporatization of SF, many of its critics asserted that the agency was still "communist." Many of these criticisms emerged from scientists living in Białowieża. The number of resident scientists grew considerably during the course of the Cold War. The Polish Academy of Sciences Mammal Research Institute and the Warsaw University

Geobotanical Station both opened research facilities in Białowieża in the 1960s. The institutes conducted research without direct relevance for timber management. This group of experts largely allied themselves with the ideals of opening to the West. For scientists, the alliance meant creating stronger international networks for their expertise. Scientists charged that SF was the "last authoritarian organization in the country." It had a "communist-like mentality" and organizational methods that stymied forest protection.

The 1990s conservation movement formed around the suspicion that foresters acted in communist ways, which were also "local" ways, and that conservationists identified with international forms and practices of nature conservation. Foresters shared "family resemblances" to Belarusian ways of social organization. For example, conservationists alleged that the system in Belarus operated based on bribery and cronyism. In Poland one could easily curry favor for better heating wood, and later, locals would thus need to "obey" foresters. One conservationist claimed that the Białowieża region was "Belarus in a little pill." Another scientist said, "There is a similarity between Poland and Belarusian totalitarianisms. There is clear evidence of links between Polish SF and the Orthodox and Belarusian activists related to the campaign against national park expansion." "Why don't we just move to Belarus!" another conservation activist complained. Collectively, these comments suggest that conditions in Belarus were not so different from conditions in the Białowieża.

By and large, foresters at the top administrative levels were still ethnic Poles. But many working for SF were also of Belarusian descent. Conservationists pointed to Belarusian "passivity" under Polish foresters. In conservationists' discourse, the Belarusian "character" of the region lent itself to this tolerance for "authoritarianism." In other words, a stereotyped Belarusian was historically conditioned to live under tsars and communists, and thus to uphold illegitimate expertise. Many conservationists pointed to the monument in front of the Orthodox church, which read, "To the heroes of socialism." As monuments to communism disappeared throughout Poland, traces of communism lingered in Białowieża. Conservationists also drew my attention to artifacts of the regional "susceptibility to communism," such as a street named after a former KPZB member.

Between 1989 and 1996, scientists managed to double the size of the Białowieża National Park. They also drafted a detailed plan for expanding the national park over the whole Polish part of the forest, including administrative structures, zones for different types of human activities, and even zones for multifunctional forestry (Gutowski 2000). Biologists attempted to create a civil society that could shake off the power of the forester. A few key biologists worked in off-hours to build local autonomy. They encouraged locals to open businesses, started independent newspapers, and ran (though unsuccessfully) for local councils, a completely new form of government for the village, which would come to be dominated by foresters and their supporters.

Neighboring Belarus's forest management would inflect on Polish foresters' expertise in this same period. After five decades as a hunting preserve for Soviet premiers (with no logging), the entire Belarusian forest became a National Park in 1991. The forest had received better protection as a hunting preserve. The Belovezhskaya Pushcha National Park opened the forest to industrial logging with some strict preserves for the oldest tree stands. They sanitized the forest against bark beetle outbreaks and removed dead wood from the forest floor. All residents of the Belovezhskaya National Park worked for the state. Many foresters and locals spoke to me about how the Polish forest could be better managed if they had the power to do what was done in Belarus. In Belarus the forester could manage the forest without interference.

A major protest against national park expansion in the year 2000 came as a surprise to many conservationists. "Locals" were projected as Belarusian (even as the population was ethnically mixed), and thus stereotyped as passive, in need of their Polish counterparts to see a kind of "Western" political freedom. Significantly, conservationists spoke of how foresters organized locals in the 2000 protest to hold signs in Polish *and* Belarusian calling conservationists "Judas traitors" for wanting to use the EU to protect the forest from logging (Fleming 2010; Franklin 2002). The EU was somehow seen as selling out Belarusian interests. Activists identifying as Belarusian (within Poland) wrote articles claiming that a huge national park would destroy the Belarusians' chances of modernizing. These Belarusian activists claimed that Belarusians in Poland were poor and needed logging to modernize themselves and the region. Many locals, Belarusian, Orthodox, as well as those who were Polish and Catholic, supported foresters.

Logging and a Belarusian or Polish identity could be one and the same in that era. It did not matter to the logging lobby or to the forester if people were Polish or Belarusian. More importantly, the foresters had to log to rid the region of its stigmata of "backwardness." Conservationists tracked the diffusion of relict communist ideas by comparing the scientific findings of Western conservation biology with those of *Polish* forestry. Logging culture associated with Belarus, in its ethnic form within Poland or within Belarus, represented a time lag.

Foresters in Poland during this period received abundant training by foreign foresters, including US State Department exchanges between the US Forest Service and Polish foresters, and a Danish-funded project (DANCEE) to collaborate with foresters on sustainable development. Ready-made models for development also came from the World Bank's Global Environment Facility. Yet these sorts of international exchanges went unnoticed by many local inhabitants, whereas locals challenged the expertise of Polish conservation biologists precisely because they were shamelessly making careers at the international level, while foresters counted as "locals" and fit in.

Among the most controversial ideas was the conservation biologist's approach to the native bark beetle. International studies validated the role of dead and dying timber in the ecological cycle of the forest (Bobiec et. al 2005). Bark beetle could multiply exponentially with dead and dying timber. The real problem was that the national park valued dead wood, and scientists fought for more dead wood in the managed forest, with the ulterior motive of maintaining large protected areas with little logging. For nature conservationists, Polish forestry (and its local cultural attentiveness to neighboring Belarus) and its professed need to log Europe's last primeval forest were simply outdated. Bark beetle were *European* and could bring back an "original" nature without causing forest die-off. Belarus was like the old communist Poland, the old Soviet Union. The way in which foresters spoke of their paternalistic mission to care for all local people, to revive the timber economy, and provide everyone with jobs as in the "good old days of communism," suggested the possibility of a future socialist state. Foresters acted against nature conservationists who scaled up to get support at the level of Europe and the international community. But the ecology of this narrative changed with the rise of Polish nationalism in the first two decades of the twenty-first century.

REINVENTING COMMUNISTS AND MARTYRS

In 2017 the bark beetle and its cyclical outbreak became a useful tool in PiS's assertion that the West wrongly persecuted Poland for a number of historical reasons. The only solution for PiS was to unify Poland by purging anyone who disagreed with them. PiS began a project of building the "Fourth Republic of Poland," against the "illegitimate" Third Republic, which had not properly reformed communist-era rules, people, and infrastructures. The beetle fit within PiS's narration about "threats" to the Polish nation, which became threats to Polish forests. The bark beetle made it easier to identify one of the current enemies of the Polish nation for PiS, *lewactwo*, or the alter-globalization left (made up of feminists, environmental activists, LGBTQ activists, vegetarians, and refugee-rights activists). This time, the old foes (communists, Bolsheviks, Soviets, Russians, German or British logging interests, or conservationists) appeared in the new face of conservationists and ecological activists. Green activists became traitors of the Polish nation, serving foreign affairs and peddling foreign values. And foresters ceased to be spoken of as "communists" in the conservationist's discourse. Foresters became "mafia-like" or "corporate."

PiS's centering of the Białowieża Forest as a national affair became a way of exposing nature conservation as a foreign plot, of showing how previous governments allowed foreign experts to weaken the expertise of Polish foresters. The bark beetle outbreak was thus not only an excuse for delineating the stranger in an otherwise ordered, Polish forest, the bark beetle also became a pretext for establishing the one and only proper history of (and for) the nation, vis-à-vis the forester by showing, through force (against nature activists) as well as through science and inventory commissioned by State Forestry, that the forester was a proper expert.

In Białowieża, State Forest's expertise was performed as historical continuity. History of the forest was not based on what was within the forest (for that would need to include old growth and bark beetles), it was built upon what was *excavated* from the forest (through foresters' management and labor). PIS wanted not only to rework nature for Poles' benefit but also to show that their work was fundamental to forest existence (Konczal 2017).

Notwithstanding, the character of Białowieża provided tools and materials for this (the bark beetle, the repeated foreign occupations, the martyrdoms vis-à-vis graves in the forest, its Belarusian otherness). Foresters under PiS adeptly calibrated local long-term opposition to nature conservationists to meet the nationalist forester myth. But this strategy began before PiS's current mandate to govern.

By about 2012, Catholic media, ultra-right-wing in its programming, took up the cause of Białowieża. Both Radio Maryja and TV Trwam broadcast religious messages that were at once rosary specific as well as antifeminist, against the Islamization of Europe, Middle Eastern refugees, and LGBTQ rights. The radio station frequently used the term "ecoterrorist" for those who opposed logging the Białowieża Forest. Jan Szyszko, both before and after he was appointed minister of environment, appeared dozens of times as a guest on Radio Maryja and TV Trwam. He and the chief of State Forests, Konrad Tomaszewski, had a close relationship with the media's founder, Tadeusz Rydzyk.

By 2013, a Hajnówka-based group calling itself "Santa, In Defense of the Białowieża Forest" organized a protest in which the first in situ signs of Polish nationalism appeared in relation to the logging cause. Their banners read, "Save the Polish forest from pseudoecologists and those who would sabotage it." Szyszko was their main speaker. Polish protesters in folk costumes (Central Polish costumes) milled about as well as Catholic and Orthodox priests. Santa's president, Walenty Wasiliuk, owned several sawmills and spoke of nature conservation as an international "Jewish" plot to undermine Polish independence.[6]

Białowieża's local attitudes to the ruling party changed quickly together with a visible rise in the nationalistic approaches among the region's inhabitants. When Szyszko was appointed to the Ministry of Environment he made it his personal mission to radically transform the ecology of the forest, through an attack on bark beetle. A few months later Szyszko dismantled the National Council on Nature Protection, most of whose members favored a national park expansion or backed the studies that native bark beetle could architecturally reconstruct a forest as a true primeval forest. Szyszko dismissed thirty-two out of thirty-nine members and constituted a new council of twenty-nine members, among them mostly foresters, agricultural scientists, and directors of botanical gardens.

In the winter of 2016 European Union bodies watched Poland, criticizing its moves to dismantle its Constitutional Court. Later, in July 2017, the European Union's Court of Justice condemned Poland for unnecessary logging that violated Natura 2000 Habitats Directive laws. In the same month, the EU gave Poland an ultimatum to drop judicial reforms that would have allowed PiS to put judges under its political control. If Poland did not comply it might lose its voting rights in the EU, the ultimate "nuclear" option. The EU stepped up its efforts to suggest that Poland's actions in Białowieża and its actions within the ruling party were antithetical to EU standards of democracy.

When nature activists from all over Poland built a forest encampment, foresters resorted to militaristic tactics to remove the protesters. They wore bulletproof vests, even though the protesters were unarmed and committed to nonviolent civil disobedience. They roughly held down activists and used knives to cut through materials that bound protestors' hands to logging equipment, prompting more than one hundred lawsuits against SF. In total volume, foresters removed 190,000 cubic meters of wood in one year, which was four times the annual average harvest, a record not met since the communist era in 1987. But it was not just nature that needed to be reordered to keep Poland and Polish forests belonging to Poland.

Belarusians increasingly became the target of local ultranationalists, who took to celebrating the "cursed soldiers" (*żołnierze wyklęci*), who were represented as ethnic Polish foresters during World War II and after. The term "cursed soldiers" refers to those who fought in Poland's Home Army (Armia Krajowa)—the dominant Polish resistance movement under the World War II occupation. The "cursed soldiers" refused to put down their arms after the Red Army's "liberation." They chose instead to fight as partisans against "illegal occupants," meaning the new Soviet-imposed rulers, and constituted anticommunist Polish resistance movements between 1944 and 1946. However, many of these soldiers died during fights with Soviets and Polish communists. Partisans associated with "cursed soldiers" were responsible for revenge killings against Belarusian inhabitants who identified Polish foresters to the Soviets during the Soviet occupation of eastern Poland. During the communist period, public mention and memorialization of these combatants were prohibited.

PiS has revived national memory of the cursed soldiers. Schools now teach this history as required curriculum. Monthly periodicals commemorate newly discovered soldiers, each with their own code name, and PiS has even suggested an official holiday to commemorate the soldiers.

The National Radical Camp (Obóz Narodowo-Radykalny, ONR), a far-right movement, has revived the memory of the cursed soldiers. Established in 1934 by anticommunists and nationalists, the interwar ONR called for the elimination of national minorities from public life. In 2012 the ONR registered legally as a common-interest association using white-power symbols at their rallies. Their local member group, Narodowa Hajnówka (National Hajnówka) linked the cursed soldiers cause with the cause of Polish foresters beleaguered by "ecoterrorists." They supported Szyszko, calling for more intense and expedited logging on their website. They attended blockades in the forest and threatened conservation activists, described on Narodowa Hajnówka's website as "enemies of the nation." Their Facebook page icon is a slashed-through hammer and sickle, and calls upon its members to form militias to track down "ecoterrorists." The fight of Narodowa Hajnówka members (who sometimes come from mixed-ethnic and mixed-religious families) with "enemies of the nation" is based on the one correct version of history and forest management. Wearing sleeve patches that say, "Death for enemies of the fatherland," they argue not only for sanitizing bark beetle but also for sanitizing the nation from "foreign factors."

In a highly politicized move at the peak of the forest conflict, foresters erected a statue to one of the cursed soldiers within the Białowieża Forest. Narodowa Hajnówka members, foresters, and those on the far right attended the statue's commemoration. The Narodowa Hajnówka members chanted in unison, "Instead of leaves hanging from the trees, let us hang the Communists." The statue's likeness was then seventeen-year-old Danuta Siedzikówna, who lived in a forester's lodge. She has become a symbol of the "cursed soldiers" movement, referred to by the code name "Inka." The Soviets captured her forester father and sent him to Siberia in 1940 and the Gestapo executed her mother for collaborating with the Home Army. Danuta, sentenced to death by a communist court, died from a bullet to the head in 1946.

She and the statue symbolize a renewed fight with anything that appears to be outside the Polish narrative of victimhood or the Polish narration of history with respect to its heritage and value. This fight includes the bark beetle—the pest inside the Polish forest and the enemy of Polish foresters—and the debate about whose ecology and science count.

Simultaneously in the PiS narrative ecoterrorists are the bad part of the EU—the worst version of a decadent secular West with loyalties to socially liberal causes rather than the nation. Ecological activists use arguments about the forest as "the heritage of humanity." They have garnered the support of the international conservation community and appealed to the European Union, its Court of Justice, and other international groups. PiS considers these instances of foreigners acquiring juridical and legal power over Poland and Polish foresters.

Beginning in 2016, Minister of Environment Szyszko commissioned new "scientific" inventories of the forest. Szyszko attempted to catalog biodiversity and history at the cost of 12 million zlotys. Biologists published a paper condemning the inventory as politicized science, without scientific merit, conducted by unqualified researchers. Foresters used their inventory as a technological fix meant to stabilize foresters' expertise. The inventory accompanied forestry treatment against bark beetles. In effect their strategy was to remove the bark beetle habitat, remove the conservation activist, remove the trees that might make the forest seem primeval, and to justify this through a forest inventory and nature that had been engineered to prove that the Polish forester was one of the four pillars of Polish society.

A FOREST CREATED BY POLES

Since its birth as a field during the Enlightenment, forestry has always involved ideologies and power. Science and techniques stood in opposition to peasants using the forest as commons. Among experts who have opposed State Forestry there is a dialectic relationship between foresters' legitimacy as experts and those who oppose their power (Sivaramakrishnan 1995; Matteson 2015). In creating their governing strategy, PiS shrewdly assessed the historical importance of Białowieża for the project of the Polish nation as a nation of forests, created by foresters, with Białowieża as the ultimate example of a Polish-created forest.

The assertion of sweeping state expertise via SF over the Białowieża Forest in 2017 enacted a relationship between the Polish forester and the body politic. As PiS's minister of environment, Jan Szyszko, annexed the forest management plan to allow massive logging, PiS drew upon two centuries of the Polish forester as a patriot, fighting for the cause of an independent Poland, against foreign rule and saboteurs. This history becomes partly juridical, partly real, and partly ideological, and it becomes hard to separate any version of nature as untransformed, as primeval and natural, without the forester.

Even as SF uses a discourse of forest protection in its fight against the bark beetle, SF's concern for the biological integrity of its "most important forest" can hardly be called a higher standard of nature protection. Applying the hubris of authoritarian tactics to changing the 2011 forest annex (which had curbed logging and earned the area selection as a UNESCO World Heritage site in 2014) and then sending highly armed forest guards to break up nonviolent forest activists demonstrates the illegitimacy of PiS state intervention in an era of EU governance. However, those same authoritarian tactics reproduced a pattern of purported respect for the forester at the local level, in its mixed ethnic form.

Looking at the interplay of cultural constructions of ethnicity and the past in recent years, including the cursed soldiers' commemoration where fascist youth groups surveilled forest activists, shows us that specific slices of the past have been central to the establishment of the Polish forester as an expert today. The circumstances under which the forest comes to be defended as a Polish logged forest does not take in the total historical complexity of this forest, nor can it. The constructedness of expertise in the *longue durée* provides a lens through which to see Białowieża as a multicultural, multi-era construction, partly logged, partly protected, and always with a surprising level of natural agency in the face of plans to create an ordered forest.

PiS narratives take the basic historical features of Polish forestry in Białowieża, and in Poland more broadly, and turn them into a coherent narrative. The rise and fall of different ideologies around SF, represented in the communist and postcommunist period, are consistent with PiS insistence that their narrative is "true" and that the communist period was a nonnative imposition. The naive idealizations of the Polish nation, as reflected in the more extreme elements of this narrative, such as the ONR and Narodowa Hajnówka, demonstrate that PiS's political projection is being accepted by groups that see the need for extra-state activity to carry out the nationalist agenda.

The attempt to know the forest through State Forestry's ongoing inventory demonstrates the use of a familiar technique in the history of forestry. When foresters attempt to thoroughly document all life in the forest, they inevitably miss some of the more important aspects of the forest's composition and history (Scott 1998). The inventory takes place in the present because forester legitimacy has been slipping due to international attention focused on the forest's unique biodiversity status. The forest cannot be a sacred grove for a larger international public and a site of Polish historical memory embodied in the forester.

The narrative in this chapter opens the opportunity to see state foresters and state experts in Eastern Europe as part of a historical shift in the configuration of power at the level of borders, ethnic groups, and supranational governance. Logging protected parts of the Białowieża Forest needed to occur on the premise not only that the state needed money but also that the Polish nation (not just a rational bureaucratic state) had its own best experts.

NOTES

1. Konrad Tomaszewski, "Leśnictwo dla Polskiej Wsi" [Forestry for the Polish countryside], May 13, 2017, https://www.youtube.com/watch?v=9u-Tf1NDZL0.

2. The term "Belarusian" arose in the nineteenth century as a linguistic category. Belarusian was spoken by the Orthodox and Catholic populations of the region. But as Russia consolidated its grip on Poland after 1864, Catholics became Poles and Orthodox became Belarusians. (Woolhiser 2003).

3. Between 1920 and 1925, the Second Republic closed 346 private Belarusian language schools, which had opened between 1918 and 1919 (Rudling 2014).

4. Per conversations with Paczoski's colleague, Andrzej Sokołowski, director the Forestry Research Institute at the time of the interview, April 14, 1999.

5. Per interview with Czesław Okołów, the former Białowieża National Park director, who was among the forester families sent to Siberia in those years, April 15, 1999.

6. "Minister Szyszko i jego zielone ludki: czy Władimir Putin [Minister Szyszko and his elves: will Vladimir Putin save the forest from logging?], http://wyborcza.pl/10,159118,22001598,minister-szyszko-i-jego-zielone-ludki-czy-wladimir-putin-uratuje.html.

REFERENCES

Bajko, Piotr. 2001. *Białowieża: zarys dziejów* [Białowieża: a historical outline]. Białystok: Orthdruk.

Bajko, Piotr. 2017. "Rozstanie z Olga Gabiec" [Parting with Olga Gabiec]. *Czasopsis*. https://czasopis.pl/rozstanie-z-olga-gabiec/.

Barton, Gregory. 2002. *Empire Forestry and the Origins of Environmentalism*. Cambridge: Cambridge University Press.

Blood, Phillip. 2010. "Securing Hitler"s *Lebensraum*: The Luftwaffe and Białowieża Forest 1942–1944." *Holocaust and Genocide Studies* 24(2) (October): 247–72. https://doi.org/10.1093/hgs/dcq024.

Bobiec, Andrzej, Jerzy Gutowski, Karol Zub, Pawel Pawlaczyk, and William F. Laudenslayer. 2005. *The Afterlife of a Tree*. Warsaw: WWF Poland. https://blogs.helsinki.fi/deadwoodmeeting/files/2016/03/Bobiec@al2005.pdf.

Broda, Józef. 2006. "Główni architekci organizacji i gospodarki lasów państwowych" [Main architects of the organization and management of state forests]. In Andrzej Szujecki, ed., *Z dziejów lasów państwowych i leśnictwa polskiego 1924–2004, Vol I: Okres międzywojenny* [The history of state forests and Polish forestry 1924–2004, vol. 1: The interwar period]. Warsaw: Centrum Informacyjne Lasów Państwowych, 113–39.

Daszkiewicz, Piotr, Tomasz Samojlik, and Bogumiła Jędrzejewska. 2012. *Puszcza Białowieska w pracach przyrodnikówi podróżników 1831–1863* [The Białowieża Primeval Forest in the works of naturalists and travelers 1831–1863]. Warsaw: Semper.

Faliński, Janusz. 1968. *Park Narodowy w Puszczy Białowieskiej* [The National Park in the Białowieża Primeval Forest]. Warsaw: Państwowe Wydawnictwo Rolnicze i Leśne.

Fleming, Michael. 2010. *Communism, Nationalism, and Ethnicity in Poland: 1944–50*. New York: Routledge.

Franklin, Stuart. 2002. "Białowieża Forest, Poland: Representation, Myth, and the Politics of Dispossession." *Environment and Planning A: Economy and Space* 34(8) (August): 1459–85. https://doi.org/10.1068/a34259.

Gellner, Ernest. 1998. *Language and Solitude: Wittgenstein, Malinowski, and the Habsburg Dilemma*. Cambridge: Cambridge University Press.

Grewe, Bernd-Stefan, and Richard Hoelzl. 2018. "Forestry in Germany, c. 1550–2000." In Oosthoek and Hoelzl, *Managing Northern Europe's Forests*, 15–65.

Grzywacz, Andrzej. 2014. "Adam Loret (1884–1940?): pierwszy dyrektor naczelny lasów państwowych" [Adam Loret (1884–1940?): first general director of the state forests]. *Studia i Materiały Ośrodka Kultury Leśnej* 13: 93–120. http://www.studia.okl.lasy.gov.pl/documents/1004479/27614097/Andrzej+Grzywacz.pdf/25ffd48f-7ce6-4705-a46c-c7c07c0c76ea.

Guha, Ramachandra. 1989. *The Unquiet Woods: Ecological Change and Peasant Resistance in the Himalaya*. Delhi: Oxford University Press.

Gutowski, Jerzy M. 2000. *Zasady funkcjonowania Białowieskiego Parku Narodowego: powiększonego na cały obszar polskiej części Puszczy Białowieskiej* [Principles of functioning of the Białowieża National Park: extended to the whole area of the Polish part of the Białowieża Forest]. Białowieża: Białowieża National Park.

Hedemann, Otto. 1939. *Dzieje Puszczy Białowieskiej w Polsce przedrozbiorowej (w okresie do 1779)* [History of the Białowieża Primeval Forest in prepartitioned Poland (up to 1779)]. Warsaw: Warsaw Księgarniaśw. Wojciecha.

Konczal, Agata Agnieszka. 2017. "Leśnicy a percepcja i kształtowanie wizerunku w przyrody w Polsce" [Foresters, their perception and their image of nature in Poland]. PhD diss., Adam Mickiewicz University, Polish Academy of Sciences.

Kossak, Simona. 2001.*The Białowieża Forest Saga*. Białystok: Muza S. A.

Krasiński, Zbigniew A. 1994. "Restitution of the European Bison in the Białowieża Reserve in the Years 1929–1952." *Parki Narodowe i Rezerwaty Przyroda* 4: 2–23.

Majchrowska, Anna. 2018. "The History of State Forests and Forestry in Poland." In Oosthoek and Hoelzl, *Managing Northern Europe's Forests*, 318–58.

Matteson, Kieko. 2015. *Forests in Revolutionary France: Conservation, Community, and Conflict, 1669–1848*. New York: Cambridge University Press.

Matthews, Andrew. 2011. *Instituting Nature: Authority, Expertise and Power in Mexican Forests.* Boston: MIT Press.

Michaluk, Dorota. 1997. *Dobra i miasteczko Narewka na tle dziejów regionu (do końca XIX wieku)* [The good little town of Narewka against the backdrop of history (to the end of the nineteenth century)]. Narewce: Urząd Gminy.

Mikusińska, Anna, Bernadeta Zawadzka, Tomasz Samojlik, Bogumiła Jędrzejewska, and Grzegorz Mikunsiński. 2013. "Quantifying Landscape Change in the Last Two Centuries in Białowieża Primeval Forest." *Applied Vegetation Science* 16(2) (April): 217–26. https://doi.org/10.1111/j.1654-109X.2012.01220.x.

Nadkarni, Mangesh Venkatesh. 1989. *The Political Economy of Forest Use and Management*. New Delhi: Sage.

Nikitiuk, Borys. 1958. *Czerwona Hajnówka* [Red Hajnówka]. Hajnówka: Komitet Powiatowy PZPR.

Okołów, Czesław. 2009. "History of Protection." In *Białowieża National Park: Know It, Understand It, Protect It*. Białowieża: Białowieża National Park.

Oosthoek, K. Jan, and Richard Hoelzl, eds. 2018. *Managing Northern Europe's Forests: Histories from the Age of Improvement to the Age of Ecology*. New York: Berghahn Books.

Peluso, Nancy Lee. 1992. *Rich Forests, Poor People: Resource Control and Resistance in Java*. Berkeley: University of California Press.

Rajan, S. Ravi. 2006. *Modernizing Nature: Forestry and Imperial Eco-Development 1800–1910*. New York: Oxford University Press.

Rudling, Per Anders. 2014. *The Rise and Fall of Belarusian Nationalism, 1906–1931*. Pittsburgh, PA: University of Pittsburgh Press.

Rykowski, Kazimierz. 2012. *Huragan w lasach: klęska czy zakłócenia rozwoju?* [Hurricanes in forests: disaster or disruption of development?]. Pisz: Instytut Badawczy Leśnictwa.

Samojlik, Tomasz. 2005. "Stanisław August Poniatowski in the Białowieża Forest (30 August–2 September 1784)." *Kwartalnik Historii Kultury Materialnej* 53: 35–52. http://www.zbs.Białowieża.pl/g2/pdf/1500.pdf.

Samojlik, Tomasz, Anastasia Fedotova, and Dries P. J. Kuijper. 2016. "Transition From Traditional to Modern Forest Management Shaped the Spatial Extent of Cattle Pasturing in Białowieża Primeval Forest in the Nineteenth and Twentieth Centuries." *Ambio: A Journal of the Human Environment* 45(8) (June): 904–18. http://doi.org/10.1007/s13280-016-0795-4.

Schama, Simon. 1995. *Landscape and Memory*. New York: A. A. Knopf.

Scott, James C. 1998. *Seeing Like a State: How Certain Schemes to Improve the Human Condition Have Failed*. New Haven, CT: Yale University Press.

Sivaramakrishnan, Kalyanakrishnan. 1995. "Colonialism and Forestry in India: Imagining the Past in Present Politics." *Comparative Studies in Society and History* 37(1) (January): 3–40. https://doi.org/10.1017/S0010417500019514.

Sivaramakrishnan, Kalyanakrishnan. 1999. *Modern Forests: Statemaking and Environmental Change in Colonial Eastern India*. Stanford, CA: Stanford University Press.

Sunseri, Thaddeus. 2012. "Exploiting the *Urwald*: German Post-Colonial Forestry in Poland and Central Africa, 1900–1960." *Past & Present* 214(1) (February): 305–42. https://doi.org/10.1093/pastj/gtr034.

Szafer, Władysław. 1920. "Plan utworzenia rezerwatu leśnego w Puszczy Białowieskiej" [Plan for creating a forest reserve in the Białowieża Primeval Forest]. *Chrońmy Przyrody Ojczysta* 6.

Szafer, Władysław. 1926. *On the Protection of Nature in Poland in the Last Five Years*. Kraków: Nakładem Państwowej Komisji Ochrony Przyrody.

Więcko, Edward. 1984. *Puszcza Bialowieska: problematyka pchrony Puszczy Białowieskiej w ujęciu historycznym i społecznym* [Białowieża Primeval Forest:protecting Białowieża in historical and social terms]. Warsaw: PWN.

Woolhiser, Curt. 2003. "Constructing National Identities in the Polish-Belarusian Borderlands." *Ab Imperio* 1: 293–346. https://muse.jhu.edu/article/559754.

10

The Political Ecology of Scientific Innovation in Russia

A Study of a Muskox Domestication Experiment in Siberia

Vladislava Vladimirova

> Comrades! Allow me to lead a toast to science, to its flourishing, to the health of the men of science. For the prosperity of science, that science that is not fenced off from the people does not keep herself away from the people, but is ready to serve the people, ready to transfer to the people all the achievements of science, which serves the people not under duress, but voluntarily, willingly (applause).
>
> —Joseph Stalin, 1938

Some stories from childhood stay with us, such as one my sister liked to tell. As a senior student in biology in Bulgaria, she obtained a job as a secretary at the Academy of Sciences. One obligation of junior staff was to keep the professor emeritus and founder of the institute away from the office on the days of international events. On such occasions, the stuffed sheep-goat, the result of the professor's research in crossbreeding in the 1960s, which was exhibited in the main hall, was taken away. Bulgarian scholars of the 1980s felt embarrassed by the ugly and short-lived "achievement" of his devotion to animal engineering, which lacked practical value. Nevertheless, no one wanted to insult the old, truly respected colossus of agricultural science in the country.

How is this story related to this chapter, which takes the reader to a different place and time—Northern Russia in 2010? I was interested in how research projects, especially in a practically oriented field like agricultural science, acquire legitimacy when they have little direct utility. Even more mysterious is how such projects and the fantasies behind them get long-standing support and hence social credibility, despite people's disbelief in their economic and ecological rationale. Although pure scientific curiosity can be a valid explanation, from the perspective of time, however, the hybrid from the opening story has rarely been mentioned in international scholarship, and thus its importance for the improvement of knowledge has been limited at best. One can argue, of course, that the negative result of an experiment also constitutes knowledge—it proves what does not work. An important consideration in formulating research questions, however, is cost. Was the expected result of the sheep-goat crossbreeding of such importance as to occupy the time and resources of a whole institute? The fate of the stuffed sheep-goat in retrospect calls into question the value of the experiment: it is forgotten together with other "curiosities" of out-of-date science and natural freaks, like the carnivorous Tibetan yak and cow hybrid, the three-headed lamb, and the fourteen-legged sheep in the rarely visited exhibition of the Agricultural Faculty. More likely then, as I show in this chapter, some scientific projects receive support and social credibility because of their political, rather than scientific value.

This chapter addresses the politics of science through the microscale analysis of a muskox domestication experiment in the village of Surinda in Central Siberia. The project is intrinsically interconnected with reindeer herding, which has played a dominant role in the history and culture of the local Evenki population. The initial idea for reacclimatization of muskox dates back to 1920. The muskox was aboriginal to Siberia until the late Pleistocene, and scholars claim that relic populations survived there long after (Lent 1999). At the beginning of the twentieth century, muskox remained only in Greenland, where their number was rapidly decreasing. The 1920s was a period of great interest not only in eugenics and species engineering but also in species (re)introduction, as part of designing new ecological habitats and landscapes. The objectives at the time were conservation, the restoration and enrichment of ecological habitats, agricultural experimentation, and research. Simultaneously, scientists searched for ways to improve the condition of the impoverished and socially troubled native populations, who had been deprived of many of their resources. V. Stefansson, an Arctic explorer, advocated domesticated muskox as a source of wool, milk, and meat, to replace the more vulnerable reindeer (Stefansson 1924). His ideas received support from powerful politicians of the day, such as Franklin D. Roosevelt, who was interested in imported species and stated that taming them had civilizational value for northernmost America (Stefansson 1924: 164). In the 1930s, the US Bureau of Biological Survey imported muskox and established an experimental herd in Alaska.

In Russia, between the 1920s and 1970s, forty-eight species were introduced, eleven of them for the first time in history. The stated objectives were similar to those in America. These projects have been discussed within the general context of establishing Soviet order, ideology, and science (Lent 1999). At the same time, equally ambitious plans for reforming, modernizing, and reimagining indigenous people of the North were realized. As a result, reindeer husbandry was singled out, and more specifically, a particular model of organization was put forward as most viable and productive, in the idiom of scientific evidence, yielding progress and rationality.

In the early 2000s, when I started research in the North, all its previous structures had significantly decayed, and new ones were not in place. As anthropologists observe, the radical transformation of the lives of hunter-gatherers and herders did not create the basis for a durable modern Arctic economy: ironically, the Soviet state provided the rural people with skills useful mostly for a life "as proletarians in a Moscow suburb" (Anderson 2004: 4).

To strengthen the recently reestablished reindeer herding enterprise in the village of Surinda, its director is working in close cooperation with the Research Institute of Agriculture and Ecology of the Arctic (the Siberian Branch of the Russian Academy of Sciences). Two joint projects received funding through municipal and federal programs: the first is the selective breeding of the aboriginal Evenki breed of reindeer, and the second—muskox domestication. Project participants, from both the scientific and herding domains, stress the potential support that the projects can bring to the enterprise through their attraction to authorities whose support and financial help is seen as essential for the restoration of reindeer husbandry. Scientific innovation has come to the fore in Russia again, and has become especially salient in state ideology. It is thus a discourse that is instrumental in state governance, agricultural science, and economy.

In this chapter, I show how a scientific project such as muskox domestication becomes a field of negotiation between the state administration and the Evenki people. The analysis below touches on areas such as the context of post-Soviet indigenous politics and economy, regional management, and the state of Russian agricultural science. I draw in this analysis from the history of the global exchange of ideas and scientific practice in the field of animal breeding and agricultural science across borders (see Dorondel and Şerban, introduction in this volume). I examine how different actors make sense of the experiment and variously frame their participation in it, based on their knowledge, social context, and political agenda. I try to understand what their stakes are, and finally, how science provides a framework and language for this encounter.

This study draws on one month of anthropological field research conducted in the Evenki region in September 2010.[1]

CULTIVATING EVENKI PEOPLE AND REINDEER IN HISTORY

In this chapter I borrow from analysis regarding how the field of science is navigated by states to control territories, nature, economies, and people (Scott 1998). I engage with anthropological research among the indigenous people of the Russian

North that critically examines the consequences of Soviet and post-Soviet state regulation of reindeer husbandry. Creative and more or less efficient ways to counteract, subvert, and resist such policies have also produced much academic work (Ssorin-Chaikov 2003). Scholars point to the common genealogies of state technologies of control and legacies of colonial encounters (Anderson 2011; Fondahl 1989; Gray 1998). However, a careful analysis of Russia's "crossbreeding" of ideas and approaches in its governing of indigenous populations is still relatively new to Siberian anthropology.

Cultivation

In this chapter, I use the metaphor of cultivation developed by David Anderson (2004) to analyze the role of science in reindeer economy. It provides for a nuanced understanding of the complex ecology between state officials, ideologists, scientists, and local people, and the natural environment. The notion attributes different levels of agency within each of these realms that shape relations and interactions. Through cultivation, Anderson points out, following Tim Ingold, indigenous people know the world, "which ties together moral, technical, rational and economic actions." Cultivation employs memory, listening, watching, and participation; living together within the environment (Anderson 2004: 2; Ingold 2000). The environment is cultivated through the practices of hunters and herders, but also by scientists and administrators (through their licensing regimes, monitoring practices and wildlife management). Modern state governance seems to have left similarly disruptive ecological footprints on human–animal relations in all regions of the North, irrespective of the different political regimes labeled democratic, socialist, or capitalist. Cultivation provides a lens for studying these similarities: "If rational state management of lands and people can be recognized as a form of cultivation, it can be perceived as form of pervasive monoculture with little tolerance for local styles" (Anderson 2004: 3).

Evenki

Evenki are one of the Numerically Small Indigenous Peoples of the North who number thirty thousand and live in Central and East Siberia. In the Evenki region of Krasnoyarsk Province, they are a titular ethnic group. Surinda is an isolated village that can only be reached by air transport; it was created in the 1950s at the location of an older trading post.

Before collectivization in the early twentieth century, Evenki were nomadic hunter-gatherers. Each household owned a small reindeer herd (five to thirty head) that was used for transportation (Vasilevich 1964). Reindeer herding had a primary role in the formation of Evenki social relations and institutions, and annual migratory cycle and land use were organized around it. Evenki migrated in small groups of three to four families, which at the end of the nineteenth century were losing their stability and exclusive kinship basis, due to intensive migration (Tugolukov 1985; Karlov 1982). While anthropologists debate the nature of property relations among Evenki, many believe that agreement between groups on migratory routes existed, in order to avoid overgrazing and competition. Twice a year groups migrating in a given area gathered to discuss questions of common importance and conduct rituals (Togulokov 1985; Karlov 1982). Finally, permanent meeting points were trading posts held by Russians, where Evenki came to exchange furs and meat for industrial goods (Alekhin 2001). As sites for the collection of yasak, a kind of state tax paid in-kind, such points were instrumental for the collection of knowledge about and imposition of control over Evenki (Slezkine 1994).

The State

However pre-Soviet statehood in the periphery is described—as barely present or as politically salient and providing symbols for the identification of indigenous people (Ssorin-Chaikov 2003: 1)—the Soviets took important steps to solidify it. One of the first steps in establishing "order" over Northern peoples and territories was the Polar Census from 1926 to 1927. The survey was designed to gather demographic and economic data on households as well as materials including photographs, maps, kinship charts, and artifacts. For the first time, the number and ages of domestic animals were carefully registered. For many indigenous people, Anderson argues, the idea of fixed households with an economic strategy within a yearly budgetary cycle must have been incomprehensible. The wealth that indigenous residents harvested "looked to them more as gifts of the land" (Anderson 2011: 17). Thus, the survey ended up fracturing kin connections, building people into places disregarding their annual movements, and tying them to inventories of property. Anderson sees the survey as a "biopolitical enterprise where people are placed into a relationship with their lands." The census provided a model of how a complex relationship might be described in the orderly language and dominant "performative" vocabulary of the regime. This architectural type of biopower, to use Anderson's terms, leaves open the possibility for unique forms of resistance, or mutual cultivation (Anderson 2011: 17).

As Gail Fondahl states, Soviet methods of governing indigenous people are not entirely novel. The three rationales behind the perpetuation of reindeer husbandry that she lists—(1) promoting the prosperity of indigenous people, (2) strengthening the northern economy by most efficiently exploiting a given resource, and (3) attending to development in the North for ideological reasons—have all been employed by other states, including the United States and Canada. States impose their institutions and "civilize" natives (i.e., destroy their own institutions and economic relations) allegedly to alleviate internal social inequality. Reindeer herding is also seen as supportive for the development of an industrial economy in the North, and as geopolitically important for legitimizing a claim that the Soviets observed the human rights of all citizens equally (Fondahl 1989).

As anthropologists show, the Soviet authorities also transformed people's relation to the state through rebuilding identities via complex ethnic, administrative, territorial, and economic engineering (see also Féaux de la Croix and Roberts, chapter 7 in this volume). A powerful tool in this process was collectivization and thorough reorganization of reindeer economy into cooperative and state-controlled enterprises, following the industry model. Animals and people were regrouped into productive units (brigades), and further divided into a taxonomy of categories (Anderson 2000; Ssorin-Chaikov 2003). The model of reindeer economy also mirrored the conditions of only one climatic zone, the tundra, where particular methods had historically evolved, such as intensive large-scale herding for meat production. Evenki herding developed in different geographic, climatic, economic, and cultural conditions. Thus, by imposing the Soviet system of productive reindeer nomadism, the authorities further disrupted Evenki social relations, including with the land. By the 1960s, the state forced the resettlement of nomadic Evenki into permanent villages. The short- and long-term problems that such ruptures produced have become the subject of many ethnographies (Vladimirova and Habeck 2018; Povoroznyuk, Habeck, and Vaté 2010).

The role of science in these transformations has been only partially investigated. The environmental historian Andy Bruno shows how the Sovietization of reindeer occurred in three areas: restructuring the economy, understanding locality, and understanding the environment. Reindeer economy was developed through scientific investigations, technological changes, education, collectivization, the restructuring of herd migration, and the establishment of collective farms (Bruno 2010: 120). This process was based on two tropes, which on the surface seem contradictory: besides being a species of economic utility that can be modeled through science, reindeer also represent an anachronistic animal, which because of its semidomesticated nature, binds people to the past and to the wilderness that it occupies. At the heart of collectivization was an effort to impose a new regulation of space since nomadism was seen as a prefeudal practice, and migratory animals posed problems for rational and predictable modern husbandry. The regime was obsessed with regulating the movement of human and animal populations. Bruno sees this as part of the Soviet competition with other states in imposing what James Scott calls a high-modernist ideology, a self-confidence about scientific and technical progress, the expansion of production, the growing satisfaction of human needs, the mastery of nature (including human nature), and, above all, the rational design of social order commensurate with the scientific understandings of natural laws (Bruno 2010: 136; Scott 1998).

THE ROLE OF REINDEER SCIENCE

Reindeer research and veterinary services at the beginning of the twentieth century were initiated both bottom up and top down, in response to anthrax epidemics (D'iachenko 1975: 3). Their development reflects the establishment of agricultural and veterinary sciences in Russia as a branch of medicine since the late eighteenth century. Medicine, as many other sciences, evolved in the changing social and economic context of the late nineteenth century, prompted by the abolition of serfdom (1861), the opening of new universities, and the wide popularization of scientific knowledge. An important characteristic of Russian medicine in the late nineteenth century was the idea of its public service (Hachten 2002). As a result, Soviet Russia inherited a tradition of physicians who were predominantly state employees emphasizing preventive medicine through sanitation and education, and community service (Graham 1993: 247). Russian medicine, like veterinary and agricultural science, combined imported knowledge with original ideas. The science of immunology and bacteriology, central to medicine, for example, evolved in cooperation with the Pasteur Institute in France in the late nineteenth century (Ulyankina 1990).

While critical study of reindeer science's genealogy and politics is still in an initial phase, its importance for the Soviet authorities in managing animal and indigenous populations is unquestionable. Yuri Slezkine provides examples of the discourse of veterinary science in the years of establishing the Soviet state in the North. Health, including that of animals, was of vital importance for political stability, productive industry, and military power (Slezkine 1994). A parallel with Bruno Latour's conclusions about the importance of immunology and medicine in colonization can be productive for analysis (Latour 1988: 142). Soviet health was grounded in the German theory of cameralism, and contributed to surveillance of the population by creating treatment and monitoring centers. The Soviet political objectives and specific categorization of home, body, life, and leisure within public and private spheres created health control through programs that were more invasive and pervasive. Soviet conceptions of dirt and hygiene, of utopian societies, perfected citizens, and state authority alongside new techniques for control, performed in the realm of science and medicine, were entangled in ethnic, class, and political categorizations that informed the construction of institutions and identities (Starks 2009: 7). Agricultural and veterinary sciences were not an exception, not least by an insistence on controlled and selective breeding, as well as taming and domestication. These specific mechanisms of biopower also framed scientists' maneuvering within a social space ruled by public and moral obligations (Hachten 2002: 208).

In 1931 the Scientific Research Institute of Reindeer Husbandry was created. Its work revolved around increasing efficiency, labor organization, capital investment, pasture use, forage and diet, immunizations and guarding of herds, herd structure, and application of modern technology. Soviet science has praised itself for its success in assessment of breeds and individual animals' quality, methods of selective breeding, and maintenance of breeds in large-scale agricultural production (Borozdin et al. 1979: 15). What distinguishes Russia from other countries is its intensive application of scientific methods in reindeer husbandry, but this can hardly be unexpected, given the minuscule scale of reindeer husbandry in other countries.

Like other modern states, the Soviet Union ideologically constructed a belief that agriculture is a production economy, and most problems that occur are amenable to technical solutions. The Soviet literature on reindeer husbandry and its poetics attribute utmost importance and authority to science (Borozdin et al. 1979: 29). My work in other parts of Russia shows that a middle-range sovkhoz (8–10 herding teams and 30,000 reindeer) would hire at least 1–2 veterinary doctors and 3–4 zoo technicians. I have also discussed the increasing complexity of reindeer herding methods and procedures described in the specialized literature (Vladimirova 2006). Such prescriptions were not easy or even possible to follow and their disciplinary role seems prominent.

Despite ideological vigor, political and financial support, the success of agricultural science in maintaining an efficient reindeer economy was limited. Reindeer husbandry in many areas was experiencing a crisis well before the dissolution of Soviet structures. Before the 1980s, failures were rarely discussed, and problems were attributed to work discipline. During perestroika, problems attracted more public attention, but they were explained as the resistance of indigenous culture to adapt, or attributed to environmental conditions. This reproduced classical tropes of thinking about reindeer and indigenous people as "uncivilized" (Vladimirova 2011). Among others, the problems included so-called nonproductive losses because of difficulties in guarding the herd, the extreme workload on herders, increase of wild reindeer, and bad work discipline. The persistent application of random innovations and rationalizations, such as mass-scale building of fences, and their subsequent rejections, brought disorientation and demoralization. Nikolai Ssorin-Chaikov coined the term "poetics of unfinished construction" to grasp the politics of replacement of one innovation with a "new" one, in order to further the notion of progress and distract attention away from the disastrous results (Ssorin-Chaikov 2003). In 1992, it finally became obvious that the main problem of reindeer organization was the rupture with the preceding Evenki herding tradition (Boiko and Kostiuk 1992: 132). These problems should not be taken as "single examples of sloppy science," but as the results of a driving ideology, a competing idiom of cultivation: the necessity for there to be a clear, rational, accountable way of planning nature's providence (Anderson 2004; Beach 2004).

The withdrawal of state support for reindeer herding in the 1990s exacerbated existing problems, bringing enterprises into deep crisis and even bankruptcy. In late Soviet time, Surindinskii Sovkhoz was specialized in selective breeding (*plemzavod*). It was presented as "one of the most important spheres of exercising traditional work habits of the indigenous population" (Boiko and Kostiuk 1992: 82). In 1992 it was transformed into a joint-stock company, and later it underwent privatization. The stock was distributed among local residents and by 1997 the number of reindeer dropped from 10,000 to 1,000. One herd leader preserved 500 reindeer belonging to 20 families (Anderson 2004: 4), and they became the basis for the present restoration of the enterprise in 2003, by the initiative of the municipality. This is attributed to the installation in Krasnoyarsk District of a new governor, who, as a high-level business manager, became upset with his team's inability to alleviate the significant unemployment, poverty, and low life expectancy in indigenous communities. Taking advice from Evenki political and civil representatives, he appointed a director to rebuild the enterprise. As an indigenous representative whose identity was deeply inscribed in this locality, he is considered the force behind slow but perceptible improvement and stabilization. "The all-terrain vehicle might be old and break often, but it goes, it goes!" expresses the trust that local Evenki have in the current director. His excellent qualities in negotiating local interests is captured in the following comment: "He respects both scientists and herders."

In an interview, the director explained that despite their support, the regional authorities do not see reindeer breeding as business that can bring profit if wise investment and management are provided. Rather, they look down upon it and subsidize it as a form of aid to a vulnerable societal group. Following the Soviet practice, Evenki are treated as a low-income population in need of patronship and aid instead of as entrepreneurs and producers. As Ssorin-Chaikov argues, the state constructs Evenki identity both as a projection of itself and as a limit. This limit is expressed within the nature–civilization dichotomy (Ssorin-Chaikov 2003: 207). Indigenous people are identified with nature and in opposition to civilization and rational economy, as the director's comments show. Within this framework, Evenki in turn use entrepreneurial creativity to cultivate state authorities in other arenas that they find more advantageous, like reindeer science and more specifically, the Evenki breed of reindeer.

MUSKOX DOMESTICATION AS A DRIVER OF REINDEER ECONOMY

Soviet scientists most probably imported the idea of muskox domestication together with the import of the first animals from Canada and Alaska, where John Teal from the Institute of Northern Agricultural Research started his famous experiments in the 1950s. The first phase of Teal's experiments showed that muskox calves feel little fear of or aggression toward humans, quickly respond to feeding and human contact, and reach almost double the size expected in the wild (Wilkinson and Teal 1984). The second stage, measuring the feasibility of large-scale breeding, showed that from ten male and twenty-four female calves in 1960, the herd totaled seventy-eight in ten years (Wilkinson 1971).

From the start, improvement of the economic and social condition of native peoples of the Arctic based on an indigenous species was an objective of muskox farming. At the same time, an inherent presumption is that scientific and managerial knowledge are a better ground for judging productivity and land use. In the project founder's words, "what is sought is an animal which can convert northern grasses into ready employment and cash incomes" (Wilkinson and Teal 1984: 39). This rhetoric points to the transformative mission of the project: spreading economic rationality that is in harmony with modern capitalist values rather than indigenous hunter-gatherers. For many politicians, however, the main objective of the project remained conservation and scientific research, rather than the development of indigenous people.

Muskox breeding experimentation revealed the huge requirement for pasturelands, and the rapid pace of their deterioration, as well as their vital significance for animal health and mortality (Lent 1999). It also showed how investment-dependent it is, the lack of prospects for self-support based on sales of its products. Muskox breeding is also heavily reliant on hay procurement, which in Arctic conditions means transport costs that are too high. After all, in the 1980s, the state of Alaska incorporated nonprofit muskox farms and set forth the following objectives: the raising and maintenance of muskoxen by improving the condition and breeding of the animals, experimentation with methods for the production of wool, and finally the increase of public knowledge about muskox through informational programs, exhibition of the animals, and tourism (Lent 1999: 239).

Importing Muskox in Russia

Despite its interest earlier in the century, Russia managed to import muskox only in the mid-1970s, bringing the species into the realm of international and national politics and diplomacy. In the early 1970s, President Nixon visited the Soviet Union; as a result, the two nations signed a series of agreements for environmental cooperation. In 1974, Canada sent ten calves to Siberia, and later Russian scholars traveled to Alaska to study muskox management and biology. In exchange, the Soviets offered the US Fish and Wildlife Service ten Siberian polecats. Even though American scholars did not plan to release polecats in the wild, in order to be polite they accepted them and stated that they needed them for experiments. It is an interesting detail that the export of muskox for USSR was exploited by regional politicians to push American legislators to allow muskox hunting in Alaska (Lent 1999).

Soviet biologists summarized the goals of the project as (1) filling an empty ecological niche and better exploiting vast areas of Arctic pastures, (2) strengthening Arctic ecosystems, and, (3) consolidating the position of the species in the world. Domestication and farming of muskox was also mentioned as a possible future task, for the production of meat, milk, and wool (Uspenskii 1984).

In 1983 the Council of Krasnoyarsk made a decision to establish the state reserve (*zakaznik*) Bikada on 870,000 hectares of land for the conservation of the muskox (Gaydin and Burmakina 2014). The muskox population has unevenly but steadily increased from 1,250 in the 1990s, to 2,000 by the early 2000s, and to 3,500 in 2003 (Sipko et al. 2007: 621). In response to its successful adaptation, in 2000 it was allowed for sport hunting, which raised criticism among biologists (Maskin 2004). Stable growth allowed for the expansion of muskox into other parts of Northern Russia, such as Sakha Yakutia, where at the end of 2012 scholars counted approximately 8,000 animals (Gaydin and Burmakina 2014). The next targets are Yamal and Nenets Autonomous Region. The existing literature explains that the objectives of such expansion are the enrichment of biodiversity and the increase of the resource basis for leisure hunting and tourism. Its proponents claim that it helps exploit little used Arctic tundra without undermining the food basis for reindeer and horse husbandry. Russian scholars predict that wild reindeer will not compete with muskox, since both species come into the same areas only in the summer, but the abundance of grazing in this season mitigates competition. Further, the authors present examples of successful coexistence in Canada (Terekhov, Popov, and Iakovlev 2015), which, however, contradict Canadian studies and the experiences of indigenous people (Nagy 2004).

Muskox Domestication

In 2010, ten muskox were transported on an airplane from Taymyr to Surinda, where they were placed in a small fenced compartment near the enterprise office building. The trial phase of the experiment was one year. If in this time some progress in their domestication was achieved, the enterprise could apply for participation in a federal program. According to the director, economists calculated that if domestic muskox are bred on a mass scale, they will bring more profit than cattle. They are better adapted to the Arctic and require less human labor to protect them from predators than reindeer. They can be a good source of meat, hides for warm clothing, and wool. The latter is considered to be the highest quality in the world (Schell 1972).

Field data from Surinda indicate no knowledge of the domestication experiments in Canada or Alaska. We sensed innovative spirit in people's attitude and they emphasized being the first in Russia to breed muskox. Employees of the enterprise as well as local residents stated they are going to use the same methods of domestication as their ancestors who have tamed reindeer. When the snow falls, the director explained, they will let the muskoxen graze together with the reindeer in the forest. "We will start taming them to let themselves be caught, just like reindeer, and make them accustomed to being near other animals and people." This is obviously seen as the first step in a process of cultivating wild animals after their capture. Stories that informants were told in their childhood described how Evenki tamed reindeer by secretly observing when a wild female would give birth and then capturing the newborn calf. When the calf could stand on its legs, they would tie it "like a yeanling." The mother always stays around the calf, so the more calves people capture, the bigger the herd they will manage to accumulate.

Transferring reindeer characteristics and herding knowledge to another species is a phenomenon observed among other Arctic people. Nuccio Mazzullo, for example, analyzing the beginning of cow breeding among indigenous Sami people in Finland, argues that cows were initially conceived of in analogy to reindeer and treated accordingly (Mazzullo 2018). An Evenki woman pointed out that as a child, her parents encouraged her to interact with a young calf, in order to teach it to trust humans and to start following her as if she were its mother. This kind of child play-acting is an important part of cultivating reindeer in pastoralism. Thus, in the Evenki understanding, an important factor for taming an animal is daily human interaction. Ethnographic accounts also describe Evenki practicing calving in fenced territories. While scholars do not explain the practical role of this custom, M. G. Turov revisits older ethnographic observations that calves born in the bush tend to be less tame and more difficult to control later. He also quotes reindeer biologists asserting that calves that were close to people right after being born perceive humans as closer to their species in their later lives (Turov 1990: 110).

Further methods of domestication, according to Evenki, involve "teaching muskox" by harnessing the females in sleds, castrating males in order to prevent territorial behavior, and using them for transport. Metal fences can be erected to prevent muskox from escaping and becoming feral. Scientists agreed that domestication requires a long time for new features to become genetically transmitted. Herders, however, emphasized continuous, lifelong cultivation of animals by their living together in constant interaction, rather than through genetic inheritance. People evoked the notion of "living" in phrases such as "living with reindeer" and "we need reindeer to live," thus reiterating the mutuality in cultivation. They contrasted this notion with other ethnic groups' exploitation-oriented views of reindeer, predominantly for transport and food. Exploitation was stated as the reason that such groups eventually lost their reindeer and would never become herders again. To be a reindeer herder, an Evenki should constantly live with the reindeer from early childhood. "If one does not work with the herd every day, lifelong . . . even the reindeer here, in the brigade, will become wild. This is what our authorities in the past disregarded. This is what the boarding school did with our kids, a huge shift. . . . If we want to improve things we need to start teaching our children again to live with reindeer." Scholarly traditions, which describe human–animal relations in pastoralism using dichotomies, would contrast here the kinds of relations envisaged by Soviet science as domination, and by Evenki as traditional knowledge, as care. Anderson and colleagues (2017) have offered a more nuanced vocabulary for studying emplaced human–animal relations, as part of particular landscapes. They use the notion of domus, to describe long-term human–animal interactions, where "many agents work together in evocative settings to create the conditions for a good life." Such vocabulary is less deterministic and better accounts for the dynamic of evolving, adaptability, and cultural peculiarity of human–reindeer relations among Evenki, where "animals and people come together and move away from each other in a type of communicative dance that acknowledges the other but also maintains the differing preferences of the various participants who share a landscape" (Anderson et al. 2017: 412). Similar mutual cultivation is inherent in expectations of how muskox will be domesticated. It would be wrong to attribute the avoidance of the discussion of genetic inheritance to inadequate education, ethnic culture, or the Soviet history of ideas about genetic transmission. Discussing the famous experiment of fox domestication in Siberia in the 1950s through insistent rigid selection of tameness, Loren Graham underlines that its leader, Dmitry Belyaev, believed that Trofim Lysenko, the doyen of the Soviet theory of the inheritance of acquired characteristics under the influence of the environment, where genetic pedigree meant nothing, was wrong. Graham also writes about his visit to Belyaev's fox farm in 1976, when one of the farm assistants explained to him that the foxes were tame "because we take such good care of them and because we love them. We constantly stroke them, supply them the best food, give all of them names, call them individually by these names, and show our affection for them. They respond by returning our love, and that love becomes hereditary" (Graham 2016: 10). Belyaev later commented on Graham's observation that it would be extreme to call his assistants followers of Lysenko. They are simply supporters of the doctrine of the inheritance of acquired characteristics by observation, leaving their abstract school knowledge of genetics disconnected from practical reality (Graham 2016). The scientists involved in the muskox experiment also seemed positive about the Evenki vision of domestication methods organized by the metaphor of "living."

The interview with the director of the Surinda enterprise also revealed a pragmatic perspective. The experiment was initiated by the regional government, which had been persuing him for two years. The director's delayed and hesitant agreement was praised and his request for the purchase of 1,500 reindeer, long ago filed with the agriculture ministry, was immediately approved. He was promised to receive the same number during the following year. "The reindeer make it worth the trouble of domesticating muskox." was his final comment on questions about the economic rationale behind the experiment. This is indicative of the political and disciplinary role that the domestication experiment has in the Northern periphery. It communicates the dominant values and practices to local indigenous populations and exerts direct and indirect pressure to accept and internalize them. Despite popular hesitation about the feasibility of the project in the local community and social media, and the list of complications in preceding experiments abroad, the reindeer-herding enterprise is coerced to participate, as this seems to be the only way to receive much needed support from the municipal authorities.

SCIENTIFIC INNOVATION AS A POLITICAL FIELD

This chapter began with a discussion of Soviet "cultivation" of reindeer herding communities through a number of social, economic, and political innovations. In the spirit of the time, one of these innovations was the development of knowledge and the political role of veterinary and *zootechnia* sciences. In the course of time, they acquired a powerful position in reindeer economy, transforming and standardizing a number of life practices, previously based in a multiplicity of reindeer herding cultures. This transformation included animal engineering as much as social engineering, in which reindeer were "cultivated" and reshaped through different procedures (Borozdin et al. 1979). Within this complex ecology, both indigenous people and animals adapted and created their ways of resistance and "cultivation" of state-envisaged routines and values (Anderson 2004), through a continuously evolving conversation with state power (Konstantinov 2015).

As anthropologists have shown, the skills and structures of Soviet reindeer husbandry in different regions could not survive without the structures imported and subsidized by the center. Thus the crisis caused by the dissolution of central control and finance was seen by some scholars as regional adaptation (Anderson 2006). As David Anderson and Hugh Beach remind us, modern states, Russia as well as Sweden, fail in their attempts at "cultivating" reindeer husbandry along scientific rationality because of blindness to its historical, cultural, and environmental contexts (Anderson 2004; Beach 2004). Rather, science provided a powerful ideology, legitimacy, and discourse for transformation and control of people and territories in the Northern peripheries. Scientific practices and rhetoric could simultaneously be reversed to accommodate and help practices of resistance to persevere.

The current attempt to rebuild Evenki reindeer herding is a complex political and economic undertaking that is navigated from different directions simultaneously: by the authorities, by Evenki ethnic politicians and reindeer experts, by scientists, and finally by local reindeer herders and community members. Some of these actors have divergent visions of reindeer herding methods, economic and political rationales, and finally future prospects. All of them are interested in different aspects of the reindeer economy complex and are thus thrust into ongoing communication and negotiation. Not all of them speak the same language or are able to understand each other (Konstantinov in this volume). In addition, different stakeholders are entangled in unequal and complex relationships that can rapidly change. In this chapter, the area of muskox domestication as a driver of reindeer herding economy was explored. Thus, the domestication experiment can be seen as a part of politics of cultivation of the Northern periphery: political, economic, and ideological.

Evenki reindeer breeders complain that the authorities do not understand their culture and its existential multisided dependence on reindeer. By their cooperation with scientists in an innovative project such as muskox domestication, they hope to raise popular respect for the enterprise in Surinda. Participation in scientific work helps make Evenki reindeer herding visible in an innovative and symbolically more prestigious way. Moreover, it can secure funding for the enterprise. It is also a way of communicating that local people hope will be audible and reach the authorities, and be better than insistence on heritage and preservation of traditions, which have been dominant idioms in other reindeer herding regions of Russia (Habeck 2005).

From the perspective of the authorities, the value of the experiment, although higher than reindeer husbandry, has not been articulated. Other than some ideologically loaded texts repeating the objectives of muskox domestication as stated abroad in the twentieth century, careful analysis of the practical issues of muskox farming are lacking. The problems that similar projects faced elsewhere have been neglected by both scientists and herders. In this sense, the economic rationale of the project is vague at best. The attraction of the project seems to be mostly in the area of scientific curiosity and the rethoric of innovation (Ábrán in this volume).

Why has science, which seemed to be heavily marginalized after the dissolution of the Soviet Union, come to take the role of a mediator between an indigenous community and politicians? This role can be seen on the level of language and communication. Science in the contemporary Russian political economy has been tasked with helping to legitimize national government. While I cannot provide here a detailed analysis of the process, I will reiterate a few points that can contextualize this study. As part of reviving its image as a responsible and respected state, Russia is eager to revaluate Soviet heritage in a number of fields, including science, and in direct relation to this study, of genetics. Loren Graham observes "the rebirth of Lysenkoism." Lysenko became influential in the 1930s, when the Communist Party proclaimed Mendelian genetics suspicious and favored Lamarckian biology (Graham 2016: 67). Lysenko received the support of the authorities, who subjected his opponents to repression (Trapezov 2009). Lysenko believed in the inheritance of acquired characteristics and he spoke of the "internalization of the environmental conditions" by organisms, which he thought was the means by which an organism acquired heredity. Thus, inheritance is tied to the materialistic view of Marxism: "The heredity of a living body is built up by the conditions of the environment over many generations. Each alternation of these conditions leads to a change of heredity, a process he called 'the assimilation of external conditions'" (Graham 2016: 92). A number of publications have appeared in the past decade (often written by nonspecialists) that aim at the rehabilitation of Lysenko, pointing to contemporary epigenetics, a branch that shows that environmental conditions can indeed influence the expression of genetic information through different inhibitors (Graham 2016). Many analysts relate this attempt to the revival of nationalism, nostalgia for Soviet times, and anti-Western sentiments. Finding in old figures like Lysenko a reason to claim superiority over Western biologists accords with these trends. This revival should thus be seen more as a political process than a scientific debate (Graham 2016: 120).

It would be a stretch to connect the proposed revival of Lysenko to muskox domestication in Surinda. None of the scholars involved hinted at such a connection or spoke of the inheritance of acquired characteristics. Both phenomena however belong to a particular political environment and indicate an ideological shift in the status of science and innovation in Russia. While such status is perhaps more discursive than inscribed in practices, it restores science as a field for communication and interaction between actors such as politicians, intellectuals, and local populations. One more example can be seen in the words of the enterprise director that the regional authorities imposed the experiment on him, in their attempt to receive resources and the approval of their superiors at the federal level.

An important tool that scientists seem to possess in addition to symbolic and institutional authority is the rhetoric of innovation. The publications of the FKNC on Evenki breed selection in Surinda present multiple examples of topics that resemble Soviet herding science, but are replete with the rhetoric of innovation. The problems being solved in this research—such as the small increase in newborn calves' body mass by the artificially induced simultaneous impregnation of female deer—also seem far from relevant for the present state of reindeer husbandry. In this context, claiming to realize the old Soviet ambition of muskox domestication sounds perhaps like a politically attractive innovation. With the rhetoric of innovation, science not only secures its financial prosperity and higher status in society. It also extends some of its historically established symbolic legitimacy to state authorities and their political and economic projects.

Through their participation in scientific innovations, Evenki people speak the language of the state in making a symbolic claim for the legitimacy of their area on the political and economic map of the country. To use the rhetoric of the Evenki municipal authorities, indigenous people are entering a "new civilization." In 2010, the municipality launched a social project funding competition under the name "Northern Village—New Civilization." The reindeer selective breeding enterprise and its participation in scientific experiments was a high point in motivating the proposal submitted by Surinda (Krasnoiarskii krai 2010). Thus, science seems to be the way to "New Civilization" for indigenous people, following a very familiar Soviet rethorical scheme.

NOTES

Epigraph: I. Stalin, "Rech Tov. Stalina na prieme v Kremle rabotnikov vyshei shkoly" [Speech by Comrade Stalin at a meeting of members of the Higher School], May 17, 1938, *Sovetskaia Etnografiia: Sbornik Statei* 1:3–4.

1. This research has been done under the program "Dynamics of Circumpolar Land Use and Ethnicity (CLUE): Social Impacts of Policy and Climate Change," funded by the National Science Foundation of the United States (grant no. 213665).

REFERENCES

Alekhin, Konstantin Anatol'evich. 2001. "Etnokul'turnaia kharakteristika lokal'nogo soobshchestva (na materiale evenkov Surindy): problemy kommunikativnoi kul'tury" [Ethnocultural characteristics of local communities (on the example of Evenki from the village of Surinda: problems of communicative culture]. Russian Academy of Sciences, Siberian Unit.

Anderson, David. 2000. *Identity and Ecology in Arctic Siberia: The Number One Reindeer Brigade.* Oxford: Oxford University Press.

Anderson, David. 2004. "Reindeer, Caribou and 'Fairy Stories' of State Power.' In Anderson and Nuttall, *Cultivating Arctic Landscapes,* 1–16.

Anderson, David. 2006. "Is Siberian Reindeer Herding in Crisis? Living with Reindeer Fifteen Years after the End of State Socialism." *Nomadic Peoples* 10: 87–104.

Anderson, David. 2011. 'The Polar Census and the Architecture of Enumeration: The 1926/27 Soviet Polar Census Expeditions.' In David Anderson, ed., *The 1926/27 Soviet Polar Census Expeditions.* New York: Berghahn Books, 1–32.

Anderson, David, and Mark Nuttall, eds. *Cultivating Arctic Landscapes: Knowing and Managing Animals in the Circumpolar North.* London: Berghahn Books.

Anderson, David G., Jan Peter Laurens Loovers, Sara Asu Schroer, and Robert P Wishart. 2017. "Architectures of Domestication: On Emplacing Human–Animal Relations in the North." *Journal of the Royal Anthropological Institute* 23: 398–416.

Beach, Hugh. 2004. "Political Ecology in Swedish Saamiland." In Anderson and Nuttall, *Cultivating Arctic Landscapes,*110–23.

Boiko, V. I., and V. G. Kostiuk. 1992. *Evenki basseina Eniseia* [The Evenki of Enisei River]. Moscow: Nauka.

Borozdin, E. K., V. A. Zabrodin, P. N. Vostriakov, N. O. D'iachenko, V. V. Kriuchkov, and V. N. Andreev. 1979. *Seevernoe olenevodstvo* [Reindeer husbandry]. Moscow: Kolos.

Bruno, Andy. 2010. "Making Reindeer Soviet: The Appropriation of an Animal on the Kola Peninsula." In Jane Costlow and Amy Nelson, eds. *Other Animals: Beyond the Human in Russian Culture and History.* Pittsburgh, PA: University of Pittsburgh Press, 117–37.

D'iachenko, N. O. 1975. "Socialisticheskie preobrazovaniia i perspektivy razvit'ia severnogo olenevodstvo" [Socialist transformations and prospects for the development of reindeer Husbandry]. *Voprosy zhivotnovodstvo na krainem severe: Nauchnye trudy SO VASKhNIL* 21: 3–18.

Fondahl, Gail. 1989. "Native Economy and Northern Development: Reindeer Husbandry in Transbaikalia." PhD diss., University of California at Berkeley.

Gaydin, S. T. , and G. A. Burmakina. 2014. "Istoria rasseleniia, akklimatizatsii i reakklomatizatsii pushnykh zverei i promyskovykh zhivotnykh v Krasnoiarskom Krae" [Hstory of settlement, acclimatization, and reacclimatization of fur-bearing animals and hunting animals in Krasnoyarsk krai]. *Vestnik KrasGAU*: 269–75.

Graham, Loren. 1993. *Science in Russia and the Soviet Union: A Short History.* Cambridge: Cambridge University Press.

Graham, Loren 2016. *Lisenko's Ghost: Epigenetics and Russia.* Cambridge, MA: Harvard University Press.

Gray, Patty. 1998. "Pursuing a Native Movement in a Russian Space: The Predicament of Indigenous Peoples in Post-Soviet Chukotka." PhD diss. University of Wisconsin–Madison.

Habeck, Joachim Otto. 2005. *What It Means to be a Herdsman.* Muenster: Lit verlag.

Hachten, Elizabeth A. 2002. "In Service to Science and Society: Scientists and the Public in Late-Nineteenth-Century Russia." *Osiris* 17: 171–209.

Ingold, Tim. 2000. *The Perception of the Environment: Essays on Livelihood, Dwelling and Skill* London: Routledge.

Karlov, Viktor Vladimirovich. 1982. *Evenki v XVII–nachale XX v. (Khoziastvo i sotsial'naia struktura)* [Evenki at the eighteenth to beginning of the twentieth century (economy and social structure)]. Moscow: Moscow State University.

Konstantinov, Yulian. 2015. *Conversations with Power: Soviet and Postsoviet Developments in the Reindeer Husbandry Part of the Kola Peninsula*. Uppsala: Uppsala University: Uppsala).

Krasnoiarskii krai. 2010. *Poiasnitel'naia zapiska: proekt "Severnoe selo—novaia civilizatsiia"* [Motivation: project "Northern Village—New Civilization"]. Krasnoyarsk krai, Evenkiiskii municipial'nyi raion, Poselon Surinda.

Latour, Bruno. 1988. *The Pasteurization of France*. Cambridge, MA: Harvard University Press.

Lent, Peter C. 1999. *Muskoxen and Their Hunters: A History*. Norman: University of Oklahoma Press.

Maskin, Viktor. 2004. "'Eksperiment' s eksperimentom" [To experiment with the experiment]. In *Zapoliarnaia Pravda*. Noril'sk.

Mazzullo, Nuccio. 2018. "Breeder's Knowledge of Cows: Cow as Reindeer." In the project *Arctic Ark: Human–Animal Adaptations to the Arctic Environment: Natural and Folk Selection Practices*. Rovaniemi, Finland.

Nagy, Murielle. 2004. "'We Did Not Want the Muskox to Increase': Inuvialuit Knowledge about Muskox and Caribou Populations on Banks Island, Canada." In Anderson and Nuttall, *Cultivating Arctic landscapes*, 93–109.

Povoroznyuk, Olga, Joachim Otto Habeck, and Virginie Vaté. 2010. "Introduction: On the Definition, Theory, and Practice of Gender Shift in the North of Russia." *Anthropology of East Europe Review* 28: 1–37.

Schell, Lillian Crowell. 1972. "The Musk Ox Underwool, Qiviut: Historical Uses and Present Utilization in an Eskimo Knitting Industry." PhD diss., University of Alaska Fairbanks.

Scott, James C. 1998. *Seeing Like a State: How Certain Schemes to Improve the Human Condition Have Failed*. New Haven, CT: Yale University Press.

Sipko, T. P., A. R. Gruzdev, S. S. Egorov, and V. G. Tikhonov. 2007. "Analiz protsessa introduktsii ovtsebyka na severe Azii" [Analysis of the Process of Introduction of Musk Ox in Northern Asia]. *Zoologicheskii Zhurnal* 86: 620–27.

Slezkine, Yuri. 1994. *Arctic Mirrors: Russia and the Small Peoples of the North*. Ithaca, NY: Cornell University Press.

Ssorin-Chaikov, Nikolai V. 2003. *The Social Life of the State in Subarctic Siberia*. Stanford, CA: Stanford University Press.

Starks, Tricia. 2009. *The Body Soviet: Propaganda, Hygiene, and the Revolutionary State*. Madison: University of Wisconsin Press.

Stefansson, V. 1924. *The Northward Course of Empire*. New York: Harcourt, Brace.

Terekhov, S. I., A. L. Popov, and F. G. Iakovlev. 2015. "Rasshirenie areala i uvelichenie chislennosti obtsebyka v tundrovoi zone iakutii" [Increasing the number and distribution of musk ox in the Yakutian tundra]. *Ekonomika Vostoka Rossii*, no. 1: 62–67.

Trapezov, O. V. 2009. "Darvinizm i uroki rosiisskoi prakticheskoi selektsii" [Darwinism and the lessons of practical selection]. *Vestnik VOGiS* 13: 249–97.

Tugolukov, V. A. 1985. *Tungusy (evenki i eveny) Srednei i Zapadnoi Sibiri* [Tungus (Evenki and Eveny) of Central and Western Siberia]. Moscow: Nauka.

Turov, M. G. 1990. *Khoziastvo Evenkov taezhnoi zony srednei Sibiri v kontse XIX–nachale XX v.* [Economy of the Evenki People in the Forest Zone of Central Siberia at the end of the nineteenth to the beginning of the twentieth century]. Irkutsk: Irkutsk University.

Ulyankina, T. 1990. "Origin and Development of Immunology in Russia." *Cellular Immunology* 126: 227–32.

Uspenskii, S. M. 1984. "Muskoxen in the USSR: Some Results and Perspectives on Their Introduction." In D. R. Klein, R. G. White, and S. Keller, eds., *Proceedings of the First International Muskox Symposium*. Fairbanks: University of Alaska, 12–14.

Vasilevich, G. M. 1964. *Tipy olenevodstva u tungusoiazychnykh narodov v sviazi s probemoi rasseleniia iz Sibiri* [Types of reindeer herding among Tungus-speaking peoples in connection with the problem of their settlement in Siberia]. Moscow: Nauka.

Vladimirova, Vladislava K. 2006. *Just Labor: Labor Ethic in a Post-Soviet Reindeer Herding Community*. Uppsala: Uppsala Universitet.

Vladimirova, Vladislava K. 2011. '"We Are Reindeer People, We Come from Reindeer': Reindeer Herding in Representations of the Sami in Russia." *Acta Borealia* 28: 89–113.

Vladimirova, Vladislava, and J. Otto Habeck. 2018. "Introduction: Feminist Approaches and the Study of Gender in Arctic Social Sciences." *Polar Geography* 41: 145–63.

Wilkinson, Paul. 1971. "The Domestication of the Musk-Ox." *Polar Record* 15: 683–90.

Wilkinson, Paul, and P. N. Teal. 1984. "The Muskox Domestication Project: An Overview and Evaluation." In D. R. Klein, R. G. White, and S. Keller, eds., *Proceedings of the First International Muskox Symposium.* Fairbanks, Alaska. Biological Paper of the University of Alaska, Special Report, 162–66.

11

Humans, Predators, and State Projects

A Look at the Lower Danube, Northwest Bulgaria

Yulian Konstantinov

In this chapter, I propose that a proper understanding of the human impact on nature will be advanced by including nonhuman animals in the reflective process. My chosen animal for performing the tasks of observation and analysis has been the golden jackal (*Canis aureus*). To explain how all this happened, I introduce the site, the local people, and the jackals themselves.

The site forms part of the Bulgarian bank of the Lower Danube, some distance downriver from the port city of Lom. A rectangular shape of roughly two by one kilometers, the riverbank forms its longer side. This faces the Romanian lowlands, about a kilometer across the water. The opposite, hinterland side of the rectangle meets a steeply rising slope. Climbing it, one sees the fertile Mizian Plain—level land, stretching all the way south—to the Balkan Mountains on the far horizon.

The ethnography for this chapter comes from many years of fieldwork at this location. My first visit was in the summer of 1976 when I pitched my tent on the riverbank. This location for the camp offered the advantage that I could communicate with local fishers and other villagers. Conversations, stories, fairy tales, fables, and anecdotes referred to events from times immemorial to the day of telling. Conversations took place while performing fishing tasks or having a break between them. During the summer, fishing was done at night, with long breaks by an open fire on the beach after each hauling in of a net. This setting for campfire talks was the main stage for telling stories of past events and commenting on current ones.

Not long after the political changes of 1989, the nightly campfire talks came to be increasingly accompanied by the howling of packs of jackals.[1] These were new voices. For my companions they stimulated a reading of recent events as the emergence of hitherto dormant evil. Where had the jackals come from? From what hiding places had they crawled forth? These questions were often asked side by side with questions about suddenly conspicuous crime. "Where had all the '(mafia) mugs' (*mutri*) been before?" Uncle Kircho, an elderly fisherman, would ask in a trembling voice.[2] "In what holes had they been hiding?"[3]

My companions were not in the mind of delving into the specialized literature on that topic, regarding neither the jackals nor what they saw as the human counterparts of the animals. Both appearances were seen as sudden eruptions with an identical source. That was "the exploding of democracy" (*kogato izbukhna demokratsiyata*), as the events of 1989 were usually called.

People's connecting in the same breath jackal and "mafia" emergence, on the one hand, and "the explosion of democracy" in 1989, on the other, suggested the coming out in the open of evil, previously suppressed. This motif organized a larger fable-like historical narrative of events from ancient times to the present day. Predators, as impersonators of evil, were prominent protagonists in this narrative.

The advance and retreat of predators formed the structuring device through which my interlocutors delineated three successive periods in recent Bulgarian history. The first was most often called "the wolves' times" (*vulchite vremena*), or "the shaggy years" (*runtavite godini*), associated with the shaggy fur of the wolf and simultaneously with violence. This stretched from the communist turnover of 1944 until about the mid-1950s.

After the "wolves' times," the stories told of a predator-free period of calm. Its beginning came with Stalin's death in 1953 and the rise of Todor Zhivkov as the supreme and long-lasting ruler of recent Bulgarian history. For the villagers the predator-free period came to a sudden and totally unexpected end with the events of 1989.

Of these new times after 1989 the jackal was forming as an emblem. Unlike the wolf of the pre-Zhivkov period, it had not yet stabilized in a set phrase. What impressed was a general perception of sudden and overwhelming emergence of a smaller, but wilier predator. As I point out further down, zoologists themselves were noting a spectacular increase in jackal population and habitat (the latter by a factor of thirty-three), and were asking in amazement, "What is going on?" (Markov 2012).

CHOOSING A PATH

It is against such a background and with such considerations in mind that I put forward my central argument. To understand the course of local events in the period from 1944 until the present in the terms of its grassroots presenters, one needs to approach it *as a whole,* uniting humans, animals, and the infrastructural projects that affected their environment.

I have found a theoretical inspiration for such an analysis in studies of *biosemiosis.* The founding premise of this growing field has been formulated as "a recognition that sign processes and the processes of life may be intimately and inseparably interconnected" (Kull et al. 2010: 168). Spelled out more extensively in an article specifically addressing jackals, Timo Maran says: "An essential characteristic of biosemiotics is that it can study semiosis without creating divides between sign processes in animals, the environment and human culture. In other words, semiotic processes in all three can be treated indiscriminately, as a whole" (Maran 2015: 68). What I have found lacking in this very attractive theoretical direction and where I have the ambition to suggest possibilities of advance is as regards two closely related aspects.

Essentially, they boil down to overcoming the limitations imposed by the fact that the voice of humans can be heard by employing, as in my case, a classical anthropological technique of participant observation. Doing the same with animals, on the one hand, or with the more diffuse entity of the environment, on the other, poses serious difficulties.

A direction for dealing with this problem has been offered by scholars in the field of human geography. In a programmatic article, Kersty Hobson (2007) makes a bold move for treating animals as *political subjects.* She uses the case of the Asiatic black bear (*Ursus Selenarctos thibetanus*), popularly known as the moon bear. When taken out of appalling farming conditions, the animals' "ability (and perhaps willingness?) to be healthy, happy and active once freed has constituted a driving force in the region's animal welfare debates" (Hobson 2007: 263). In other words, the liberated moon bears' enjoyment of newly acquired conditions of life constitute semiotic production that drives a particular political process forward. In such a political process, Hobson claims, we can argue for political subjecthood on the part of the liberated bears. Their communication of enjoyment by a variety of expressive means can be seen as the making of a political statement and not just because of favorably changed living conditions. Running somewhat ahead, I will use the same argument to claim political subjecthood for jackals in a situation that bears a resemblance to the case described by Hobson. In a more general way, I will use the insight offered by Hobson for "uncovering" the nonhuman part in the indiscriminate biosemiotics whole, to paraphrase Kull and colleagues (2010: 168).

I note, at this point, that notwithstanding the emphasis on holism in the biosemitic theoretical position, the nonhuman part can be seen as a semiotic subject in the treatment of the above authors, most persuasively on the level of the organic cell. When it comes to higher organisms, this component part of biosemiosis is overshadowed by the human one. Such is the case, in which the jackal is luckily in focus in the already mentioned article by Maran (2015). In it, we can hear humans trying to make sense of suddenly emerging jackals as far north as Estonia, but a discussion of a possible presence of semiotic activity on the part of the jackals themselves, suggesting how they may look at a novel situation is an avenue that has been chosen not to be explored.

Given such difficulties, my proposal builds on the unity between humans, animals, and the environment evident in the historical accounts of my campfire companions. As mentioned above, predators—or their absence—tend to symbolize that unity as an oscillation between advancing or receding evil. As it happened in the particular case, evil had radically advanced in the first communist decade—in the "wolves' times." Ambitious infrastructural projects had nearly "liquidated it" by various removals, bringing about in its place emblematic creations: an irrigation system and an orchard. Then all of it collapsed, and, in the words of my local companions, "turned into a jungle" (*tuk sega e dzhungla*). It is in this "jungle" that the jackals have come to reign.

Prompted by this fable-like telling of recent history, my general thesis will seek theoretical support from studies in the subfields of *oral memory* and *oral history*. Here I lean on insights like that of Alessandro Portelli, who says: "History is an invention which reality supplies with raw materials. It is not, however, an arbitrary invention, and the interest it arouses is rooted in the interests of the teller. . . . This is why 'wrong' tales are so very valuable. They allow us to recognize the interests of the tellers" (Portelli 1991: 2).

I will show in what way the tales of my tellers are "wrong." Compared with mainstream history, their periodization and evaluation of the periods follow different principles and terms. The "wrongness" of these comes from the above-stressed preoccupation not with facts as such, but with their perceptual manipulation according to the *moral* evaluations of the tellers (Ábrán in this volume; Özkan in this volume). This epistemological issue connects with the history of fears and anxieties (Jackson and Everts 2010; Bourke 2005), or, more generally, with a history of emotions (Rosenwein 2010). It, however, does not suggest in either its human or nonhuman expressions a lack of intentionality of which proponents of the "affective turn" have been accused (Leys 2011: 469). In the case at hand, the "wrongness" of oral historical accounts reveals a particular intentionality. The latter can be seen as using embodying images—of predators (wolves, jackals), or the environment (orchard, "jungle")—as political statements in which a separation between anxiety or emotion, on the one hand, and rational appraisal, on the other, would be artificial and irrelevant. This calls for "thick description of life experience of the kind that are familiar to anthropologists" (Leys 2011: 471). In the ethnographic part of the chapter, I attempt to include my nonhuman subjects—the jackals—in such an appealing program.

Another direction that needs to be considered, given the material at hand, leads to the extensive field, known under the increasingly controversial term "postsocialism" (Humphrey 2002). Two central findings associated with this body of studies must be mentioned here. One concerns the "nostalgia" for a Soviet/state socialist past, noted by researchers of the former state socialist universe—from one end of its map to the other, and certainly for this region (Dorondel 2016; Creed 2002; Verdery 1996). The other side of the same coin is experiencing "postsocialism" as a "fall" into a harsh and lawless reality. Or, as David Kideckel aptly sums it up, "post-Socialism hurts" (2008: 6).

Both observations of postsocialist reality are abundantly applicable to my micro case. In it, where the oral version of history differs from that of the scholarly one is in the stress the first one puts on subjective evaluation and ordering of past events. For the academic mind, the "wrongness" of the oral history version comes from the pronounced relativity and subjectivity of the backward-looking perspective. For the oral history tellers, the key period of "evil removal" was when there was radical intrusion in environmental affairs—that is, of the "nature-fighting" projects of state socialism. This coincided with their youth, while, in a fateful and tragic contrast, their advanced years in the new postsocialist period came to depend either on low-paid manual labor or miserable old-age pensions. Their evaluation of both the liquidation of predators during "communist times," and the impacting of the environment in a radical way that gave its stamp to that age, would therefore predictably differ from today's "mainstream" and "green" appraisal of the same developments, which sees the impact of "communist times" as ruthless and senseless destruction (Blavascunas in this volume). For my tellers though, these were the good times, and for them the liquidation of wolves and swamps, as well as later major anthropogenic intrusions, were steps toward a hopeful future. By contrast, the "re-wildening" (*podivyavane*), in their words, of society and nature after the political changes of 1989 has confirmed a deep-seated sense of doom.

Below I proceed with the ethnographic part of this chapter.

PROJECTS AND PREDATORS

From the establishing of the communist regime in 1944, and until Stalin's death in 1953, there stretched in local memory a period marked by violence. A wave of extrajudicial killings of political opponents swept over the whole country. Within days of the September 9, 1944, communist takeover, approximately thirty thousand people were killed or had disappeared.

Along the Danube, the theme of the wolf-as-a-dreaded-menace was additionally enhanced by recurrent awakenings of ancient fears of wolves crossing over from the Carpathian Mountains whenever the Danube froze in winter. Such was the winter of 1952–1953, when fears of human and nonhuman wolves merged in the violence and fears of late Stalinism in these parts.

Violence continued with the nationalization of commercial and industrial property in 1948, and the beginning of collectivization of agricultural land. As is well-known, that latter campaign was voluntary in word, but enforced in practice. In stories about it, the image of the wolf comes out strongly. The animal served as a *poetic* vehicle for telling about unbridled violence and desire for revenge—of the newly empowered as regarded their predecessors. These latter, branded as fierce enemies of collectivization, formed a "wolf" metaphorical trope in the ideological poetics of the new times. The "wolflike" formerly rich peasants (*kulatsi*) and current enemies of collectivization had to be rounded up and exterminated as bloodthirsty, cruel, and destructive "elements." Similar measures, taken, some two decades before that, during the mass collectivizing drives in the Soviet Union (1928–1932), found here their local application.

It is in such ways that the wolves' unfortunate fate of being dreaded and intensely hated creatures since times long past, came to be burdened with new metaphorical extensions. As shown above, these served to construct the image of the ideological enemy. Further than that however, the wolf symbolized the part of the natural environment that, according to the dominant ideological accents of the time, had to be removed as subversive and destructive. In the same way that this applied to human "enemy elements" (*vrazheski elementi*), as the phrase had it, it also applied to environmental features (like swamps). These latter were deemed not only useless for socialist agriculture but also harmful, being breeding grounds for all manner of "pests." The environment was reclassified into useful and harmful parts—both animate (including human), and inanimate. In this new inventory of being, the wolf remained as an apex symbol of harmfulness and put a mark on the period.

REMOVING HARM AND SANITIZING EXISTENCE

The ideological trope of liquidating human as well as nonhuman "enemy elements" in a spirit of revenge was gradually superseded by a more generalized ideological accent.[4] That was removing *all* harmfulness from society and nature and thus sanitizing the totality of existence.

This new trope followed Soviet "cultural revolutionary" slogans of the 1930s "war against backwardness" (Slezkine 1994: 219–20; Vladimirova in this volume), and the drive for removal of obstacles to progress. The doctrine, demanding the removal of harmfulness in the name of improved existence, is well traceable to ideas of European modernity for social and environmental intervention and engineering (Hoffmann 2000: 246; Şerban and Dorondel in this volume).

Sanitizing the environment along the Lower Danube targeted swamps in the first instance, but also other more generally found features of the rural landscape. Among them were the physical boundaries between private plots of land (*sinuri*). As in the case of swamps, the logic of their removal was twofold. On the one hand, both swamps and plot borders obstructed large-scale agriculture. Their removal turned wasteland into a newly opened agricultural resource. Alongside this motive was also the benefit that breeding grounds for pests would be annihilated. In the case of plot borders, the animals that developed the larger ideological trope—of the benefit of annihilating private property—were snakes and lizards, lurking in the tangled growth of the private plot marking strips. In other similar cases in the rest of the country, the "pest breeding" motive concerned other "subversive critters" (Harraway 2008). Putting mountain caves to a useful purpose (as cheese fermenting cellars) or blasting mountain gorges to create new road and railway passes celebrated the annihilation of bats and owls.[5]

The liquidation of predators as well as humans branded as "enemy elements" can thus be read as a variant of an "illiberal modernity" in its propensity for "gardening the body politic" (Bauman 1989). Running ahead, I note here that Zygmunt Bauman's metaphor fits very well with the embodied symbolism of my case. As we will see further down, after the extermination of the "enemy elements," and side by side with them, of the wolves, a co-operative orchard was planted, which came to symbolize the order and relative prosperity of "developed socialism." The explosion of jackals alongside the destruction of the orchard after 1989 resurrected the "predators as criminals" trope.

In the case of "my" site, the will of the socialist state to cleanse the environment demanded the draining of the local swamp.[6] The goal to be achieved—mainly by pick and shovel—was proclaimed as "gaining from nature thousands of hectares of arable land." Such were the slogans of the day. A similarly important goal, and one connected with a deep-seated dread of malaria, was the extermination of blood-sucking insects.

Accordingly, the denudation of the marshland between the two villages in my research site was carried out in the space of two years (1948–1949). An earthen dike was built to separate the marshland from the Danube. It was followed by digging a system of canals, with a pumping station to take the water back into the river.

By the late 1960s the project was quietly abandoned. It had become evident, early on, that earthen dikes could not stop the river from seeping in. By the time of writing (the fall of 2020), the draining of the swamp belongs to distant memory. The pumping station is a long-gone ruin, overgrown with weeds. The network of canals can still be seen, but the land is underwater once again.

The beginning of the post-Stalinist period in Bulgaria (April 1956), proclaimed as "developed/mature socialism" in accordance with the new Soviet doctrine of Leonid Brezhnev (Thompson 1988), brought about a second project of scale. It reflected a will not so much to destroy the existing, but to improve on it by technological means. In the local case, the aim was to moor a floating pumping station by the Danube bank and pump water up into an extensive system of canals, and from them to the fields of the Mizian Plain.

By the early 1970s, the new irrigation system was completed. At that time, two of my present fishing companions—Ivan and Slavi—had found jobs in the facilities after completion of their military service. Ivan became a mechanic at the floating pump, and Slavi worked at the new electric transformer station by the port. The jobs were modestly paid, but carried the prestige of not having to do physical work in the fields. They had free time in abundance and it was mostly spent fishing and socializing around campfires. I return below to the memories of this life and what happened with it.

In the early 1970s another new project of scale was begun. That was the turning of fallow land by the river into a peach and apricot orchard of the local co-operative farm. The plan was to uproot all bushes and trees, which had grown on the rich soil in the past twenty years or so, and plant peach and apricot saplings. Connecting dirt roads intersected the rectangular shape of "the massif" (*masiva*), as it came to be called, some three by one kilometers in size. Enforced-concrete reservoirs were built at regular intervals to hold the spraying liquid. Additionally, a second pumping station was installed close to the riverbank. This smaller pump was to provide water for the orchard—for irrigation and spraying purposes.

In a few years, the orchard was already increasing the income of the co-operative farm. The great pride and satisfaction of local leadership and farmers came from the fact that the produce was not only readily absorbed by the local urban market but also went to Austria and West Germany. This was proudly announced in the local and national press, on TV, and even in a film newsreel. The peaches and apricots were bringing in hard foreign currency.

THE JUNGLE

By the time of writing, which follows these events by some forty years, all of this is distant memory. The floating station stopped functioning in the mid-1990s, and sometime later was floated downriver to be dismantled and sold for scrap. The pipe is still there, rusting away. The canals, for which so many tons of earth had been excavated and removed by huge trucks, and into which other tons of reinforced concrete went, are now empty and overgrown.

The peach and apricot trees are long gone. Several decades of "democracy," as villagers sneeringly call the postsocialist period, beginning with the liquidation of the co-operative farm in 1990, saw to it that left without care, the fragile fruit trees quickly degraded. Bushes and wild trees overgrew them together with the connecting roads. "It is a jungle here now" (*Sega tuk e dzhungla*), became an often heard phrase.

Ivan and Slavi are still living in their home village—among the few remaining there. By the riverbank, where the pumping station once stood, a big private vineyard of two hundred hectares came into being in the early 1990s—right after the demise of the co-operative. Its ownership is shrouded in mystery. According to rumor shared around the campfires, the first owner was a Ukrainian "mafia guy." He got the land, according to the same rumor, at some ridiculously low price by bribing the local authorities. The story went that the deal was made possible by connections of the shady Ukrainian businessman with a "very high-level politician" from Sofia. Eventually, that politician found a way to remove the Ukrainian and appropriate the estate. In recent years, the estate is rumored to be in the hands of a "Chinese," but no one has ever seen that even more mysterious person.

With the demise of the irrigation system and the resulting loss of the few available jobs, Ivan and Slavi, one after the other, began selling their acquisitions from the "orchard" period. They did this to make ends meet, and it was very painful, especially for Ivan. His proudest possession was a Soviet Volga limousine. In former socialist times, he had to wait fifteen years for his turn to buy it from the state retail system. The family cow was then sold, then the few sheep, then another proud possession that could fetch a price: an electric grinder for sun-dried paprika. In Slavi's household a similar stripping to the bone occurred.

When all this money was gone, and no more likely to come in, both of them became farm hands in the vineyard estate of mysterious ownership. "They pay us only twenty levs a day [10 euros], but that's only at peak times," Ivan told me last summer. "During the rest of the time there are simply no jobs around." It is as if Kideckel's words "post-Socialism hurts" (2008: 6) were written exactly for this case—one of a myriad others.

In sum, the time of relative prosperity had gone, together with youth. The lull of developed socialism was over.

THE JACKALS

The turning of the orchard into a jungle, and the shady appearance of the "mafia vineyard" coincided with the unexpected appearance of jackals. The howling of their packs, roaming over the former orchard, became a nightly event.

At the fringe of the orchard, on its southern side, a cluster of small vineyards had been steadily expanding during the socialist period. Of an average size of a *dekar* (one-fourth of an acre) they were known as plots for personal use (*partseli za lichno polzvane*). The swift decay of the co-operative orchard was soon followed by their abandoning.

The villagers who were tending them had built shacks and even small huts at one end of their plots. On each of my yearly visits after 1989, I could see the huts and shacks disappearing one after the other. To my questions about the exodus, the typical answer was that tending the plots did not make sense anymore. "Why should I be hoeing and spraying (the vines) month after month?" a friend in the village asked rhetorically. "You see what it is like now after this democracy came. I dig like mad and then when the grapes ripen, burglars come and take everything. There is no law and order here now, bandits have sprung up as if from nowhere and the plots are theirs and the jackals'. The only difference is they [the burglars] are not howling."

Against this background, a point often made in conversations about the abandoning of the small vineyards, inevitably leading to bitter comments about current poverty and lawlessness, was that criminals and jackals descended upon the villagers as if from nowhere. "Where were they before? In what dens were they hiding? From where did they crawl out?"—these were the questions often asked in reference to both the legion of high and low "mafia bandits," and to the animals themselves. Jackals had thus become part of local political life. Different from politically disengaged "objective" naturalists' talk, to which I come in a moment, jackals, in the oral literature of the place, came to be directly linked to political change as their living symbols.

Hunters' Jackal Talk

This is the situation as seen from a local human perspective. From this angle, the explosion of jackals, revealed by nightly vocal performances—or "concerts" (*kontserti*) as the local fishers joked—was built on the store of associations that had been accumulated around the animal. That was uniformly negative, profiling the creature as a rapacious, wily predator.

Local hunters were influential in sustaining that image. A particular theme was recurrent: jackals eating newborn roe deer. "They eat all the newborn (*priploda*) and that is why you can't see any roe [deer] as before," Ivo, a local hunter would often repeat. "Where there are jackals there is nothing."

The cue for the telling of the newborn roe deer story was, as a rule, the question about why "game" (*polezen divech*) such as roe deer, hare, pheasant, and the like, as opposed to "vermin" (*vreden divech*) had so much diminished after 1989, while jackals and birds of prey had become so abundant. Had not the game animals mobilized "traditional" means of survival in the new "jungle" conditions?

The usual explanation fused all "predatory" features of the postsocialist situation into an image of all-encompassing destruction. In the words of Dimo (age forty-two), a local hunter, "it is not only the jackals. There was order before. You could see roe deer in herds in the fields, you could see them by the roadside. Red deer also! But then came this democracy and now everyone has a rifle, the rich shoot from fancy jeeps, everyone poaches, and the deer have disappeared. Where are the fields, where is the orchard? It is a jungle there now, it is a jungle all around, there are jackals all around" (July 2017).

In analyses of this type, the roe deer would be singled out as the most vulnerable to predatory pressure, and, in addition, at the most vulnerable part of their life cycle: when giving birth. Hunters tell of jackals surrounding a pregnant roe deer and waiting for her to give birth. Then they devour the newborn fawn and also the mother if they can get her. From there on the theme is generalized to include human actors also: the newly sprung "mafia," "criminally enriched" class. Their possession of high-power rifles, fancy SUVs, and all sorts of electronic equipment for night shooting is a constant refrain at hunting gatherings. What is bemoaned even more comes from the conviction that these "predators" are in cahoots with local strongmen, or "players" (*igrachi*) as they are called, and are thus above the law. A paradox of the situation is that in the egalitarian spirit of after-hunt communal dinners, urban hunters—many of whom themselves belong to the much maligned newly rich and are eagerly invited to be part of village hunting teams—talk as if they are excluded from the circle of "predators."

I have often asked if anyone had ever seen with their own eyes the devouringof-the-newborn-calf scene. No one had seen it, but the answer has always been: "It is known to be so" (*znae se che e taka*).

One can hardly argue on such a basis, but that is beside the point. The point made in hunters' talk is that the law of the jungle has replaced previous order (the orchard) in which the disempowered could feel secure. Jackals eating a newborn fawn in the orchard-turned-jungle, on the stage of hunting campfire talks, can be interpreted as the poetic rendering of the new times.

The "Wrong" and the "Right" Jackal Talk

Against such a background, the central accents of what may be called "folk jackal talk" need to be profiled. This will help us to understand how this particular oral history is being made.

Compared with naturalists' accounts, the central accent is about the suddenness and unexpectedness of emergence. That is, how did it happen that the jackals increased so suddenly in numbers, and spread over the whole country? How was it that they suddenly sprang out as if from nowhere?

Answers to the question are rather different in both discourses. Zoologists say that the increase in density and spread of the jackal population in Bulgaria was spectacular indeed, but not sudden. The whole picture here is impassionate and rather different.

SCIENTIFIC JACKAL TALK

According to learned opinion, the golden jackals' traditional habitat is big and diverse, due to the ecological plasticity of the species. The habitat stretches from East Africa to the Middle East and farther to South Asia. In Europe, the jackal is mainly found in the Balkans, with the greatest presence in Bulgaria of over thirty thousand animals, according to Georgie Markov (2012: 67) and Stoyan Stoyanov (2012: 50).[7]

The same authors claim that an increase in numbers of animals and, particularly, in expansion of the habitat in Bulgaria were already evident by the early 1960s. The main reason is thought to be the declaring of the jackal as a protected species in 1962. This ban was revoked in 1984, but despite that, the density of the population continued to grow steadily. Judging by the number of jackals shot by hunters, a further spectacular rise in density began in the early 2000s. Thus, while the number of harvested animals was around five thousand annually from the late 1980s to about 2002, by 2009 that number had sharply risen to over twenty-five thousand (Stoyanov 2012: 50). The result is the present situation in which despite such a serious annual cull, Bulgaria has the greatest jackal population in Europe.

Besides the increase in density, the jackal's habitat expanded more spectacularly. Nationally, it grew by a factor of thirty-three over the period noted above. At present, almost anywhere you go in the rural parts of the country, you can hear packs of jackals howling after sunset.

The jackal expansion gradually crossed national borders, creeping northward. Since about the 1980s, the jackal populations of Romania (Banea et al. 2012), and Hungary (Tóth et al. 2009) began to rise. Jackals started being seen in Poland (Kowalczyk, Kołodziej-Sobocińska, and Wójcik 2015), and eventually as far north as Estonia (Maran 2015). A general rise was observed in many other parts of Europe (Krofel et al. 2017; Arnold et al. 2012).

Something was indeed happening and the zoologists began asking themselves what it was. Explanations differed. As far as Bulgaria was concerned, apart from the 1962–1984 ban on jackal shooting, the general rise of hunting, in particular boar hunting, was mentioned as a possible cause. The logic was that boar hunters would throw away freshly skinned hides and offal, which the jackals would utilize (Markov 2012: 67). Others saw the jackals as taking the upper hand vis-à-vis their direct competitor—the fox. An advantage they had was in their larger portfolio of land-use tactics, among which, besides what foxes were capable of, the jackal's hierarchical sociality allowed, critically, group ("pack") hunting (cf. Tsunoda et al. 2017). A critical cause of jackal expansion in Bulgaria and northward is connected with the extermination of wolves. While the wolves had been around until about the late 1950s, they suppressed jackals, but once they were removed, the smaller predators regained their numbers and distribution.

In naturalists' description this is called "mesopredator release" (Krofel et al. 2017). In lay talk however, an accent was put on what was believed to be a fact: there were no jackals during the socialist period, they came only after "the explosion of democracy."

As already mentioned above, in naturalists' opinion this is no more than folk belief—jackals were already on the rise by the early 1960s. The important point is elsewhere. In folk discussions of the matter, talking about jackals is to a significant extent talking about politics. Seen from such an angle, in folk jackal talk the animal is a symbol of criminality and insecurity, similar to the wolf's symbolization of terror of the postwar Stalinist period.

In this way, when compared to folk explanations of expansion, a common line can be seen to unite the variety of scientific hypotheses: the absence of *political* factors among the drivers. Politics does not figure in academic talk to the extent that one is left with the impression that it may not even be considered. Naturalists' accounts thus differ epistemologically from lay ones. Jackal expansion is treated as belonging to different temporalities, reflected in "scientific" versus "folk" representations. While the naturalists' version registers a temporally extended span of developments, the folk versions resort to temporal compression. A long, drawn-out process is compressed into a spectacular happening, poetically highlighting the work of higher, fatal forces.

Emergence, as opposed to a cause-and-effect processual structuration of events, is the better tool as regards the ideological objective of lay talk about change. In this specific case, jackals appear as if out of nowhere in the early 1990s. This emergence is directly connected with the political changes in the country. The groundwork for a *politically* charged message is thus that the systemic change of 1989 ("the explosion of democracy") produced the jackals in a general upsurge of predation. A reading of democracy/capitalism as a state of affairs in which the powerless are left at the mercy of the powerful—as in the scene of jackals surrounding a pregnant roe deer—is not difficult to make.

Romanian researchers doing fieldwork in roughly the same area across the river report a variant of politically related jackal emergence. Local fishers linked the emergence of the jackal with political events in the Middle East and the ensuing wars, such as in 1991 in Kuwait and thereafter. According to them, deterioration in the Middle East drove the animals to Romania and Bulgaria (Stefan Dorondel and Stelu Şerban, personal communication).

FROM THE JACKALS' POINT OF VIEW

This particular aspect of the situation has not been examined so far. Maran's (2015) study of the arrival of the jackal in Estonia, mentioned above, examines human cultural responses to the event, but stops short of investigating the jackals themselves acting as semiotic subjects over and above the way they are perceived by humans.

As already mentioned, Hobson (2007) has already attempted to include animals in the political process as subjects. Following her approach, I tried to observe and record instances of jackal behavior that crossed the boundary between adaptation to new environmental conditions and spoke of more—that is, intentional semiotic agency. In what follows, I illustrate what I have experienced with jackals in the particular locality of my case, which I argue reveals such a crossing.

A principal part of such work is a slow building up of a relationship of solidarity and trust. In my case, a background feature of positive effect was the fact of living in a tent, alone, in a remote location "worked" nightly by a jackal family group, and for a sufficiently long period. Needless to say, any form of aggressive behavior toward jackals or other animals was strictly avoided.

In my pilot field study in the summer of 2017, and the more extensive annual five months of studies during the following three years, this setting was additionally improved by creating sites that invited common use: a waste heap (the "table"), and the cutting of paths through the "jungle." These established a regular pattern of commensality, as well as what can be called "co-passage." In the summer of 2018 I began to use a camera trap trained on the "table." This latter was developed as a bait site by a nightly offering of an average of three kilograms of food—mainly water-soaked muesli, fish, fish innards, leftovers from meals, and bread. The camera recorded the jackals use of the bait site, as well as their behavior in regard to both conspecific audible presence (of other neighboring jackal groups) and humans.

In the process of establishing and sustaining a relationship on this basis, the acts of semiotic production began to be observed in proportion to individuation of the various members of the group. It fairly soon became obvious who was the mother in the group. Her deep-throated voice would be the first during the night, she was the first to come down to the "table," and was the one to pass nearest by my tent—her bigger tracks could be seen perfectly imprinted on the mud by the water's edge. In broad daylight, I saw her twice: she was of an impressive size for a jackal.

I called her Dzheki (Jacky) and would address her by this name. Dzheki would announce the coming of the group at night with a characteristic howl (cf. Gese and Ruff 1998: 1042). After her, other voices would be heard, until, at the end of the line there were the screaming, giggling howls of the pups.

At some distance, one could hear humans talking around a fire. At night, human talk could be heard even from the Romanian bank across the kilometer-wide river. Often the pack from Romania would begin howling first, then Dzheki's bunch would answer, and a third neighboring pack, positioned to the west of the camp, would join in. The howling would thus travel across and downriver first, then do a circle to the west and upriver. The cross-river howling session would last for about half an hour, marking respective rights and positioning over a territory of roughly three kilometers in diameter.

When the first pack began howling, humans stooped around a distant campfire would stop talking. As a rule, they would stand hushed for a few minutes, listening to the "concert." I felt strongly that Dzheki and her pack were aware of the impression they were making on humans.

From my vantage point by the tent, even now I can still see Dzheki howling toward the distant huddled shapes round the reddish glow of fire. There is enough starlight on the water to see her in profile. My presence seems to be recognized not as that of targeted audience, but of provider, and thus, closer to an inner circle. A sign of this may be seen in the fact that no food, except what was left on the "table" would be touched, not even fish hanging out to dry and within easy reach. These encounters left a strong impression with me that jackals were acting as political agents, among other existential business.

Two features of political behavior can be pointed out. During daylight, the pack was resting and invisible. But the traces of their presence were abundantly evident. The whole of the former orchard was crisscrossed by jackal paths with numerous resting places under trees. A message of repossession and reassertion could be read in it.

During the night the message would be driven home more pointedly by the howling described above. The functional purpose of howling is well described in the ethological literature. It is discussed primarily as "inter-pack communication, favouring territory maintenance and inducing the avoidance of one pack by another" (Gazzola et al. 2002: 157; Giannatos et al. 2005). This opinion thus reiterates the hypothesis that "howling functions primarily as a mechanism to advertise territorial occupancy and the location of dominant animals, so that confrontations with neighboring pairs and transients can be reduced" (Jaeger, Pandit, and Haque 1996: 773; Harrington and Mech 1979).

Nowhere in this literature of bioacoustic studies, however, have I ever found the nonhuman–human communicational aspect mentioned. At the risk of being accused of the sin of anthropomorphism (Root-Bernstein et al. 2013: 1578), I still maintain that the howling sessions I had witnessed accounted for human presence within their communicational purpose.[8]

I thus argue that the nightly howling ritual included humans among its participants. It thus extended in scope and purpose the usual communicative content. Although my studies of this topic are only at their beginning, I would still venture to suggest that a message of territorial repossession and reassertion of rights was communicated. It could be read as a celebration of reclaiming the formerly human-dominated environment, subjected to human engineering and "gardening" (Bauman 1989). In this sense, it was "true" that the jackals' population explosion came with that of "democracy." It was thus the case, that the nightly howling ritual, communicating a message of reassertion, stimulated a specific human and nonhuman way of recording history.

I have tried to show in this chapter the advantages offered by employing a biosemiosis-stimulated approach in a study of environmental change.

For the local people who constituted my circle of informants, the key referential period was, unsurprisingly, that of childhood and youth. This period coincided with what Soviet and Bulgarian historiography had defined as "developed" or "mature socialism." My informants remembered it as a time of security and professional realization. Their appraisal of the period between the late 1950s and the late 1980s was based on involvement in infrastructural projects from their creation to their demise.

It so happened that their appraisal of environmental engineering came to deviate radically from today's mainstream look at the environmental projects of state socialism. This gives reason to value the opinion of grassroots actors as involved practitioners and "experts" in presenting their own version of history.

A further conclusion is that the extended depth offered by a holistic approach—both synchronically and diachronically—demands the inclusion of nonhuman actors in the oral historical record. This enriches our understanding of historical events by providing a plurality of interconnected "gazes" (to use John Urry's felicitous metaphor [2002]). Among specific problems the approach offers solutions for is "illiberal modernities": Were they imported or autochthonous developments? The approach offered here places the problem on the plane of human–nonhuman relationships in a *longue durée* perspective and thus as a matter of common European heritage.

NOTES

The field studies from which this chapter draws have been self-funded, but they could not have happened without the support and concrete help of my family and a close circle of friends: both from Sofia and the nearby Danube villages. Thank you Maria, Nikola, Bogdan, Assen, Rossen, Ganka, and Ivo—yet the list goes far beyond these names. I owe a particular debt of gratitude to Familia Nikola Ltd. for supplying me free of charge with muesli for the jackals ever since 2017. The help and academic advice of Mira Kovacheva of Sofia University was decisive in the first editing of this text and the adopting of a biosemiotic approach. Another great debt of gratitude goes to Nikolai Spasov of the National Natural History Museum and to his doctoral student Ilya Acosta-Pankov for their specialized ethological advice and concrete help in the summer of 2020. Needless to say, the responsibility for the interpretation of jackal behavior rests solely with me.

1. On November 10, 1989, the long-ruling dictator Todor Zhivkov was removed from office in an inner Communist Party reshuffle. The momentous event marked the beginning of the postsocialist period in Bulgaria.

2. All equivalents of English words and expressions given in brackets are in Bulgarian.

3. All personal names are fictitious.

4. On the rhetoric and practice of "cleansing populations" and "removal of undesirable elements" see Ross (2011: 39–40).

5. For glorifying human destruction of "undesirable" parts of nature, see Elin Pelin's (1973 [1947]) short story "Znam i moga" [I know and I can]. For the historical context, see Metodiev (2017).

6. For the history of draining, see "Tsibursko blato" [Tsibar swamp], http://www.birdsinbulgaria.org/ovm.php.

7. For an alternative view of the numbers and dynamics of expansion, see Spassov and Acosta-Pankov (2019).

8. Compare with the communication of "indifference," as in Matei Candea's (2010) famous meerkat study.

REFERENCES

Arnold, Janosch, Anna Humer, Miklos Heltai, Dumitru Murariu, Nikolai Spassov, and Klaus Hacklaender. 2012. "Current Status and Distribution of Golden Jackals Canis aureus in Europe." *Mammal Review* 42(1): 1–11.

Banea, Ovidiu, Miha Krofel, Jaroslav Červinka, Petre Gargarea, and László Szabó. 2012. "New Records, First Estimates of Densities and Questions of Applied Ecology for Jackals in Danube Delta Biosphere Reserve and Hunting Terrains from Romania." *Acta Zoologica Bulgarica* 64(4): 353–66.

Bauman, Zygmunt. 1989. *Modernity and the Holocaust*. Ithaca, NY: Cornell University Press.

Bourke, Joanna. 2005. *A Cultural History of Fear*. London: Virago.

Candea, Matei 2010. "'I Fell in Love with Carlos the Meerkat': Engagement and Detachment in Human–Animal Relations." *American Ethnologist* 37(2): 241–58.

Creed, Gerald. 2002. "(Consumer) Paradise Lost: Capitalist Dynamics and Disenchantment in Rural Bulgaria." *Anthropology of East Europe Review* 20(2):119–25.

Dorondel, Stefan. 2016. *Disrupted Landscapes: State, Peasants and the Politics of Land in Postsocialist Romania*. New York: Berghahn Books.

Gazzola, Andrea, Elisa Avanzinelli, Lorenza Mauri, Massimo Scandura, and Marco Apollonio. 2002. "Temporal Changes of Howling in South European Wolf Packs." *Italian Journal of Zoology* 69(2): 157–61.

Gese, Eric, and Robert Ruff. 1998. "Howling by Coyotes (*Canis latrans*): Variation among Social Classes, Seasons, and Pack Sizes." *Canadian Journal of Zoology* 76(6): 1037–43.

Giannatos, Giorgos, Yiannis Marinos, Panagiota Maragou, and Giorgos Catsadorakis. 2005. "The Status of the Golden Jackal (Canis aureus L.) in Greece." *Belgian Journal of Zoology* 135(2): 145–49.

Harraway, Donna. 2008. *When Species Meet*. Minneapolis and London: University of Minnesota.

Harrington, Fred, and L. David Mech. 1979. "Wolf Howling and its Role in Territory Maintenance." *Behaviour* 68(3–4): 207–49.

Hobson, Kersty. 2007. "Political Animals? On Animals as Subjects in an Enlarged Political Geography." *Political Geography* 26: 250–67.

Hoffmann, David. 2000. "European Modernity and Soviet Socialism." In David Hoffmann and Yanni Kotsonis, eds., *Russian Modernity: Politics, Knowledge and Practices, 1800–1950*. New York: Springer, 245–60.

Humphrey, Caroline. 2002. "Does the Category 'Postsocialist' Still Make Sense?" In Chris M. Hann, ed., *Postsocialism: Ideals, Ideologies, and Practices in Eurasia*. London: Routledge, 12–15.

Jackson, Peter, and Jonathan Everts. 2010. "Anxiety as Social Practice." *Environment and Planning A* 42(11): 2791–806.

Jaeger, Michael M., Rajat K. Pandit, and Emdadul Haque. 1996. "Seasonal Differences in Territorial Behavior by Golden Jackals in Bangladesh: Howling versus Confrontation." *Journal of Mammalogy* 77(3): 768–75.

Kideckel, David. 2008. *Getting by in Postsocialist Romania: Labor, the Body, and Working Class*. Bloomington: Indiana University Press.

Kowalczyk, Rafał, Marta Kołodziej-Sobocińska, and Jan M. Wójcik. 2015. "Range Expansion of the Golden Jackal (*Canis aureus*) into Poland: First Records." *Mamm Res* 60: 411–14.

Krofel, Miha, Giorgos Giannatos, Duško Ćirović, Stoyan Stoyanov, and Thomas M. Newsome. 2017. "Golden Jackal Expansion in Europe: A Case of Mesopredator Release Triggered by Continent-wide Wolf Persecution?" *Hystrix, the Italian Journal of Mammalogy* 28(1): 9–15.

Kull, Kalevi, Terrence Deacon, Claus Emmeche, Jesper Hoffmeyer, and Frederik Stjernfelt. 2010. "Theses on Biosemiotics: Prolegomena to a Theoretical Biology." *Biological Theory* 4(2): 167–73.

Leys, Ruth. 2011. "The Turn to Affect: A Critique." *Critical Inquiry* 37(3): 434–72.

Maran, Timo. 2015. "Emergence of the 'Howling Foxes': A Semiotic Analysis of Initial Interpretations of the Golden Jackal (*Canis aureus*) in Estonia." *Biosemiotics* 8(3): 463–82.

Markov, Georgi. 2012. "Golden Jackal (*Canis aureus L.*) in Bulgaria: What Is Going On?" *Acta Zoological Bulgarica* 4: 67–71.

Metodiev, Metodi. 2017. "The Writers, the Conflicts and Power in Bulgaria and Czechoslovakia 1948–1968." PhD diss., University of Glasgow.

Pelin, Elin. 1973 [1947]. *Znam i moga* [I Know and I can]. Sofia: Narodna mladezh.

Portelli, Alessandro. 1991. *The Death of Luigi Trastulli, and Other Stories: Form and Meaning in Oral History*. Albany: State University of New York Press.

Root-Bernstein, M., L. Douglas, A. Smith, and D.Verissimo. 2013. "Anthropomorphized Species as Tools for Conservation: Utility beyond Prosocial, Intelligent and Suffering Species." *Biodiversity and Conservation* 22(8): 1577–89.

Rosenwein, Barbara. 2010. "Problems and Methods in the History of Emotions." *Passions in Context* 1(1): 1–32.

Ross, Thomas. 2011. "Wolves in Wolves' Clothing: The Role of the Chechen Mafia in the Formation of an Independent Chechen Republic." PhD diss., Naval Postgraduate School, Monterey.

Slezkine, Yuri. 1994. *Arctic Mirrors: Russia and the Small Peoples of the North*. Ithaca, NY: Cornell University Press.

Spassov, Nikolai, and Ilya Acosta-Pankov. 2019. "Dispersal History of the Golden Jackal (*Canis aureus moreoticus* Geoffroy, 1835) in Europe and Possible Causes of Its Recent Population Explosion." *Biodiversity Data Journal* 7: e34825; https://www.doi.org/10.3897/BDJ.7.e34825

Stoyanov, Stoyan. 2012. "Golden Jackal (*Canis aureus*) in Bulgaria: Current Status, Distribution, Demography and Diet." Paper presented at the international symposium on hunting "Modern Aspects of Sustainable Management of Game Population." Zemun-Belgrade, Serbia, June 22–24.

Thompson, Terry L. 1988. "Developed Socialism: Brezhnev's Contribution to Soviet Ideology." In T. Thompson and R. Sheldon, eds., *Soviet Society and Culture: Essays in Honor of Vera S. Dunham*. Boulder, CO: Westview, 206–36.

Tóth, Tamás, László Krecsák, Eleonóra Szűcs, Miklós Heltai, and György Huszár. 2009. "Records of the Golden Jackal (*Canis aureus* Linnaeus, 1758) in Hungary from the 1800s until 2007, based on a Literature Survey." *North-Western Journal of Zoology* 5(2): 386–405.

Tsunoda, H., E. G. Raichev, C. Newman, R. Masuda, D. M. Georgiev, and Y. Kaneko. 2017. "Food Niche Segregation between Sympatric Golden Jackals and Red Foxes in Central Bulgaria." *Journal of Zoology*. https://www.doi.org/10.1111/jzo.12464.

Urry, John. 2002 [1990]. *The Tourist Gaze*. London: Sage.

Verdery, Katherine. 1996. *What Was Socialism and What Comes Next?* Princeton, NJ: Princeton University Press.

EPILOGUE

A New Ecological Order at a European Margin

Stefan Dorondel and Helmuth Trischler

Following the conceptual trajectory that history has seen not one single modernity, but multiple modernities (Eisenstadt 2002), this volume has discussed regional modernities (Sivaramakrishnan and Agrawal 2003) in relation to nature as they emerged and developed in Eastern Europe during the past 170 years. Eastern Europe had—and still has—an in-between status: it is rarely included in conceptions of "Europe" (which is tacitly understood as Western Europe, a provider of civilizational ideas), and yet it is also not a colony per se and is thus excluded from colonial and postcolonial discourse.[1] This in-betweenness has rendered the region an epistemological incertitude (Todorova 2009), as well as a shadowy presence in the international literature concerning the relationship between society and nature. By exploring the modern constitution of this relationship, we have attempted to bring the analysis of Eastern European statehood, economic development, and nature into mainstream discussions.

Since the creation of Eastern European nation-states, a process starting roughly in the second half of the nineteenth century, a general obsession has haunted this part of the world regarding how to find a way out of the European periphery. Small countries or regional empires, such as the tsarist, Ottoman, and Habsburg, all strove to catch up with "the West" (Todorova 2009). One of the obstacles that had to be overcome was the natural environment and its inherent resistance to the new capitalist frame of higher productivity. This was the time that Marc Levinson

(2020: 24) has called "the first globalization"—the worldwide rise and spread of industrial capitalism. And it was the time in which, according to Levinson, governments cautiously started to allow market forces to shape national economies.

The volume has focused specifically on the ways nature was mobilized by the developmental policies of Eastern Europe's nation-states in their struggle to escape the periphery. The main state actors shaping the process were planners, experts, and bureaucrats. The contributions to this volume have shown that Eastern European countries' national project borrowed from the ideas, technologies, and experts of Western countries, but also from other Eastern countries that perceived themselves to be part of the European periphery. In their infancy, East European states looked to their neighbors for ideas of what "nature" is and how it can be put to work for the national good. But there were sometimes clashes between Western experts' ideas and those of their regional counterparts. Often, Western plans and technologies were too expensive or contradicted Eastern countries' ways of seeing nature. It was the role of national experts and planners to adapt Western proposals to fit local environmental and social realities. The volume's findings concur with other works that have shown East European countries' intersectional approach to development. For instance, as Anna Mazanik (2020) has convincingly demonstrated, the urban modernization of Moscow at the end of the nineteenth century was a constant process of exchanging ideas, plans, expertise, and technologies. Urbanization resulted from the intersection of many models including local ones. The Moscow abattoir (built between 1886 and 1888) was drawn up by Russian engineers and architects using French and German structural models, sterilization machines invented in Belgium, refrigerators created according to the Hamburg slaughterhouse model, and an English sewage system (Mazanik 2020). This example epitomizes the web of technologies, ideas, and plans between local and Western European models. The East European periphery, as several authors have demonstrated in the volume, was actively searching for new ideas for development and for new ways to integrate regional environments into an international (or global) economy. In so doing, states were represented by three types of actors—planners, experts, and bureaucrats—who were, on different scales and levels, the architects of the new (modern) states in Eastern Europe.

However, as several contributions to the volume have attested, the "periphery" was also at times the vanguard of development and technology: areas that had once felt the need to "catch up" with the West became a source of inspiration for others. Thus, the transfer of techniques and ideas in reshaping nature is traceable not only on the West–East axis but also between neighboring countries. Again, our findings back up recent research on knowledge transfer between Western and Eastern countries (Gantner, Hein-Kircher, and Kochadel 2020). Moreover, East

European regional modernities often presented *themselves* as "backward" in order to push for external expertise, technologies, and ideas (Todorova 2009; Mazanik 2020; Gantner, Hein-Kircher, and Hochadel 2020). Our findings counter the idea that Western countries exported knowledge and technology to more "backward" peripheral countries. The case of the socialist period further questions the accepted vectors of knowledge and expertise transfer. Socialist economies, and not only in Soviet Russia, exported knowledge and technology to Western capitalist countries and even more to Middle East and African countries (Reynolds 2017).

In contrast to the existing literature exploring issues such as traveling experts, technologies, and ideas of "modernization" across the European continent (e.g., Kohlrausch and Trischler 2014; Kalinovsky 2018; Gantner, Hein-Kircher, and Hochadel 2020), our volume has taken a slightly different path to understanding the position of Eastern states vis-à-vis the global circulation of ideas about nature and development. Our volume finds that the Eastern states' view that the natural environment needs "corrections" to serve developmental purposes led to a plurality of nature's meanings that traversed political regimes from the 1850s to postsocialism. Governing the natural environment required coming to terms with the divide between the desired environment and the sometimes intractable reality. Forests, for instance, had a positive connotation and proved valuable for both capitalist and socialist economies. Even though forests still needed "corrections" to serve utilitarian national purposes—corrections achieved through rational exploitation, encouragement of monoculture, calculations of the volume harvested, and bureaucratization—they were more desired than wetlands, for instance. Forests were deeply embedded in national identities and in state making. As Hande Özkan shows (chapter 8 in this volume), the designation of the forest as national patrimony and its anthropomorphization as the lung of the nation speak for the imagined value of forests and their central role for the nation-state. Wetlands, on the contrary, were understood as undesirable environments, uncertain and liminal spaces that did not fit into the new capitalist economies. The socialist states had a similar negative image of wetlands. Their inner amphibious nature, with intermittences between water and land, unproductive sand, and square kilometers of seemingly useless reed and other weeds made them not only economic but also cultural wastelands in the eyes of the state: wetlands did not contribute productively to the formation of national identities.

Accordingly, the animal world was separated into unwanted critters—pests that are perceived by planners and experts as useless for development—and desirable animals. Goats, for instance, were perceived as pests that, through their feeding habits, impede the regeneration of the forest. Other undesired animals were associated with the breakdown of the socialist regime and the rise of the new neoliberal political-economic elite. The example of the jackal (*Canis aureus*) is illuminating. Having first arrived in Eastern Europe in recent decades, the jackal has come to symbolize the locus of imagined continuities and ruptures between a ruined socialist landscape and a grim economic neoliberal present. Well-adapted to the patchy decollectivized landscape, the jackal is imagined as embodying the capitalist socioeconomic and political order. Contrary to goats and jackals, other animals are seen as desirable for their economic potential—as is the case of the muskox or the reindeer. The attempt to domesticate the muskox in Siberia was part of a larger plan to use the arctic landscape to improve local social and economic conditions. In this process, the state created a new power language based on science and ideology that categorized natural resources as "good" and "bad." This categorization is made entirely through the eyes of the state and serves state interests rather than those of the local populations.

Yet the two natures of nature are not fixed categories. A weed, for instance, which would be treated as pest, and thus included in the "bad nature" category, can be transformed through scientific discoveries into a medicinal plant. Its (potentially global) economic value transforms its status. Thus, the categorization of "good" and "bad" nature depends not only on scientific knowledge but also on the economic demands of the market. To continue with the above example of plants, a rural economy may use some of the weeds as medical remedies. But the modern state would still perceive these plants as weeds as long as there is no higher economic value to add to the national economy, revealing a split between local and national categorizations of natural resources. Likewise, the view of wetlands was also intertwined with market value: transforming wetlands into agricultural land transforms it from "useless" to "valuable." The dream of quitting the periphery thus depends on changing both the economy and the natural environment.

Correcting the nature of nature and the nature of the economy also means the transformation of political subjects. Rearranging the natural world to fit the dream of development requires rearranging social relations. Building dams and reclaiming land from wetlands is not just a matter of harnessing nature. It is also a matter of the reorganization of the national territory and a tool of social engineering. Correcting natural environments to fit into developmental frames should be seen as a state project in which the nature of land, rivers, economy, and people is deeply transformed. It is the reorganization of the natural world and of human society and economy, a high-modernist act of "improving the human condition" (Scott 1998) through science, technology, and expertise. The missionaries of such transformations were the planners, experts, and bureaucrats. They are the "priests" of a new ecological order.

There are remarkable continuities and obvious ruptures in following the state's correction of natural environments in Eastern Europe over the past two centuries. Despite the ideological differences of the various political regimes that succeeded in the region, the contributions in this volume revealed multiple continuities in human–nature relationships, the agents of change, and the tools used in manipulating environments. To explore continuities and detect ruptures most of the authors have followed their respective topics along at least two different political regimes. The actors that promoted the transformation of environments remained broadly the same from the emergence of the nation-state and the regional empires heading for "modernization" until the rise of a neoliberal order: planners, experts, and bureaucrats. Their tools also remained the same: science, technology, and plans for the reorganization of the territory. Even the socialist regimes that some scholars perceive as a rupture with the preceding capitalist regimes engaged experts trained before the socialists came to power and encouraged them to continue their work and implement their visions for transforming environments. Following the evolution of large-scale environmental development projects from presocialist/pre-Soviet times to their socialist/Soviet realization is illuminating. Certainly, ideology played a fundamental role in differentiating the two regimes, but something even more fundamental separates them: the power of the socialist/Soviet regimes to mobilize bureaucratic structures and expertise for large-scale projects for transforming the natural world. Where presocialist/pre-Soviet regimes failed, socialist/Soviet regimes prevailed. The continuities and ruptures can only be identified by examining, as most contributors to this book have done, a given topic through at least two consecutive political regimes. And this constitutes a methodological innovation that the volume advances.

We conclude with the words of a Danish engineering student (which also opened the volume) that expressed the pipe dream of the nineteenth century. He strongly believed that by harnessing the gifts of nonhuman nature humans will "control and fix the great mechanism of the world" (Pontoppidan 2013, 2:439). While his idealism referred to a Europe conceived of as Western Europe, this volume has explored the neglected story of East Europe's contribution to such high-modernist fantasies.

NOTES

1. Sharad Chari and Katherine Verdery (2009: 12) suggest that former socialist countries were "under a form of colonial domination" in their relations with the Soviet Union (cf. Kalinovsky 2018). Sonja Schmid (2011) argues something similar with regard to the case of "nuclear colonization."

REFERENCES

Chari, Sharad, and Katherine Verdery. 2009. "Thinking between the Posts: Postcolonialism, Postsocialism, and Ethnography after the Cold War." *Comparative Studies in Society and History* 51(1): 6–34.

Eisenstadt, Shmuel N., ed. 2002. *Multiple Modernities*. New Brunswick, NJ: Transaction.

Gantner, Eszter, Heidi Hein-Kircher, and Oliver Hochadel, eds. 2020. *Interurban Knowledge Exchange in Southern and Eastern Europe, 1870–1950*. New York: Routledge.

Kalinovsky, Artemy M. 2018. *Laboratory of Socialist Development Cold War Politics and Decolonization in Soviet Tajikistan*. Ithaca, NY: Cornell University Press.

Kohlrausch, Martin, and Helmuth Trischler. 2014. *Building Europe on Expertise: Innovators, Organizers, Networkers*. Basingstoke: Palgrave Macmillan.

Levinson, Marc. 2020. *Outside the Box: How Globalization Changed from Moving Stuff to Spreading Ideas*. Princeton, NJ: Princeton University Press.

Mazanik, Anna. 2020. "Learning from Smaller Cities: Moscow in the International Urban Networks, 1870–1910." In E. Gantner, H. Hein-Kircher, and O. Hochadel, eds, *Interurban Knowledge Exchange in Southern and Eastern Europe*, 1870–1950. London: Routledge, 119–33.

Pontoppidan, Henrik. 2013. *Per cel Norocos* [The lucky Per]. 2 vols. Romanian translation from Danish by I. Soare and Al. Budişteanu. Bucharest: Litera.

Reynolds, Nancy Y. 2017. "City of the High Dam: Aswan and the Promise of Postcolonialism in Egypt." *City and Society* 29(1): 213–35.

Schmid, Sonja D. 2011. "Nuclear Colonization? Soviet Technopolitics in the Second World." In G. Hecht, ed., *Entangled Geographies: Empire and Technopolitics in the Global Cold War*. Cambridge, MA: MIT Press, 125–54.

Scott, James C. 1998. *Seeing Like a State: How Certain Schemes for Improving Human Condition Have Failed*. New Haven, CT: Yale University Press.

Sivaramakrishnan, K., and Arun Agrawal. 2003. "Regional Modernities in Stories and Practices of Development." In K. Sivaramakrishnan and A. Agrawal, eds., *Regional Modernities: The Cultural Politics of Development in India*. Stanford, CA: Stanford University Press, 1–62.

Todorova, Maria. 2009. *Imagining the Balkans*. Oxford: Oxford University Press.

CONTRIBUTORS

ÁGOTA ÁBRÁN is a research fellow at the Zalău Museum of History and Art, Romania. She finished her doctoral degree in social anthropology in 2018 at the University of Aberdeen (Scotland), with a dissertation titled "The Ingredients of Plant Medicines: On the Process of Plants Becoming Medicinal in Romania."

EUNICE BLAVASCUNAS is an assistant professor of anthropology and environmental studies at Whitman College (Walla Walla, Washington). She writes about conservation conflicts in East Central Europe and the politics of memory and modernity in relation to resource institutions. She holds a PhD in anthropology from the University of California, Santa Cruz. She is the author of *Foresters, Borders and Bark Beetles: The Future of Europe's Last Primeval Forest* (Indiana University Press, 2020). Her articles have appeared in the *Journal of Political Ecology*, *Historische Anthropologie*, and in several contributed volumes, including Stephen Brain and Viktor Pál, eds., *Environmentalism under Authoritarian Regimes* (Routledge, 2019) and Dolly Jørgensen, Finn Arne Jørgensen, and Sara B. Pritchard, eds., *New Natures: Joining Environmental History with Science Technology Studies* (University of Pittsburgh Press, 2013).

DRAGANA ĆOROVIĆ is an associate professor at the University of Belgrade, Faculty of Forestry, Department of Landscape Architecture and Horticulture. She was trained as an architect at the University of Belgrade, Faculty of Architecture and holds a PhD in the history and theory of architecture from the same institution. She is the author of *A Garden City in Belgrade* (in Serbian, 2009), and of a chapter in Jelena Bogdanovic, Lilien Filipovitch Robinson, and Igor Marjanovic, eds., *On the Very Edge: Modernism and Modernity in the Arts and Architecture of Interwar Serbia, 1918–1941* (Leuven University Press, 2014).

STEFAN DORONDEL is a senior researcher at the Francisc I. Rainer Institute of Anthropology of the Romanian Academy and is affiliated with the Institute for South East European Studies, Bucharest. He holds a PhD in history and ethnology from the Lucian Blaga University, Sibiu (Romania) and a PhD in agricultural economics from Humboldt University Berlin (Germany). He is the author of *Disrupted Landscapes: State, Peasants and the Politics of Land in Postsocialist Romania* (Berghahn Books, 2016) and coauthor (with T. Sikor, J. Stahl, and P. X. To) of *When Things Become Property: Land Reform, Authority and Value in Postsocialist Europe and Asia* (Berghahn Books, 2017).

JEANNE FÉAUX DE LA CROIX is a social anthropologist leading research on societies and environments in Central Asia at the University of Tübingen (Germany). She is the author of *Iconic Places in Central Asia: The Moral Geography of Pastures, Dams and Holy Sites* (Transcript Publishing, 2017). She is the coeditor with David Gullette of *Everyday Energy Politics in Central Asia and the Caucasus: Citizens' Needs, Entitlements and Struggles for Access* (Routledge 2016) and publishes on arts-based and collaborative research, labor, and aging.

JIŘÍ JANÁČ holds a postdoctoral position at the Institute of Contemporary History of the Czech Academy of Sciences in Prague (Czech Republic). His primary interests include the history of technocratic governance, state socialism, and environment (water). He is the author of *European Coasts of Bohemia: Negotiating the Danube–Oder–Elbe Canal in a Troubled Twentieth Century* (Amsterdam University Press, 2012).

YULIAN KONSTANTINOV is chair of the Bulgarian Society of Regional Cultural Studies (BSRCS). Mainly known for his work with reindeer herders of the Russian Far North (*Conversations with Power: Soviet and Post-Soviet Developments in the Reindeer Husbandry Part of the Kola Peninsula* [Acta Universitatis Upsaliensis, 2015], http://uu.diva-portal.org/smash/record.jsf?pid=diva2:865695), his current research is on human–animal interaction in the Lower Danube.

ANNA OLENENKO is an associate professor of the Faculty of Arts at Khortytsia National Academy, Zaporizhzhia (Ukraine). Her research interests are related to the environmental history of Ukraine, especially the steppe region. She investigates nature transformations in the nineteenth and twentieth centuries, and historical memory about them. She has presented the results of her research at conferences in Ukraine, Russia, Estonia, Poland, China, and Canada.

HANDE ÖZKAN is an associate professor of anthropology at Transylvania University in Lexington, Kentucky. She holds a PhD in anthropology from Yale University. She also has previous graduate training in history from Boğaziçi University in Turkey and combines ethnographic inquiry with a historical approach in her research and teaching. Her research focuses on anthropologies of the environment, state making, subject formation, and cultural representations. Her work has been published in the *International Journal of Middle East Studies* and in academic journals in Turkey. She is currently working on a manuscript on Turkish forestry.

FLORA ROBERTS is a lecturer in environmental history at Cardiff University, in Wales. She was previously a member of the Junior Research Group on Society and Environment in Central Asia, led by Jeanne Féaux de la Croix at the University of Tübingen. She has a BA from Oxford University and an MA and PhD in Soviet history from the University of Chicago. Her research on the social and environmental history of Soviet Central Asia has been published in *Ab Imperio, Central Asian Survey*, and *Global Environment*. Her current research project, "A Sea for the Valley," is an environmental history of the Ferghana Valley in the modern period.

STELU ŞERBAN, a sociologist at the Institute for South East European Studies, Bucharest, is interested in postsocialist transformations in Southeast Europe, everyday life in rural societies, ethnicity, and political ecology. He is the author of *Elites, Parties, and the Political Spectrum in Interwar Romania* (in Romanian, Bucharest, 2006), and coeditor (with Stefan Dorondel) of the special issue "At the Margins of History: The Agrarian Question in Southeast Europe," *Martor. The Anthropology Review of the Romanian Peasant Museum* (Bucharest, 2014). He has published in international peer-reviewed journals such as the *Journal of Balkan and Near East Studies, Environment and History, Canadian Journal of Development Studies*, and *Nationalities Papers.*

HELMUTH TRISCHLER is head of research at the Deutsches Museum, Munich, and a professor of modern history at Ludwig Maximilian University, Munich. In 2009, he and Christof Mauch cofounded the Rachel Carson Center for Environment and Society, one of the leading international centers for the growing field of environmental humanities. His main research interests are innovation cultures in international comparison; science, technology, and European integration; evidence practices in science, technology, and society; cooperation and competition in the sciences; and environmental history. He has conceptualized the world's first major exhibition on the Anthropocene (Welcome to the Anthropocene: The Earth in Our Hands), on display at the Deutsches Museum in 2014–2016. His recent books include: with Martin Kolrausch, *Building Europe on Expertise: Innovators, Organizers, Networkers* (Palgrave Macmillan, 2014); with Nina Möllers and Christian Schwägerl, *Welcome to the Anthropocene: The Earth in Our Hands*

(Deutsches Museum Verlag, 2015); and, with Ruth Oldenziel, *Cycling and Recycling: Histories of Sustainable Practices* (Berghahn, 2016).

GEORGE L. VLACHOS is an environmental historian and currently a research associate at the National Hellenic Research Foundation, Institute for Historical Research. He holds a PhD from the Department of History and Civilization of the European University Institute. His research focuses on the implications of large-scale technical works on the ethnic, social, and economic coherence of rural communities in modern Europe. This work has also led him to investigate the way history and environment are used and represented in the realm of videogames.

VLADISLAVA VLADIMIROVA is an associate professor at the Institute of Russian and Eurasian Studies and the Department of Cultural Anthropology and Ethnology, Uppsala University. She has conducted research in the Russian North on topics related to indigenous peoples, reindeer herding, law and ethnicity, intangible heritage, gender, and nature conservation. Her latest works include a coedited issue of the journal *Polar Geography* on "Gender and Feminist Approaches in the Arctic" (2018). She has also published papers in the *Journal of Political Ecology*, *Journal of Rural Studies*, and *Ecological Studies*.

INDEX

Note: Page numbers in *italics* indicate figures.

Ábrán, Ágota, 7, 17
Acemoglu, Daron, 68
Adonis Cooperative in Cluj, 98
Adrianople, Treaty of, 11
Advisory Board for Ensuring Water Purity, 148n13
afforestation, 67–68, 190, 196n6; act in Dnipro wetlands, 73; experimentation in, 73–74; of southern Ukraine steppes, 75
Agrawal, Arun, 6, 189
Agricultural Bank of Bulgaria, 118
Agricultural Bank of Greece, 59
Agricultural Educational Institution, 93
Agricultural School of Athens, 49
Agronomic Research Institution, 104
Aimatovo lake, 54
Akinfiev, Ivan, 74
Alexander II (Emperor of Russia), 204–5
Alexander III (Emperor of Russia), 205
Amu Darya River, 159
Anderson, David, 223, 224, 233, 234
András, Faragó, 100
Angelov, Boris, 117
anthropomorphized representations: of nature/landscape, 179, 181, 192; of trees of Asia Minor, 182–86
Antipa, Grigore, 114–15
Aral Sea, 161, 162, 171
"archival record" of Asia Minor, 180, 181, 182
Armia Krajowa, 213
Artzan lake, 54
Ash, Mitchell, 111
Asia Minor, 179, 184; binary categories, 180, 185; Euro-American gaze, ambivalences of, 186–89; imagining nation with goat and axe, 189–93; modernity of, 180, 181; national forestry policies, ambivalences of, 193–94; nationalism of, 180, 181; naturalized people and anthropomorphized trees, 182–86; Orientalism of, 179–80, 181, 183; representations of natural landscapes, 18; scholarship on colonial modernity, 181; uncertain narratives in twenty-first century, 195–96. *See also* Turkey
Asia Minor in Ruins (Ximinez), 179, 182
Asiatic black bear (Ursus Selenarctos thibetanus), 242
Association of Engineers and Architects of Bulgaria, 122
Association of Engineers of Serbia, 35
Atay, Falih Rıfkı, 191
Austria, 9; canal project, 113; debate between hydraulic engineers and biologists, 115; irrigation systems in, 122
Austro-Hungarian Empire, 6–9
axes in Turkish forestry, 190, 195; as obstacles in Republican Turkey's green civilizing mission, 192; representation of, 191–92

Baghdad Railway project, 12
Bakiyev, Kurmanbek, 169
Balkan Wars, 49
Barca, Stefania, 4
bark beetles (*Ips typographus*) outbreak, 201–2, 211–12
Barton, David, 180
Bartovský, Josef, 150n45

Beach, Hugh, 234
Belarus/Belarusians, 203, 205, 207, 210, 213, 217n2; forest management, 210; language, 206; otherness, 212; political system in, 209; stereotyped, 209
Belgrade: borders/boundaries of, 31–38, *37*; environmental transformation, 31–38; fortress, 32, *33*; increase in area of territory, 33; Josimović's "green belt" plan for, 35, 36; new landscape, 39–40; as part of Ottoman Empire, 31–32; population growth of, 42n9; sewerage system in, 40; urban development of, 40–41; water supply system in, 40. *See also* Serbia; spatial transformation of Belgrade's environment
Belgrade Cadastre, 29, 38
Belomor canal tragedy, 156
Belovezhskaya Pushcha National Park, 210; expansion, protest against, 212–13
Belyaev, Dmitry, 232
Benjamin, Walter, 195
Berdyansk model of forest plantation, 73
Bhabha, Homi, 181
Białowieża Forest, 18, 200–201; bark beetles outbreak in, 201–2, 211–12; under communist Poland, 207–8; ethnohistory of, 202–3; Poland's illegal logging into, 201, 202; as Polish-created forest, 215–16; Poniatowski's efforts to rationalize forest management, 203–4; primeval of, 201, 202; under Romanov empire, 203–5; during world wars and industrial forestry, 205–7. *See also* Polish foresters in Białowieża Forest
Bicaz hydropower plant, 120
biosemiosis-stimulated approach, 255
biosemiosis, 242, 243
Bistriţa River, 120
Blavascunas, Eunice, 18, 189
"bovine mystique," 197n10
Brezhnev, Leonid, 247
Bricogne, Louis, 187–88
Bridge, Gavin, 4
Brincken, Juliusz, 204
British and European Timber Trust, 194
British Century European Timber Corporation, 206
Bruno, Andy, 225
Bulgaria, 6, 49, 111, 112, 118–19; autonomous principality, 9; immigration conditions with Greece, 62n18; irrigation and electricity systems in, 120, 121–22; levees construction in Belene-Svishtov area, 117–18; jackals, 18, 190, 240–43, 244, 246, 249–55; map of Vidin, *118*; nationalist vision of technocrats and experts, 113–14; political instability after WWI, 116–17; river's "modernization" projects in, 113; sluggishness in transformations, 10; state institutions, creation of, 114; studies of political elites, 8; water syndicates, 120–23; western planners role in reshaping, 12
Bulgarian guerrillas. See *comitadjis*
Bulíček project, 140
bureaucrats, 19, 260, 262, 263; as agents of change, 5, 15–16; as carriers and implementers of changes, 5

capitalism, 66, 92, 253; edges of, 90, 93, 105–6; emergence of, 66; generative friction of, 90; industrial, 3–4, 260; liberal, 136; modern, 31
Carroll, Patrick, 111
Catherine II (Empress of Russia), 71, 75, 203–4
Central Water Management Administration, 142, 144, 145
cereal trade in Danubian region, 113
chifliks, 61n12; estates, 52–53, 54, 59; ownership model, 52; system, 48
Chiriţescu-Arva, Marin, 96, 97, 100, 103
chorbadzii in Bulgaria, 8
"civilizational act" of Russia, 66
"civilizing mission," 105, 130, 136, 192
Cluj Research Station, 96, 97, 104
colonization, 105, 201: in Giannitsa plain, 53–57; interwar Bulgaria, 116–17; policy of Russian government, 74; of steppes, 67
comitadjis, 48, 49
Commission for Nature Protection and Water Management of Czechoslovakia, 135, 136

common good of "forest protection," 201, 202
Conquest of America, The (Todorov), 184
continuity, idea of, 131
Cooper, Frederick, 181
Cornies, Johann, 73
Ćorović, Dragana, 16
Cossacks, 67, 81n2
Cowentz, Hugo, 205
criminalization, 190, 191
Csedő, Károly, 94, 98, 103
ČSZAV. *See* Czechoslovak Academy of Agricultural Sciences
"cultural engineers," 134
"cursed soldiers" movement (*żołnierze wyklęci*), 213–14
cypress plant, 185–86
Czechoslovak Academy of Agricultural Sciences (ČSZAV), 142
Czechoslovak Academy of Sciences, 142
Czechoslovak Communist Party, 141, 145
Czechoslovakia, 9, 135–39; 1901 Canal Act of, 132; agricultural lobby in, 134–35; Central Water Management Administration, 142, 144–45; economic crisis in, 136; environmental change in, 17–18; fund acts in, 135, 136; Great Plan for Transformation of Nature in, 139–42, 146; hydraulic mission, 132–35; Masaryk Academy of Labour in, 135, 136, 142; public water management projects in, 136; State Planning Office in, 141; state water management policy in, 135; SVP, 140–41, 143–44; Water Management Act in, 141, 150n41; water resource management system in, 131, 143
Czech Republic: capitalist transformations in, 10; evolution of volume of reservoir water, *147*

dam(s): "biographies," 157; dam-building struggles in Tajikistan and Kyrgyzstan, 167–69; dam-inauguration campaign of Franco, 56; "détente," 171; Egyptian Aswan High Dam, 158; Farhad dam project, 167; Gabčíkovo dam, 10; Iron Gates dam, 10, 11; Kairakkum dam project, 165–67; Ka Nurek dam, 168; khovka dam, 69, 78–79; Naryn dam projects, 169; Roghun dam project, 168–69, 171; Syr Darya dam projects, 162, 165. *See also* reservoirs
Danube–Black Sea connection project, 113
Danube–Oder–Elbe Canal construction, 132, 136, 137
Danube River, 32, 33, 110
Danubian Principalities, 11
deforestation: anthropogenic, 188; Bricogne's global approach to, 188; effect in Dnipro wetlands, 72; at European scale, 201; negative consequences for steppe arable farming, 74
Delta, Penelope, 61n5
Demetriadis, Vassilis, 48
Die Technische Hochschule Berlin. *See* Technical University, Berlin
Die Wasserbewegung im Dammkorper, 117
Digitalis Cooperative in Orăștie, Hunedoara County, 98, 100
Digitalis purpurea (Lady's glove), 100
Dinulescu, Constantin, 124n9
Directorate of Land Improvement of Romania, 118, 119
Dniester River, 65
Dnipropetrovsk, 79
Dnipro wetlands, 66–67; allocated to collective and state farms, 77; development policies, 77–78; patchwork landscape in, 70, 71; and Russian imperial quest, 69–76; in Southern Ukraine, 17
Dokuchaev, V. V., 74
Dorondel, Stefan, 15, 17, 18, 189
Draft Water Management Program in the Kingdom of Bohemia, A (Klír), 133
drought, 68, 135, 136, 139, 148n12

Eastern Europe: capitalist transformations of, 8–10; crushing of socialist regimes in, 15; developmental discourses and practices, 7; failure of guest experts, 13; formation of local technical cadres, 13–14; growth of urban populations, 19n7; interest in economic development, 5–6;

Eastern Europe (*cont.*): inviting foreign experts and private companies, 12–13; nation-states role in creating expertise, 10–11; new nation-states liberation, 7; regional history in modern period, 4; impact of socialism in, 6–7; transfer of technical expertise with Soviet Union, 14–15; transforming nature in, 4–5; Western European expertise contribution to, 11–12
ECD. *See* European Commission of Danube (ECD)
ecology: complexities of, 223; of the Danube, 115; of Dnipro wetlands, 69–70; of transformed Belgrade, 41–42
EGE. *See* Erdélyi Gazdasági Egylet (EGE)
Egypt, 19n3; Egyptian Aswan High Dam, 12, 158; hydraulic works and damming of Nile in, 7
Eidgenössische Technische Hochschule. *See* Swiss Federal Institute of Technology
Ekaterinoslav province, 67
electrification, 78, 119, 120, 122; in Belgrade, 40–41; of Russia, 161–62
EMGE. *See* Erdélyi Magyar Gazdasági Egylet (EMGE)
environment(al): biosemiotics study on, 242; engineering projects, 163; issues, 7; knowledge and ideas transfer about, 15; natural, 9, 13, 33, 68, 223; reclassification, 245; transformation in Belgrade, 29, 30–31, 34, 41; transformation in North America and Western Europe, 4; urban, 32, 41
Erdélyi Gazda, 93, 94, 99
Erdélyi Gazdasági Egylet (EGE), 93–95
Erdélyi Magyar Gazdasági Egylet (EMGE), 100
Euro-American gaze in Asia Minor, ambivalences of, 186–89
European Bank for Reconstruction and Development, 171
European Commission (EC), 201, 202
European Commission of Danube (ECD), 11, 110, 112, 122
European Court of Justice, 200, 202
Evenki people, 223–24, 228; herding, 225, 228; "New Civilization" for, 236; reindeer breeders, 234; reindeer herding of, 223–24, 234; relation to state, 225
experimental pharmacology, 94
experts/expertise: as agents of change, 5, 15–16; agricultural, 59; of Belgrade, 30–31; as carriers and implementers of changes, 5; contribution with states in Lower Danube, 110–12; forest. *See* Polish foresters in Białowieża Forest; hydraulic, 17; international, 172; transferring to developed countries, 10–15; US, 77, 158
"explosion of democracy" (*kogato izbukhna democratsiyata*), 241, 252, 253

Farhad dam project, 167
Farhad hydropower plant in Uzbek SSR, 162–65
Féaux de la Croix, Jeanne, 7, 18
Ferguson, James, 194, 197n10
fishing in Dnipro River, 70, *71*
Fligelman, Ya'akov, 166
Fondahl, Gail, 224
foreign experts in Serbia, 38, 39
foresters, 7, 13, 16, 186, 190, 193, 196, 209; acted against nature conservationists, 211; in Białowieża, 18; claiming expertise as common good of "forest protection," 202; French, 187, 188; German, 188, 205; Italian, 191; under PiS, 212; Polish *See* Polish foresters in Białowieża Forest; Prussian, 204; representations of Asia Minor's forestry, 187–89; in research institutions, 196n9; Russian, 75; Forest Protection Committee (FPC), 72–73, 74
forestry, 4; ambivalences of national forestry policies, 193–94; artificial work of, 4; in Asia Minor, 189; forestry-engineering, 142; German, 73, 76; Polish, 18, 171, 216; scientific, 201–2; in Turkey, 186–87, 189
forest(s): communities, 193; degradation, 194; exploitation, 4, 13, 19n2; harvesting, 14; plantation model, 73; transformation of, 4; unauthorized felling of, 72; villagers in Turkey, 195–96, 197n11; wetland vegetation as, 74. *See also* Białowieża Forest

Fornari, V., 117
Foundation Company, 54, 55, 57
FPC. *See* Forest Protection Committee (FPC)
Franco, Francisco, 56

Gabčíkovo dam, 10
GAS. *See* Greek Agricultural Society (GAS)
gemeinschaft, 203
General Water Management Plan of Protectorate, 137
Gerchyk, I., 72
German foresters of Białowieża Forest, 188, 205
German forestry, 73, 76
German School of Forestry, 4
Giannitsa Lake, 47; agricultural station in, 52; *chiflik* estates around, 52–53; draining of, 54; Greek occupation of, 49–51; land reclamation project in, 54, 55, 57–61; malaria morbidity statistics around, 51; in nationalist interest, 48
Gidófalvy, István, 100, 101
"gigantomania" of Roghun, 168
Glavinić, Konstantin Kosta, 39
goats in Turkish forestry, 184, 190; banning of, 190; as "infidels," 191, 192; as obstacles in Republican Turkey's green civilizing mission, 192; role in forest degradation, 194; role in preventing forest fires, 196n9; as safety net for peasants, 194; social significance for rural poor, 188; as "traitors," 191
GOELRO plan, 80
Goering, Hermann, 208
golden jackal (*Canis aureus*), 18, 240, 249; folk jackal talk, 253; hunters' jackal talk, 250–51; from jackals' point of view, 253–56; political subjecthood for, 242; scientific jackal talk, 251–53
Goody, Jack, 6
Gorky, Maxim, 80, 156
Government Committee for Improving Agriculture, Water Management, and Forestry (ZVL), 141, 143
Graham, Loren, 232, 233, 235
Grand Ethiopian Renaissance Dam, 157
Great Ferghana Canal, 159
Great Plan for Transformation of Nature in Czechoslovakia, 139–42, 146
Greceanu, Nicolae, 116
Greece/Greek, 48, 52, 54, 58; as autonomous territory, 9; Central Greece, 61n7; failure of guest experts in, 13; immigration conditions with Bulgaria, 62n18; occupation of Giannitsa Lake in Macedonia, 49–51; socialist political economic model, 19n8; utilizing "agricultural capital" in Giannitsa plain, 53; western planners role in reshaping, 12. *See also* southern Macedonian swamp
Greek Agricultural Society (GAS), 51–52
"Greekness"; of Orthodox refugees, 57; of southern Macedonia, 47
"green belt" plan, 35, 36
Grinţescu, G. P., 97, 100, 103

Haeckel, Ernst, 114
Hann, Chris, 6
Harvey, David, 99, 105
Haupt, Heinz-Gerhard, 15
Haussoullier, Bernard, 179, 182
Hawley, Walter, 183–84, 186
Hellenism, 196n2
Heske, Franz, 188–89
high modernism, 89, 92; at edge of capitalism, 105–6
"high modernist state," 10
Hitler, Adolf, 207
Hobson, Kersty, 242, 243, 253
Horticulture Society (Russia), 73
Hull, Matthew, 180
human impact on nature, observations of, 240; biosemiotics, 242–43; golden jackal as research animal, 240, 241; Jackal's Talk around Lower Danube, 251–56; Lower Danube as research site, 240–41; postsocialist reality, 243–44
Hungarian Hydraulic Service, 114
Hungary, 9, 113; capitalist transformations in, 10; medicinal plant research station, 90; role in industrial development, 15
hydraulic engineering: conception, 142; projects, 131, 157; radical, 142; works in Kingdom of Bohemia, 133

hydraulic engineers, 115, 134, 140, 141
hydraulic mission, 111, 119, 130, 131. *See also* Czech (Oslovak) hydraulic mission
hydrocratic/hydrocracy, 130; community of "water experts," 134; Czech, 132, 137; emergence and activities of, 130; hydrocrats acted as "system builders," 131; inclusive approach to water management, 144–45; institutionally fragmented, 132–33; relationship with state, 131; views about water resource, 131
hydrocrats, 130, 131, 137, 139; acted as "system builders," 131; Czech, 133, 142, 146; Romanian, 113
Hydroenergoproekt, 79
Hydrological and Hydrotechnical Research Institute. *See* Hydrological Research Institute
hydrological cycle, 131, 133, 136
hydrological drought, 148n12
Hydrological Research Institute, 135, 138, 142
hydropower systems: Bicaz hydropower plant, 120; in Bulgaria, 120, 121, 122; in Czechoslovakia, 141, 144–45; Farhad hydropower plant, 162–65; in Romania, 120, 121, 122; Soviet's big dam projects in Central Asia for, 158–61

IDC. *See* International Danube Commission (IDC)
Ilinden Uprising of 1903, 48
IMF. *See* International Monetary Fund (IMF)
Imperial Agricultural Society of Southern Russia, 73–74
Imperial Russian scientific knowledge, 67
Imperial Russian territory reorganization, 66
IMRO. *See* Internal Macedonian Revolutionary Organization (IMRO)
industrialization, 4, 15; agricultural industrialization program, 96–97; of Central Asia, 162; industrial capitalism, 3, 4, 260; industrial forestry in Białowieża Forest, 205–7; socialist, 141
"industrialized river" as "organic machine," 134, 148n8
"infrastructural modernization" project, 10
infrastructure: agricultural, 77; Belgrade communal, 38–41; in Eastern European countries, 7, 10; material, 111; national, 119; projects, 5, 12, 15, 18, 242, 243, 255; system of Belgrade, 30–31; urban infrastructure of Budapest, 12; water infrastructure projects, 18, 144, 146, 156, 157, 159
Ingold, Tim, 223
Interfax, 167
internal colonization, 105
Internal Macedonian Revolutionary Organization (IMRO), 48
International Danube Commission (IDC), 112
internationalization, 11
International Monetary Fund (IMF), 10, 12
International Union for Conservation of Nature, 206
Iron Gates dam, 10, 11
irrigation systems, 9, 12, 68, 76–77, 134, 147, 247; in Bulgaria, 120, 121–22; in Romania, 119–20, 122; Soviet big dam projects in Central Asia for, 158–61
Iteanu, A., 103

Janáč, Jiří, 17
Jelavich, Barbara, 19n5
Jelavich, Charles, 19n5
Josephson, Paul, 9
Josimović, Emilijan, 34–35, 36, 40

Kairakkum dam project, 165–67
Kaiser's Military Forest Administration, 205
Kakhovka hydroelectric power station, 69, 78–79
Kalemegdan (in Belgrade), 32, 33, 35
Kalemegdan Park, 40
Kálmán, Bodor, 100
Kambar-Ata 1 and 2 plants, 169
Karakis, Iosif, 163
Karavidas, Konstantinos, 47, 56, 57
Karpiński, Jan Jerzy, 206–7
Kazakhstan, water source of, 159
Kemal, Mustafa, 182
Kholoussy, Hanan, 19n3
Khorpushtak journal (The Hedgehog), 159, *160*

Kideckel, David, 244, 249
Kiselev, Pavel, 73
Klír, Antonín, 133–37, 150n45
Kneipp, Sebastian, 94
Kocka, Jürgen, 15
Kohlrausch, Martin, 14
Kolozsvár medicinal plant research station, 94
Konstantinov, Yulian, 18, 190
Kopp, Elemér, 93, 94, 97, 103
Krasnoyarsk, 223, 228, 230
Krause, Herr, 188
Krivy Rih, 79
Kroeber, Ludwig, 94
Kull, Kalevi, 243
Künksatan, Ahmet, 194
Kurdish population, 184
Kyrgyzstan, 158, 172; contemporary dam-building struggles in, 167–69; electricity supply from "water towers" of dam, 159, 161; inhabitants in, 172n2; political histories, 158–59; stop-and-go dam building in, 169–70; water infrastructure projects in, 18

labor force, 58, 162, 208
Lamartine, Alphonse de, 31
land reclamation: in Bulgaria, 117; project in Giannitsa plain, 54, 55, 57–60; in Romania, 116
land transformation of Belgrade, 31–38
"large technological systems," 148n4
Lasy Państwowe. *See* Polish State Forests (SF)
Latour, Bruno, 91, 226
Law and Justice Party government (PiS), 200; bark beetle outbreak and, 211; centering of Białowieża Forest, 212; historical features of Polish forestry in Białowieża, 216; narrative ecoterrorists, 214–15; politicized projection of Polish forester, 201
League of Nations, 13, 117
Lengyel, Emil, 184, 189, 190
Lenin, Vladimir, 161
Leonida, Dimitrie, 120
levee construction: in Bulgaria, 116–19; along Lower Danube, 13, 113, 114–15; in Romania, 115–16
lewactwo, 211
Lindley Brothers, 12
Lisitsin, P., 72
logging in Białowieża Forest, 18, 200, 201, 204, 205, 207, 210–16
Loret, Adam, 206
Lower Danube, 11–12, 240; attempts to restore areas of, 15; construction of levees along, 13; international context and models of river modernization, 112–14; international status of, 123; irrigation and hydropower systems around, 119–23; levees, politics, and development around, 114–19; link between experts on river dynamics and state, 110–12; in national political imagination, 111; orchard turning into jungle in, 248–49; as "organic machine," 123; politics of damming, 17; projects and predators, 245; removal of harmfulness along, 246–48; sanitizing totality of existence along, 246–48; stories about predators in, 240–41; technological "assault" on, 109; wolf-as-a-dreaded-menace along, 245
Lucky Per (Pontoppidan), 3
Lysenko, Trofim, 232, 233, 235

Macedonia(n): colonization in Giannitsa plain, 53–57; nationalisms, 47–49; Slavic grassroots cultural and economic institution in, 56–57; Struggle, 61n5. *See also* southern Macedonian swamp
Macedonian Strugglers, 49
Macedonia newspaper, 55
Macfarlane, Charles, 185
MacLean, Gerald M., 196n1
mahalle (neighborhood or residential quarter), 32
Makedonomachoi. See Macedonian Strugglers
malaria morbidity statistics around Giannitsa Lake, 51
Maran, Timo, 242, 243
Marder, Michael, 102
Markov, Georgie, 251
marshlands, 46, 60, 114; drainage of, 13; transformation of, 3–4
Masaryk Academy of Labour in Czechoslovakia, 135, 136, 142

Mazzullo, Nuccio, 231
medicinalness of plants, 102–3
medicinal plants in Transylvania: Agronomic Research Institution, 104; Cluj Research Station, 96, 97; commodity chains on global market, 92; conducting experiments with medicinal plants, 103; evolution of pharmacology in field of, 93–94; industrial promotions of scientists for, 95–98; legitimacy of, 94; movement to global market, 89–90; noble cause of, 95; research stations for, 90–91, 93; value of, 103, 105; "wild," 106n1. *See also* Páter, Béla
Menga, Filippo, 111
mesopredator release, 252
Michurin, Ivan, 79
Middle Danube, 110, 112
Mihál, Juraj, 145
Mikhail, Alan, 60
Mirziyoyev, Shavkat, 167
Mississippi River, 77
Mitchell, Timothy, 7, 30
Mizian Plain, 240, 247
modernist nationalism, 189
modernity/modernities/modernization, 46; of Asia Minor, 180, 181; of Belgrade, 30–31, 41–42; illiberal, 256; metanoia, 7; organized, 130; representatives of, 46–47
Moldo-Walach boyars, 8
Molle, François, 130
Mollinga, Peter P., 130
Montrichard, D. de, 188
Moore, Jason W., 92, 111
Moscow Agricultural Institute, 68, 80
Mueggler, Erik, 180
Muhlis, Esat, 194
muskox domestication experiment in Siberia, 6, 18, 221, 222, 230–33; as driver of reindeer economy, 229; importing in Russia, 229–30; scientific innovation as political field, 233–36
Mussolini's reclamation of Pontine Marshes, 56

Nabiyev, Rahmon, 171
Narodowa Hajnówka, 214, 216
Naryn dam projects, 169
nation-state, 6–9, 19, 111, 183, 259–60; emergence in Eastern Europe, 10–11, 15; nationalizing and strengthening educational institutions, 14; political model, 8
nationalism, 172, 180, 181; amphibious, 47–49; impact on rural populations, 61n2; modernist, 189; Polish, 211, 212; ultra-right-wing, 200
National Liberal Party of Romania, 114
National Peasants' Party of Romania, 114, 120
national projects, 5–6; carriers, 5; "regional modernities" concept, 6, 7; transferring knowledge, experts, and ideas about nature, 10–15
National Radical Camp. *See* Obóz Narodowo-Radykalny (ONR)
Natura 2000 Habitats Directive laws, 202, 213
naturalized people in Asia Minor, 182–86
Nature and Empire in Ottoman Egypt (Mikhail), 60
Navigația Fluvială Română. *See* Romanian River Navigation Service
networks of translations, 98; gatherers and cultivators, 102; high-modernist intelligentsia, 100–101; village intelligentsia, 98–100, 101
neudobnaia zemlia. See wasteland
NGOs. *See* nongovernmental organizations (NGOs)
Nixon, Richard, 230
nongovernmental organizations (NGOs), 15
nonproductive losses, 227
nonsubmersible levees, 117; in Bulgaria, 116; in Romania, 115, 116
"Northern Village—New Civilization" project, 236
Novotný, Jan, 150n45
Nurek dam, 168

Obóz Narodowo-Radykalny (ONR), 214, 216
Obrenović, Miloš, 33
Old Greece, 49, 51, 61n7
Olenenko, Anna, 17
ONR. *See* Obóz Narodowo-Radykalny (ONR)

On the Usefulness of Our Plants (Teleki), 100
Operation Barbarossa, 207
Orava Reservoir, 142
"organic machine," 148n8; Danube as, 123; "industrialized river" as, 134
Orientalism: of Asia Minor, 179–80, 181; nesting, 181; of Turkey, 184
Orman ve Av journal, 188, 193
Ottoman Empire, 6–8, 29, 180; Baghdad Railway project in, 12; Belgrade as part of, 31–32; dissolution of, 180; economic and military uncertainty process in, 8–9; first Balkan War against, 49; initiating modernizing reforms in Asia Minor, 186; management of water irrigation in, 60
Ottoman Imperial School of Naval Engineering, 13
Otunbayeva, Roza, 173n19
Özkan, Hande, 18

Paczówski. Józef, 206
Pagalos, Theodoros, 54, 55, 56
Palamiotis, Georgios, 50–51
Papageorgiou, Pilavios, 50
Papen, Uta, 180
Paris, Treaty of, 112
Parisian École nationale des ponts et chaussées, 13
Pašić, Nikola, 40
Páter, Béla, 89, 90; creating networks of translations, 98–102; encouraging idea of producer and seller cooperatives, 97–98; finding "noble cause" of medicinal plants, 95; founded first medicinal plant research station in Europe, 93; organization of wild-plant gathering, 97, 98; position at Agricultural Educational Institution, 95–96; promotion for medicinal Transylvanian plants, 1033; views about research station goal, 102
"Páter perception," 96
Patterson, Orlando, 196n3
Pavel, Dorin, 120, 124n10
Peloponnesus, 50, 61n7
Perera, Suvendrini, 191
Peter I (Emperor of Russia), 72
Peterson, Maya, 77
Pharmaceutical College and School (Romania), 97
pharmacology, 91; evolution in medicinal plants, 93–94; experimental, 94
PiS. *See* Law and Justice Party government (PiS)
planners, 66, 99, 113, 260, 262; as agents of change, 5, 7, 15–16; Bulgarian, 122; as carriers and implementers of changes, 5; in Romania, 119, 120; Soviet, 161, 170, 171
Poland, 6, 9; capitalist transformations in, 10; Home Army, 213; Polytechnic University in Warsaw, 14; role of forester in Białowieża, 18; transformation of political elite in, 8; violating Natura 2000 Habitats Directive laws, 213
Polar Census, 224
Polish Academy of Sciences Mammal Research Institute, 208
Polish foresters in Białowieża Forest, 200–201, 206–7, 216; common good of "forest protection," 202; under PiS, 212; PiS's politicized projection of, 201; political conflicts, 202–3, 214; postsocialist forest politics, 208–11; reinventing communists and martyrs, 211–15. *See also* Białowieża Forest
Polish forestry, 171, 189, 201, 208, 210, 211, 216
Polish Nature Conservation Council, 206
Polish–Soviet War, 205
Polish State Forests (SF), 200, 201, 208, 209, 216
Polytechnic University in Warsaw, 14
Polytechnic University of Zurich, 120
Poniatowski, Stanisław August, 203
Po River, 77
Portelli, Alessandro, 243
postsocialism, 244, 248–49
Potlog, Alexe S., 96, 97, 104
"poverty-stricken" village economies, 95
Pravilova, Ekaterina, 68
predators, 241, 250; advance and retreat of, 241; liquidation of, 246; projects and, 245; symbolization of, 243
Protectorate of Bohemia and Moravia, 137
Pugliese, Joseph, 191

Rachtina settlement, 56
Rarincesco, I. R., 124n9
"re-wildening" (*podivyavane*) of society and nature, 244
rectification, 109, 112, 133
"regional modernities" concept, 6, 7
Regulation of Belgrade Town, 40
Regulation on Preservation of Forests (1888), 72, 73
reindeer herding, 221, 222; for development of industrial economy, 225; of Evenki people, 223–24, 234; projects, 222; role of reindeer science, 226–29; Soviet cultivation of, 223, 232, 233–34. *See also* muskox domestication experiment in Siberia
Reitter, Ferenc, 12
"renting cooperatives" in capitalist state society, 96
Report by the World Commission on Dams, The, 173n20
Republican People's Party, 194
Research Institute of Agriculture and Ecology of Arctic, 222
reservoirs, 143; Kakhovka dam, 78–79; Orava Reservoir, 142; in Vltava and Berounka drainage basins, 137; Zlatna Panega reservoir, 121–22. *See also* dam(s)
Riabinin, I. V., 75
Roberts, Flora, 7, 18, 166
Robinson, James, 68
Roghun dam project in Tajikistan, 168–69, 171
Romania, 6, 111, 112, 114; debate between experts about economic development, 114–15; following "white coal" strategy, 114; hydropower systems, 120, *121*, 122; international support for levees construction, 116; irrigation systems, 119–20, 122; nationalist vision of technocrats and experts, 113–14; obtaining Western knowledge and experience, 123; river's "modernization" projects in, 113; Romanian technocracy, 119; sluggishness in transformations, 10; water syndicate creation, 116. *See also* Transylvania
Romanian Hydraulic Service, 113, 115
Romanian Medicinal Plant Conference, 97–98
Romanian political elites, studies of, 8
Romanian River Navigation Service, 113
Romaniţa Cooperative in Bucharest, 98
Roosevelt, Franklin D., 157, 221
Roşu, Vasile, 113, 114
RusHydro, 169
Russo-Turkish War, 9
Rydzyk, Tadeusz, 212

Said, Edward, 179, 181, 184
Saligny, Anghel, 114–15, 119
Salih, M. Ali, 193
Sarafanov, I., 79
Sava River, 33
Schmied, Ignaz, 117
Schulz, Hugo, 94
scientific forestry, 201; role in healing natural landscape, 191–92; system in Turkey, 186; techniques of, 202
Scientific Research Institute of Reindeer Husbandry, 227
scientism, 130–31, 133
Scott, James C., 10, 92, 99, 105, 112, 148n5, 225
Second Republic of Poland, closing of, 217n3
Secrets of the Swamp (Delta), 61n5
Selezhinskii, Iosiph, 74
"self-assigned cultural role," 99
Şerban, Stelu, 15, 17, 189
Serbia, 6; as autonomous territory, 9; capitalist development, 36–37; change in technical and engineering culture, 38; experts in, 38–39; first railway line in, 42n12; modernization, 31; Principality of, 29, 33–34, 36; under Ottoman rule, 42n5. *See also* spatial transformation of Belgrade's environment
Serbian-French Society, 40
Serbian technical journal. See *Srpski tehnički list*
Serviciul Hidraulic Român. *See* Romanian Hydraulic Service

SF. *See* Polish State Forests (SF)
"shaggy years" (*runtavite godini*), 241
Shenrok, Ernest, 74
Siedzikówna, Danuta, 214
Sivaramakrishnan, K., 6, 189
Skarzhynsky, V. P., 73
Slánský, Rudolf, 141
"Slavicness" of Macedonia, 47–48
Slezkine, Yuri, 226
Smetana, Jan, 135, 142, 146, 151n63
Smith, J. V. C., 184, 186
Smrkovský, Josef, 141
social death, 183, 196n3
social engineering, 156, 157; goals, 161–62; Kairakkum dam project, 165–67; nature and society for victory at Farhad, 162–65
socialism, 6–7, 9, 15, 139, 142, 146; as "infrastructural modernization" project, 10; mature, 255; state, 146. *See also* postsocialism
socialist technological havoc, 15
social transformations, 156
Société générale des industries agricoles, 116
Société serbe-française. See Serbian-French Society
Society of Serbian Engineers and Architects, 38
Sokołowski, Andrzej, 208
southern Macedonian swamp, 46–47, 60–61; amphibious nationalisms in, 47–49; colonizing process in, 53–57; general area of hydraulic works, 55; Giannitsa Lake, 47, 48; Greek agronomists, societies, and peasants role in, 49–53; subjugation mechanisms, 57–60
Southern Regional Ameliorative Organization, 65
southern Ukrainian wetlands, 65; administrative bureaucracy and propaganda machine, 80–81; colonization of steppes, 67, 81n7; Dnipro wetlands and Russian imperial quest, 69–76; end of wetlands, 76–80; mounds of sand in, 75; during Soviet period, 66–67, 68; during tsarist era, 67, 68–69; visions for wetlands, 80–81
sovereign (Foucault), 33
Soviet big dam projects, 157, 158; comparing generations of Central Asian dams, 170–72; dam-building struggles in Tajikistan and Kyrgyzstan, 167–69; diverse engineering goals of, 157–58; electrification and reshaping spaces by, 161–62; engineering Inland Sea at Kairakkum, 165–67; engineering nature and society for victory at Farhad, 162–65; family traits in "social life" of, 157–58; stop-and-go dam building in Kyrgyzstan, 169–70; tapping rivers for irrigation and hydropower in Central Asia, 158–61
Soviet Socialist Republics (SSRs): Kyrgyz, 158; Tajik, 158; Uzbek, 163, 167
Soviet Union: Białowieża Forest under, 201, 202, 207; First Five-Year Plan of Soviet Union, 138, 140, 141, 150n40, 161; importing muskox in, 229–30; influence in Eastern European countries after WWII, 9–10; introduction of domesticated species in, 221–22; pre-Soviet statehood, 224–26; role of reindeer science in, 226–29; transfer of technical expertise with Eastern Europe, 14–15
Sovkhoz, Surindinskii, 228
spatial transformation of Belgrade's environment, 29; communal infrastructure for, 38–41; emerging ecology in transformed landscape, 41–42; experts for, 30–31; landownership and establishment of land registry, 29; landscape, 30–32; modernization in, 30–31
Srpski tehnički list, 38
Ssorin-Chaikov, Nikolai, 227, 228
Stalin, Joseph, 220
Stalin Plan. *See* Great Plan for Transformation of Nature in Czechoslovakia
Stamboliyski, Aleksandar, 116
Stark, David, 15
State Defense Committee (GOKO), 163
State Hydrological Institute, 142
State Water Management Plan (SVP), 140–41, 143–44

"statewide water balance" model, 139
Stefanovik-Skilitsis, P., 61n9
Stefansson, V., 221
steppes of southern Ukraine, 66, 71, 75; afforestation of, 75; targeted as climatic conditions, 76; treeless landscape of, 71–72; tsarist colonization of, 67; wetlands transformation in, 81
Stoler, Ann Laura, 182
Štoll, Čestmír, 142
subjugation mechanisms in Giannitsa plain, 57–58; distribution patterns of reclaimed lands, 59–60; labor force, 58; refugee groups evolved into agricultural cooperatives, 58–59
Šumadija, Serbia, 31
Sungar, Şakir, 192
Surinda, Siberia, 224, 236
Surman, Jan, 111
SVP. *See* State Water Management Plan (SVP)
Swiss Federal Institute of Technology, 40
Swyngedouw, Erik, 111
Syr Darya River, 159; dam projects in, 162, 165
system builders, 131, 148n4
Szafer, Władysław, 206
Szyszko, Jan, 212–15

Tajik Civil War, 168
Tajikistan, 158, 172, 172n2; contemporary dam-building struggles in, 167–69; diverse engineering goals of, 157–58; electricity supply from "water towers" of dam, 159, 161; freedom of speech in, 173n18; political histories, 158–59; Roghun dam project in, 168–69, 171; water infrastructure projects in, 18
TALCO aluminum plant, 168
Tallián, Béla, 93
Teal, John, 229
Technical Faculty in Belgrade, 39
Technical Forestry High School in Białowieża, 208
Technical University, Berlin, 39
technocratic/technocrats, 9, 16, 31, 46, 47, 119, 133–36; failing in Macedonia, 52; Ottoman, 186; Romanian and Bulgarian, 113; system, 143, 145
technoeconomic power, 7
technostate apparatus, 18
Teleki, Árvéd, 100
Teleki, Domokos, 93
Tennessee Valley Authority project, 157
Theotokis, G., 61n9
Thomas, Nicholas, 182
Three Gorges project of China, 157
Tisza–Adriatic Sea canal project, 113
Todorova, Maria, 180–81, 189
Todorov, Tzvetan, 184
Tomaszewski, Konrad, 200, 201, 212
translation of plants and people: networks of translations, 98–102; people into collectors and cultivators, 90–92, 99, 104; weeds into commodities, 90–92, 99, 105; weeds into medicinal plants, 104
Transylvania, 9; high modernism at edge of capitalism, 105–6; medicinal plant research station in, 90–91, 93; medicinal plants movement from, 89–90; moving from "edge of capitalism," 93; place in global market, 92; plants as medicines in, 102–5; as raw-material exporter, 93–95. *See also* Romania
Transylvanian Agricultural Association. *See* Erdélyi Gazdasági Egylet (EGE)
Transylvanian Hungarian Agricultural Association. *See* Erdélyi Magyar Gazdasági Egylet (EMGE)
Trischler, Helmuth, 18, 118
Tsankov, Aleksandar, 116
Tschirch, Alexander, 94
Tsing, Anna, 90
Turkestan, 159
Turkey, 9; Macfarlane's dissatisfaction with trees in, 185; modernist nationalism in, 189; naturalized people in, 184; Orientalism of, 184; scientific forestry system, 186, 189; socialist political economic model, 19n8; Turkish nationalist movement, 49; Turkish political elites, studies of, 8; wet–dry binary role in, 185, 196n4. *See also* Asia Minor

Turov, M. G., 231
Twain, Mark, 183
Tyzenhaus, Antoni, 203

Ukraine, 6. *See also* southern Ukrainian wetlands; steppes of southern Ukraine
Úlehla, Vladimír, 149n29
UNESCO, 200, 201, 202
Universul newspaper, 96
Upper Danube, 110, 112
urbanization, 4; of Belgrade, 41–42; East European countries, 260; along Lower Danube, 11
Ústřední správa vodního hospodářství (ÚSVH). *See* Central Water Management Administration
Uzbekistan: Farhad hydropower plant in, 162–65; water source of, 159

Váh Cascade, 142
Varoš u šancu. *See* City within the Moat
Vaskov, M., 79
Versailles, Treaty of, 205
veterinary services in Soviet Russia, 226–27
Vidraşcu, Ion, 115, 120
Villoresi, Luigi, 119
Vlachos, George, 17
Vladimirova, Vladislava, 18
Vltava Cascade, 142
Vltava-Danube Canal construction, 132
Vollard, Ambroise, 109
Votruba, Ladislav, 151n48

Walachia, 11
Warsaw University Geobotanical Station, 208–9
weeds, common (Hu. *közönséges dudva*), 101, 105; translation into commodities, 90–92, 99, 105; translation into medicinal plants, 104
Weinberger, Hans, 118
Wenzel, Hermann, 188
Western Belarusian Communist Party (KPZB), 205–6
wet–dry binary in Turkish forestry, 185, 196n4
wetland(s), 65–66, 75, 80; forests protection activities, 74; ideal, 66; ideas about transformation, 67–68; transformation of, 3–4, 116; Yaroshevsky's proposal for, 76–77. *See also* southern Ukrainian wetlands
"white coal" strategy of Romania, 114
Wilcocks, William, 12, 119
Wittfogel, Karl August, 130
"wolves' times" (*vulchite vremena*), 241
woodlands in Dnipro wetlands, 68, 70, 71–72
World Bank, 10, 12, 169; Global Environment Facility, 210

Ximinez, Saturnino, 179, 182

Yalta agreements, 207
Yaroshevsky, Aleksandr, 65, 76
"Yavaşyavaş," 50, 61n8
Yörük, 188
Young Turk Revolution (1908), 49

zadruga, 42n3, 57
Zaimis, A., 61n9
Zaporizhia, 78, 79
Zhivkov, Todor, 241, 256n1
Zimmler, Emil, 136, 139, 149n17
Zlatna Panega reservoir, 121
ZVL. *See* Government Committee for Improving Agriculture, Water Management, and Forestry (ZVL)